KUMON MATH WORKBOOKS

Intro to Geometry

Workbook I

Table of Contents

KUMON

1 Triangles & Quadrilaterals Review

Level ☆

Date / /

Name

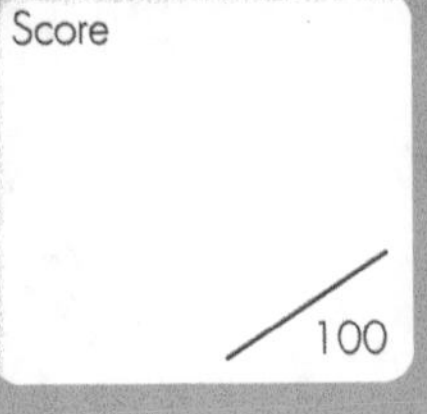

Score /100

■ The Answer Key is on page 88.

Don't forget!

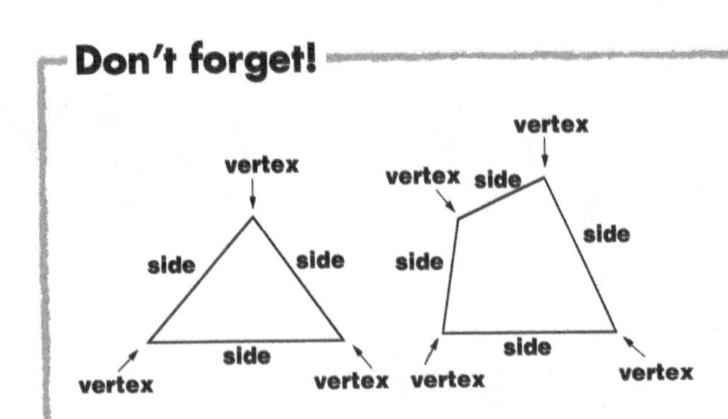

● The straight lines that make up triangles and quadrilaterals are called **sides**. Each point where those lines come together is called a **vertex**. The plural of vertex is **vertices**.

● A triangle has three vertices, a quadrilateral has four, a pentagon has five, and a hexagon has six and so on.

1 **Sort the shapes pictured here into the categories below.** 10 points per question

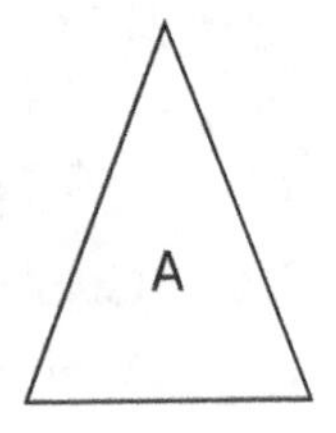

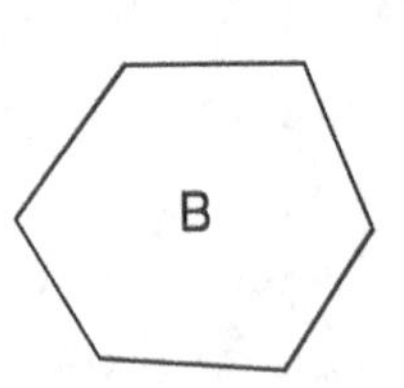

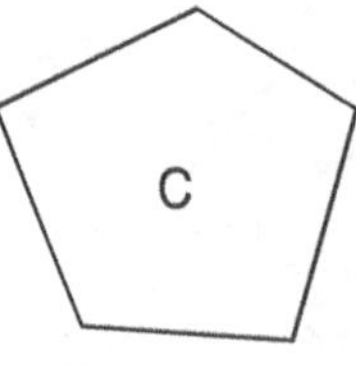

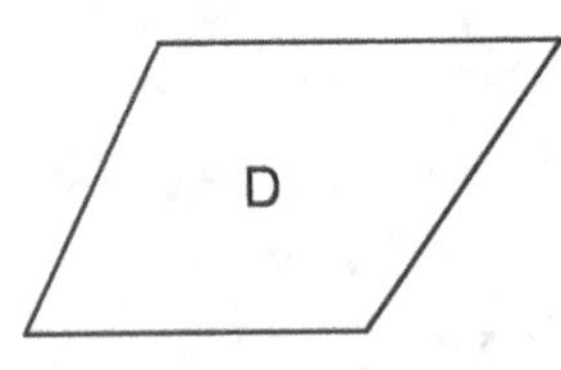

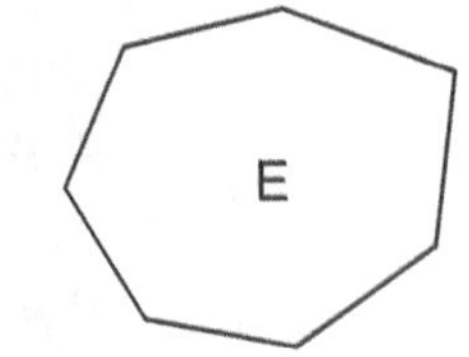

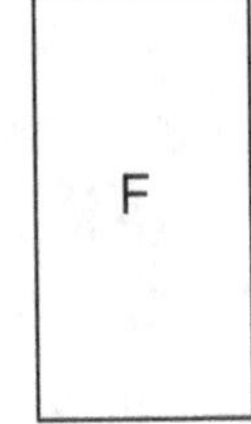

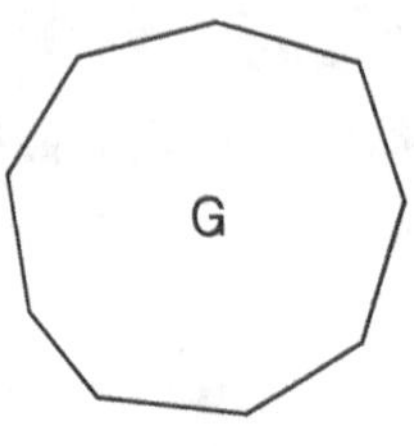

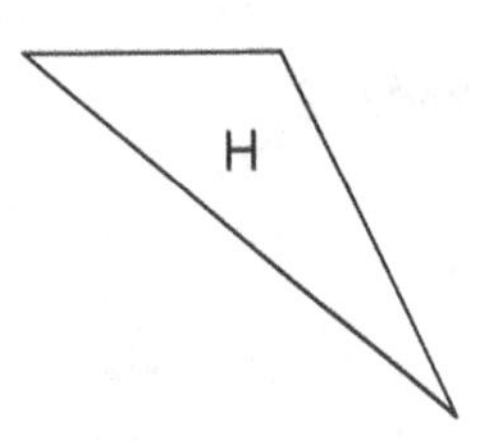

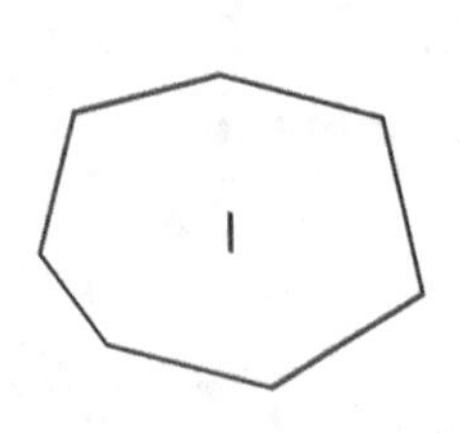

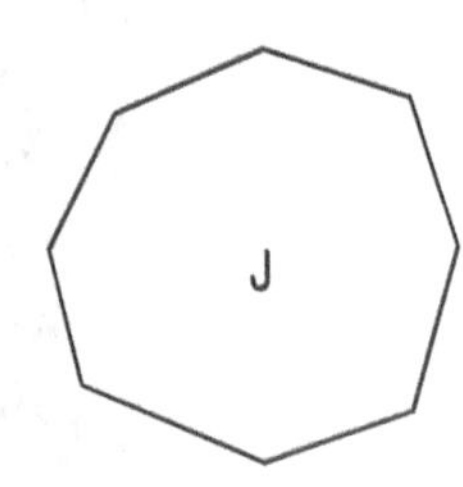

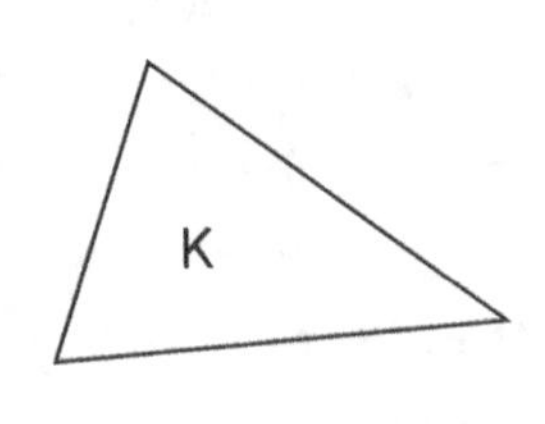

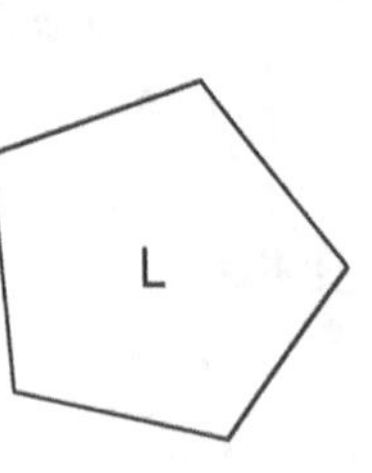

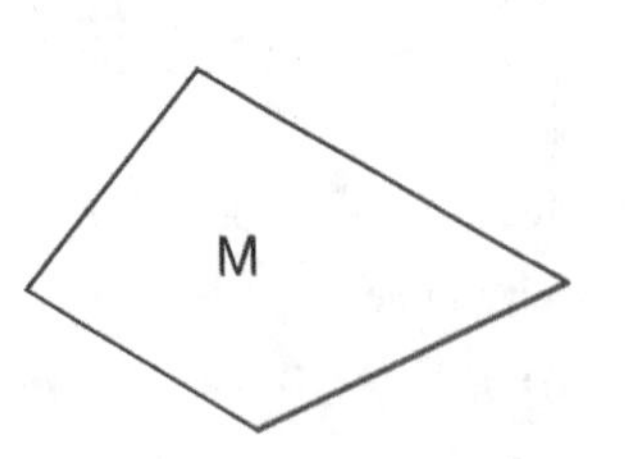

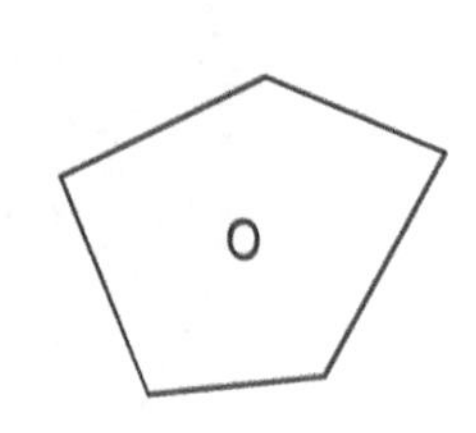

(1) Triangle 〈Ans.〉 , ,

(2) Quadrilateral 〈Ans.〉 , ,

(3) Pentagon 〈Ans.〉 , ,

(4) Hexagon 〈Ans.〉 ,

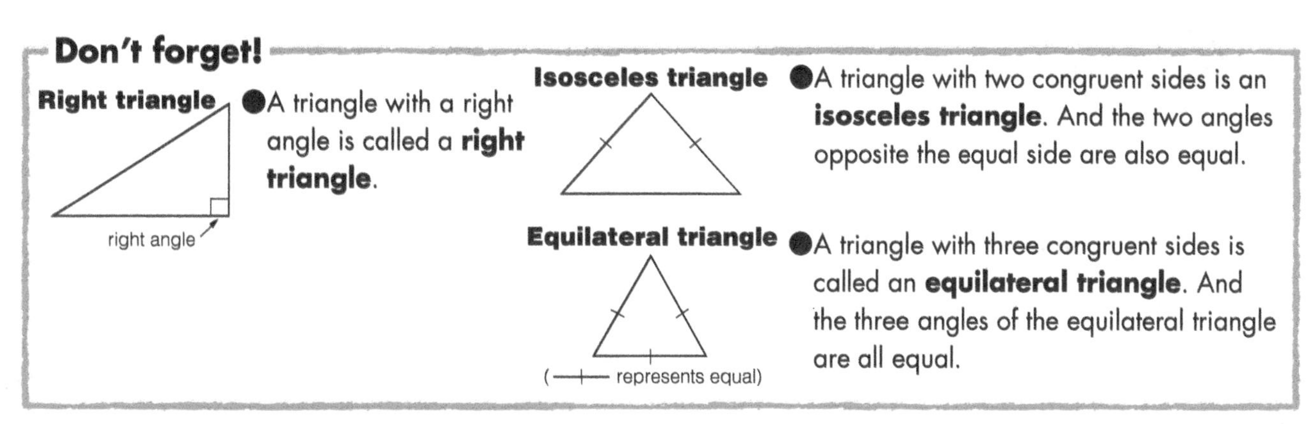

●A triangle with a right angle is called a **right triangle**.

●A triangle with two congruent sides is an **isosceles triangle**. And the two angles opposite the equal side are also equal.

●A triangle with three congruent sides is called an **equilateral triangle**. And the three angles of the equilateral triangle are all equal.

2 **Sort the triangles pictured here into the categories below.** 10 points per question

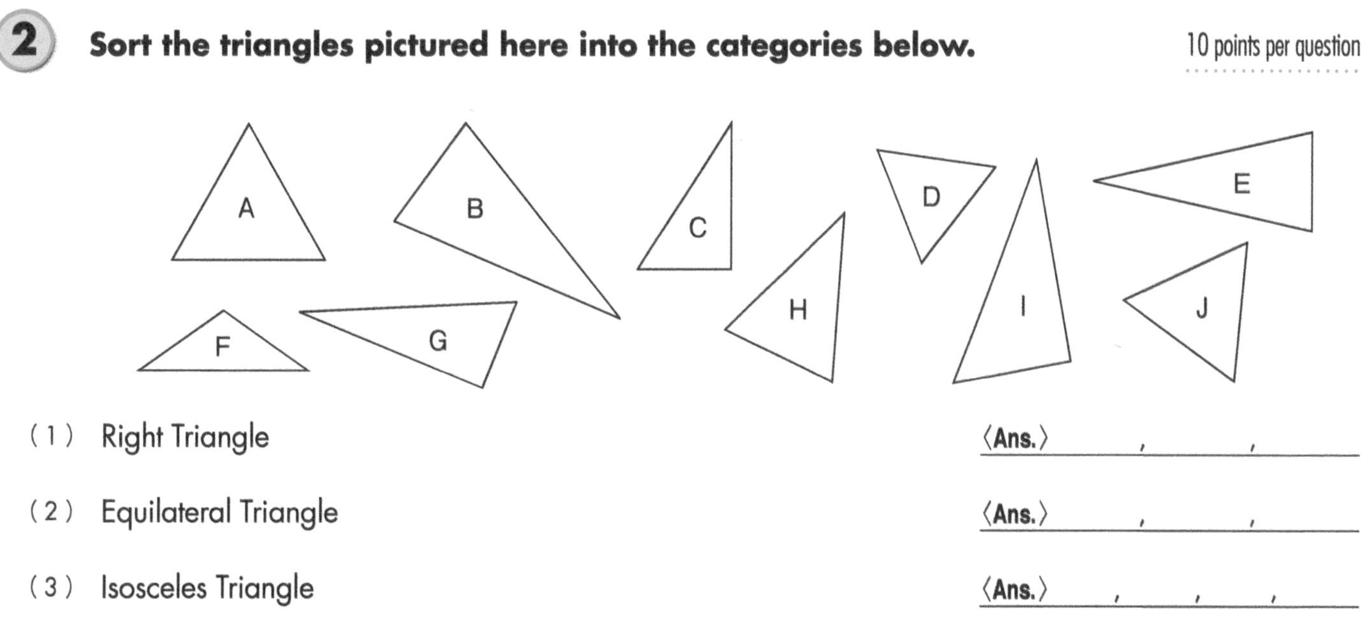

(1) Right Triangle 〈Ans.〉 , ,

(2) Equilateral Triangle 〈Ans.〉 , ,

(3) Isosceles Triangle 〈Ans.〉 , , ,

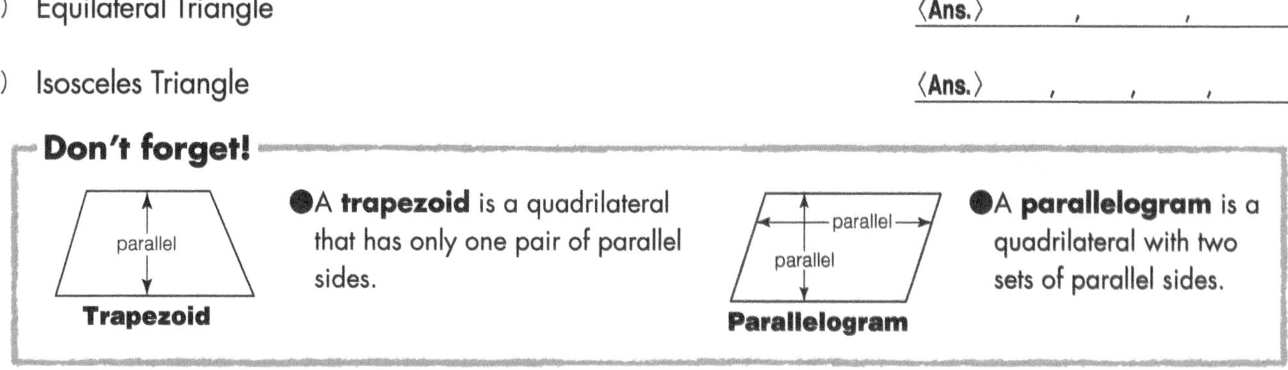

●A **trapezoid** is a quadrilateral that has only one pair of parallel sides.

●A **parallelogram** is a quadrilateral with two sets of parallel sides.

3 **Sort the quadrilaterals pictured here into the categories below.** 15 points per question

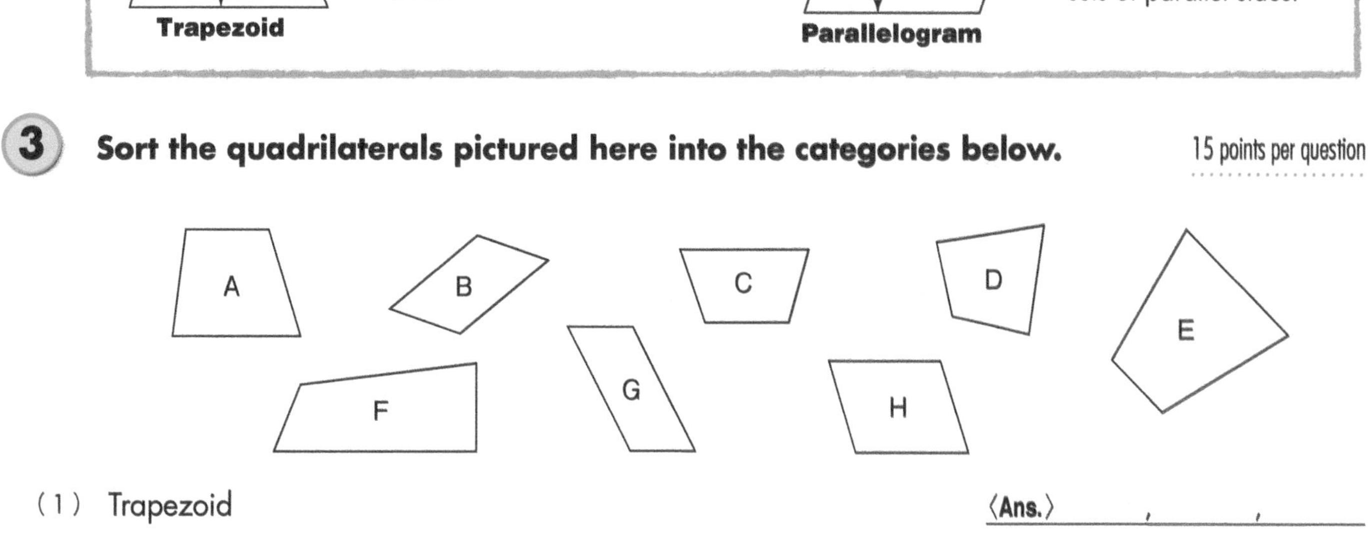

(1) Trapezoid 〈Ans.〉 , ,

(2) Parallelogram 〈Ans.〉 , ,

2 Quadrilaterals Review

Level ☆

Date / /   Name

Score /100

■ The Answer Key is on page 88.

Don't forget!

● A quadrilateral with four right angles is a **rectangle**.

● A **rhombus** is a parallelogram with four congruent sides.

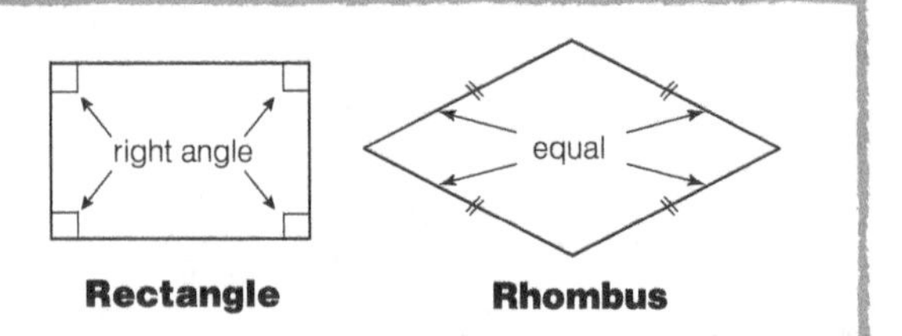

1 **Sort the quadrilaterals pictured here into the categories below.** 8 points per question

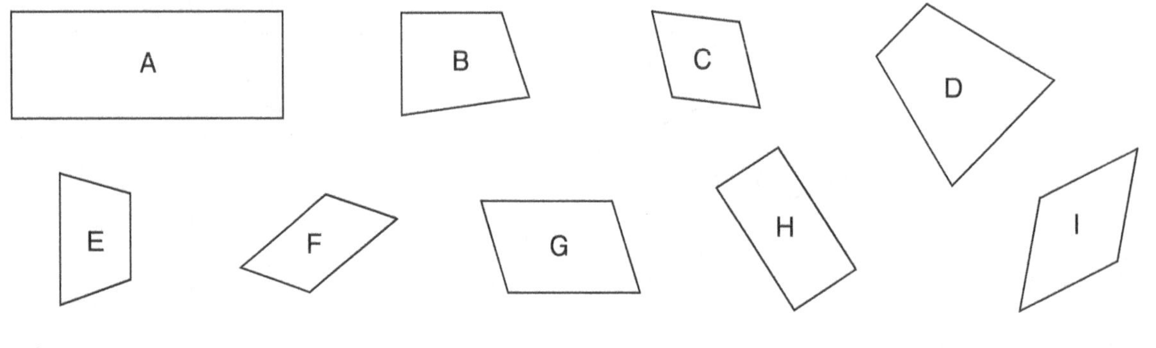

(1) Rectangle 〈Ans.〉 ,

(2) Rhombus 〈Ans.〉 ,

Don't forget!

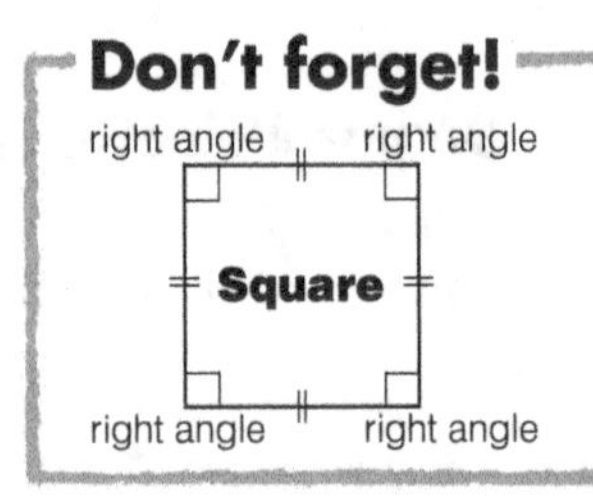

● A quadrilateral with four equal sides (also known as four congruent sides), and four right angles is a **square**.

2 **Choose the squares from the parallelograms below.** 8 points for completion

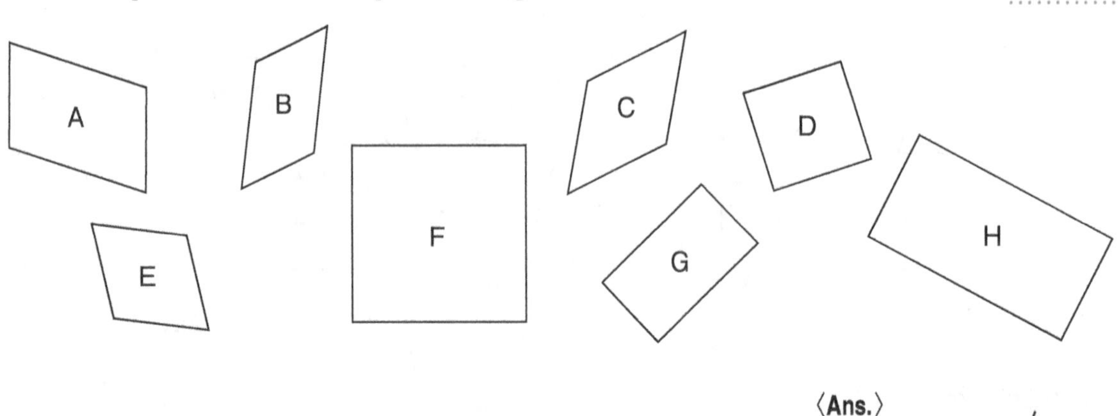

〈Ans.〉 ,

3 Use the shapes pictured here to answer the questions below.

9 points per question

A〔Square〕 B〔Rectangle〕 C〔Trapezoid〕 D〔Parallelogram〕 E〔Rhombus〕

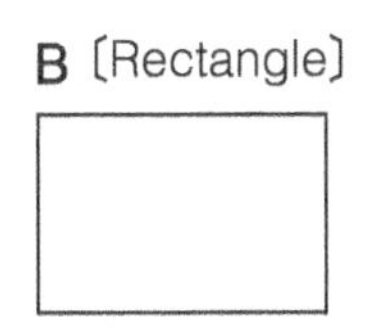
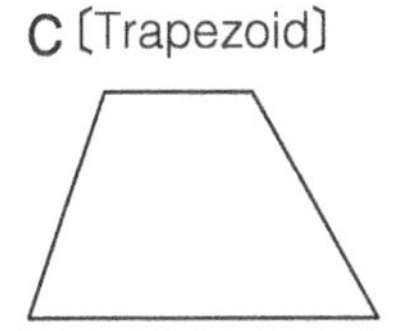
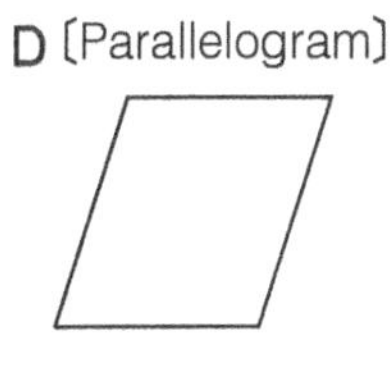
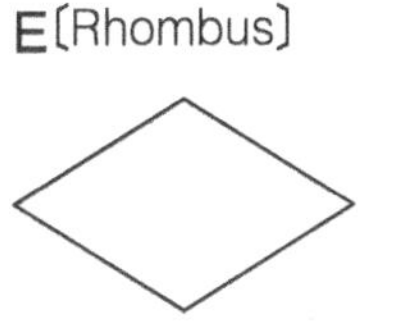

(1) Which quadrilaterals have pairs of congruent opposite sides? 〈Ans.〉 ___ , ___ , ___ , ___

(2) Which quadrilaterals have four congruent angles? 〈Ans.〉 ___ , ___

(3) Which quadrilaterals have pairs of parallel opposite sides? 〈Ans.〉 ___ , ___ , ___ , ___

(4) Which quadrilaterals have four congruent sides? 〈Ans.〉 ___ , ___

4 Write the name of the quadrilateral using the description below.

8 points per question

(1) A quadrilateral that has only one pair of parallel sides. 〈Ans.〉 ______

(2) A quadrilateral with two sets of parallel sides. 〈Ans.〉 ______

(3) A quadrilateral with four right angles. 〈Ans.〉 ______

(4) A quadrilateral with four equal sides. 〈Ans.〉 ______

(5) A quadrilateral with four equal sides and four right angles. 〈Ans.〉 ______

Classifying Quadrilaterals

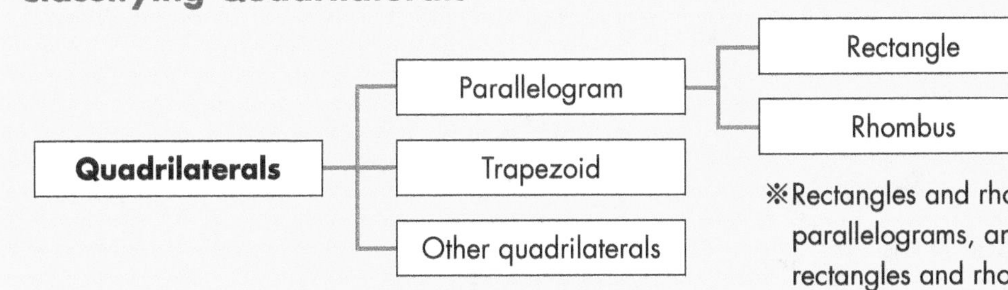

※Rectangles and rhombuses are special types of parallelograms, and, squares is special types of rectangles and rhombuses.

3 Polygons Review

Level ☆

Date / / Name

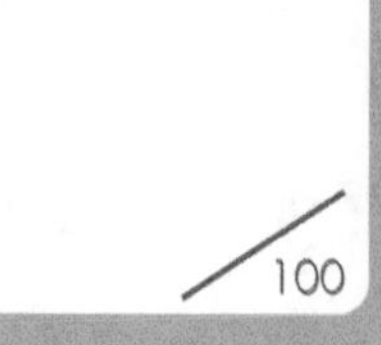

■ The Answer Key is on page 88.

Don't forget!

● A closed figure made of three or more sides that are straight lines is called a **polygon**. A polygon consists of the same number of vertices, angles, and sides. For example, a pentagon consists of five vertices, five angles, and five sides, and so on. A polygon in which all the sides are equal and all the angles are equal is called a **regular polygon**. The polygons on the right are regular polygons.

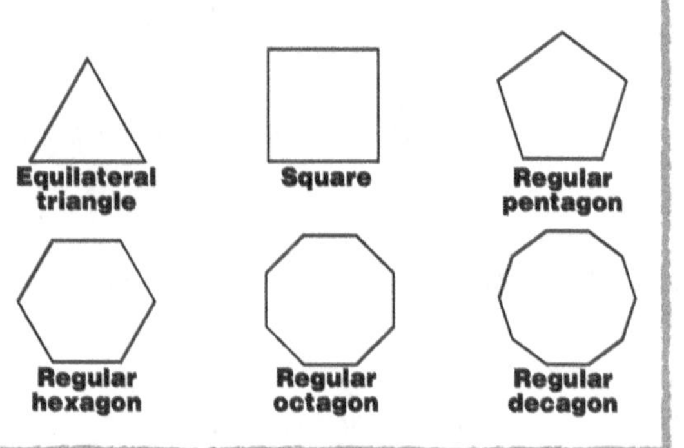

1 Write the name of each regular polygon below.

5 points per question

(1) 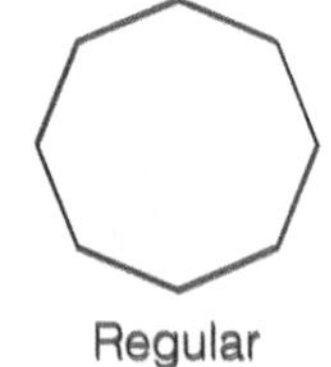 Regular

〈Ans.〉

(2) 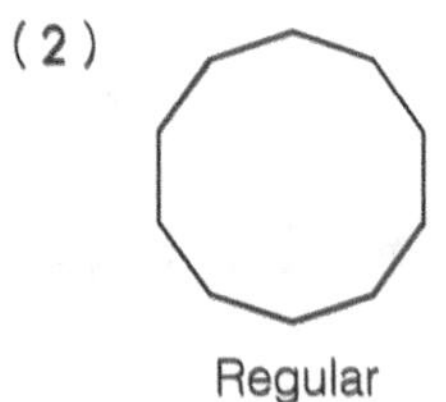 Regular

〈Ans.〉

(3) 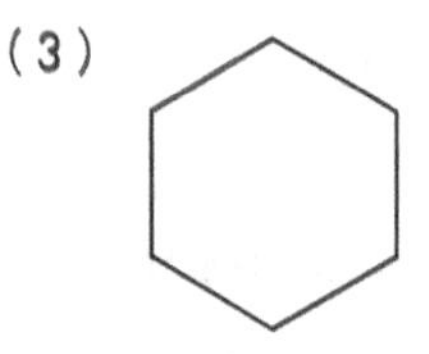 Regular

〈Ans.〉

(4) 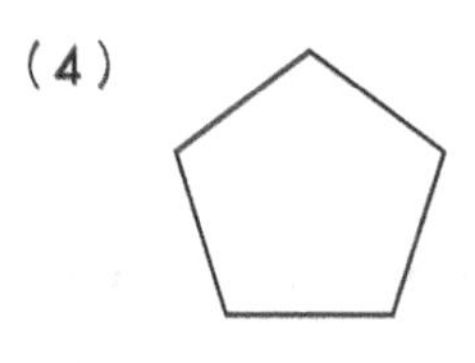 Regular

〈Ans.〉

(5)  Equilateral

〈Ans.〉

2 Write ○ under the regular polygons and × under the irregular polygons.

5 points per question

(1) 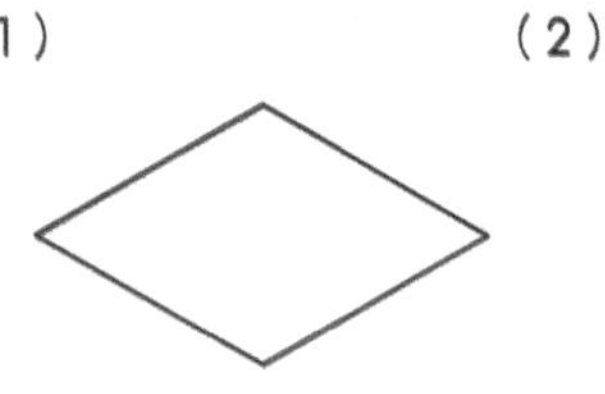

〈Ans.〉

(2) 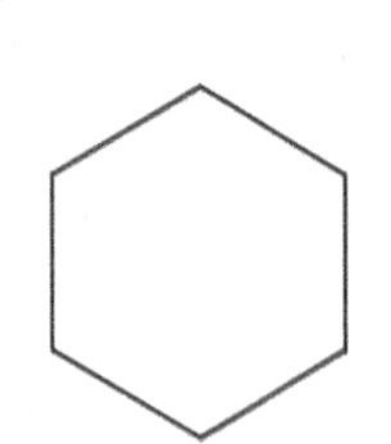

〈Ans.〉

(3) 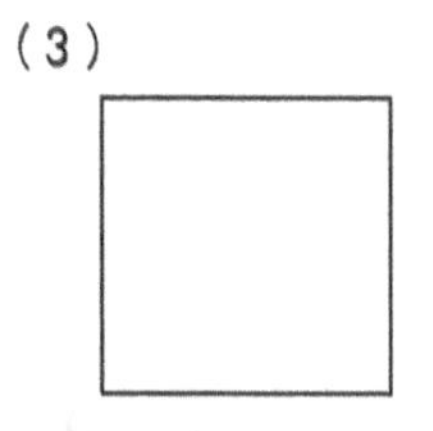

〈Ans.〉

(4) 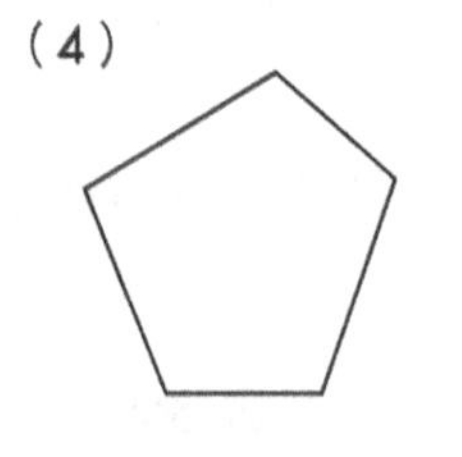

〈Ans.〉

(5) 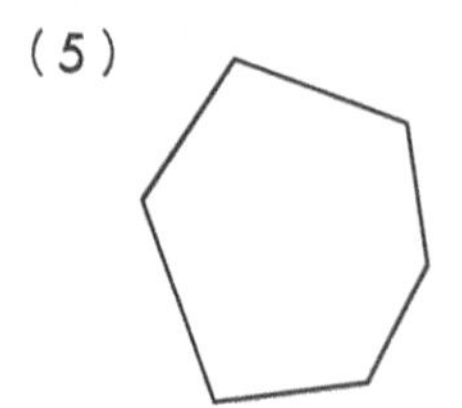

〈Ans.〉

(6) 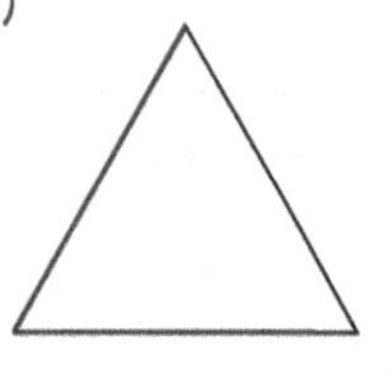

〈Ans.〉

(7) 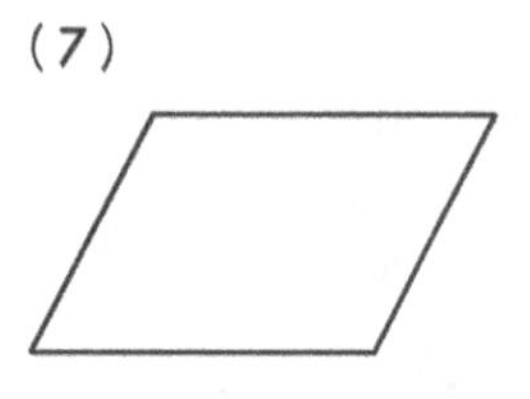

〈Ans.〉

(8) 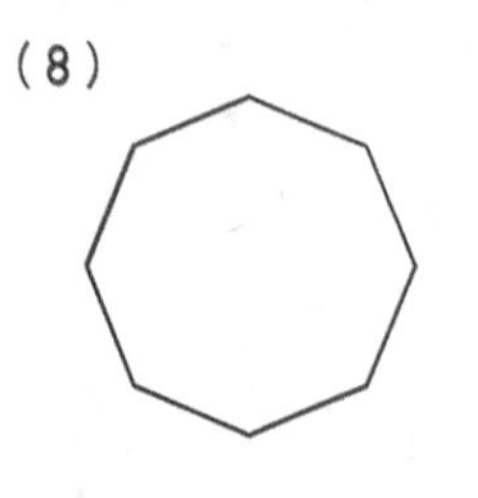

〈Ans.〉

(9) 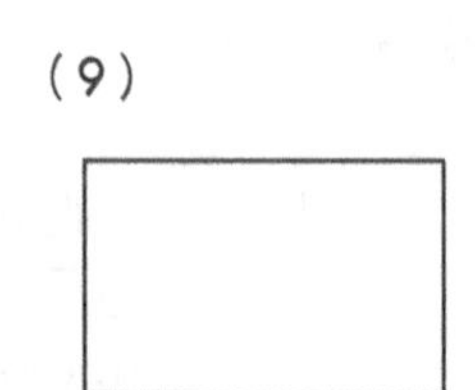

〈Ans.〉

(10) 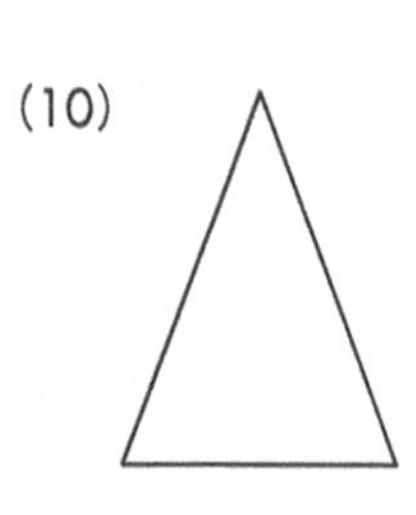

〈Ans.〉

3 **Each polygon below is regular and is created by dividing a circle into equal parts. Find the measure of angle A for each polygon.**

5 points per question

(1)

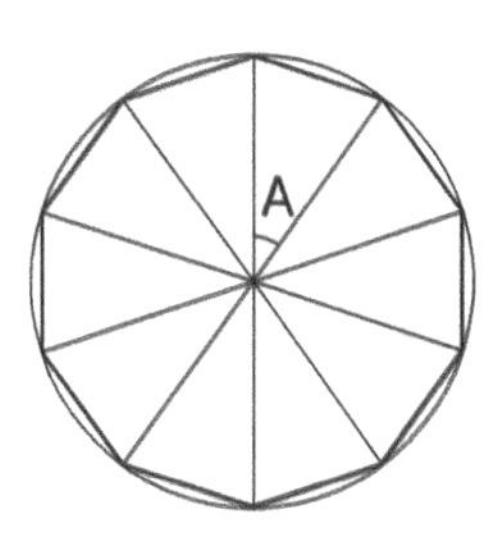

〈Ans.〉

(2)

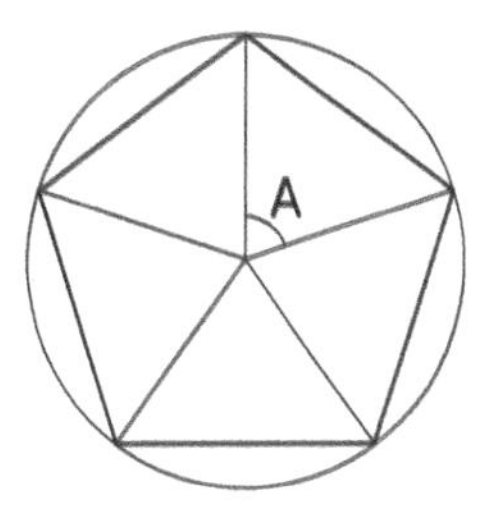

〈Ans.〉

4 **Use the example below to draw each regular polygon.**

5 points per question

Example

〔Square〕

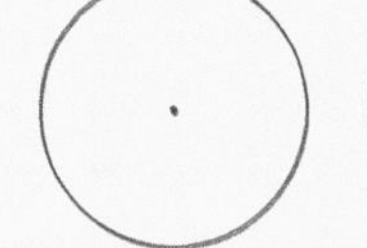

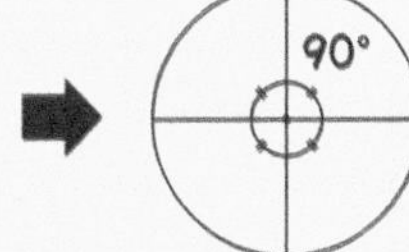

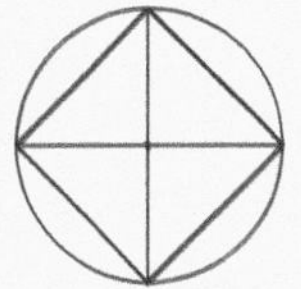

Draw a circle. Draw lines through the center point to divide the circle into 90° sections. Connect the ends of lines using straight lines.

(1) 〔Equilateral triangle〕

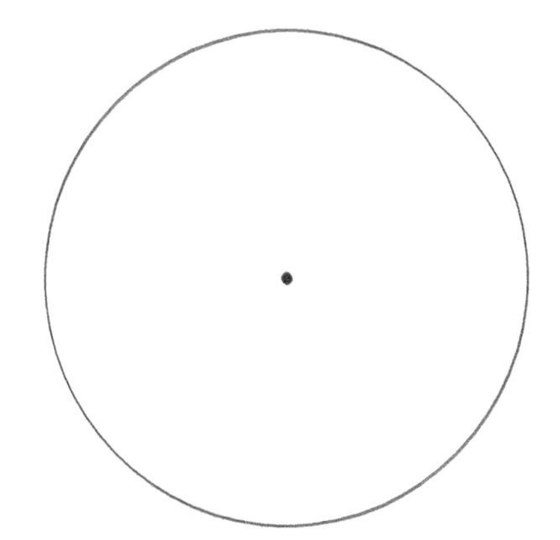

(2) 〔Regular hexagon〕

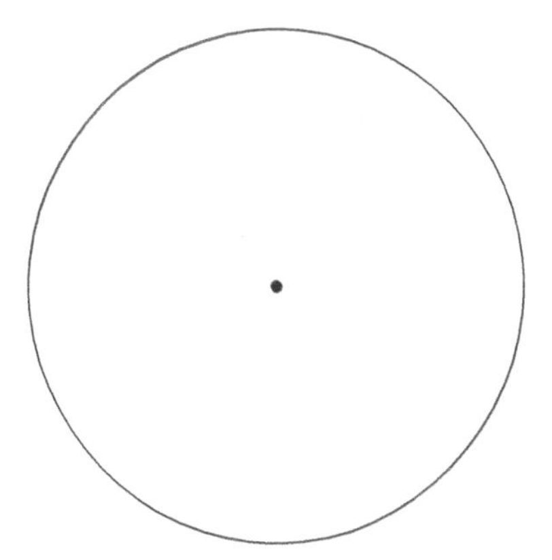

(3) 〔Regular octagon〕

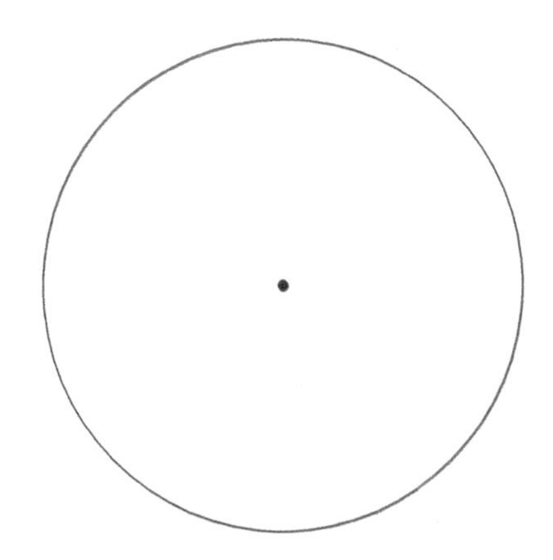

Angles Review 1

Level ☆ 

Date / / Name

Score 

■ The Answer Key is on page 88.

Don't forget!

Acute angle < 90°

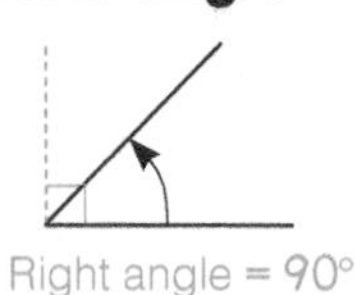

Obtuse angle > 90° < 180°

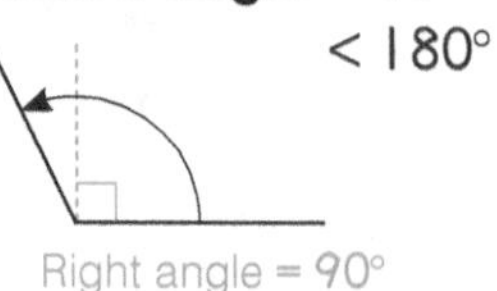

Straight angle = 180°

Round angle = 360°

360°

1 **There are two kinds of triangle rulers. Answer the measure of each angle. (A protractor may be used.)** 10 points for completion

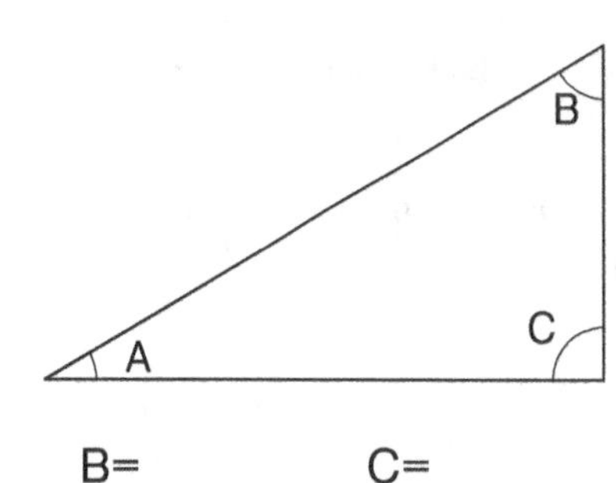

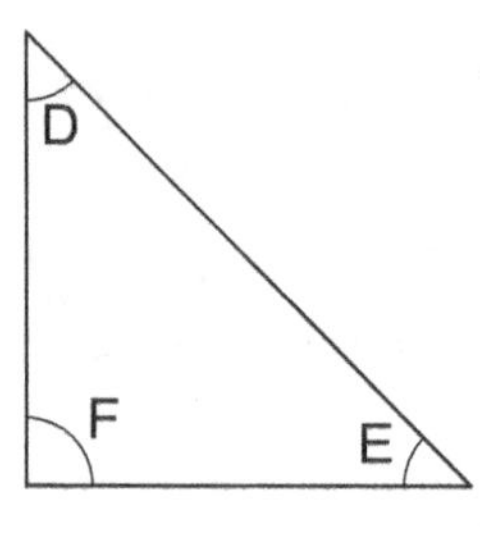

⟨Ans.⟩ A= B= C= D= E= F=

2 **Find the measure of angle A in each illustration below.** 6 points per question

(1)

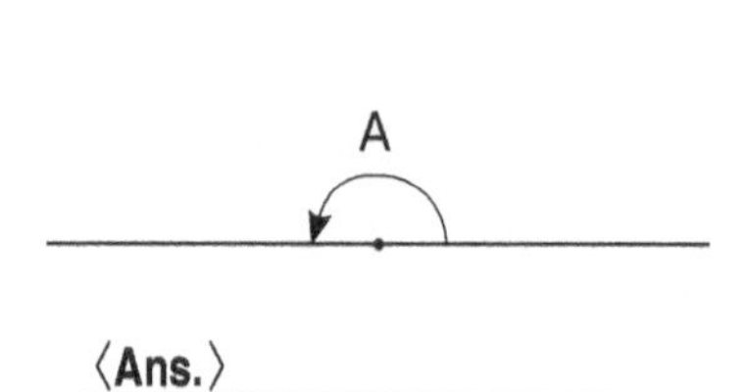

⟨Ans.⟩

(2)

A

⟨Ans.⟩

(3)

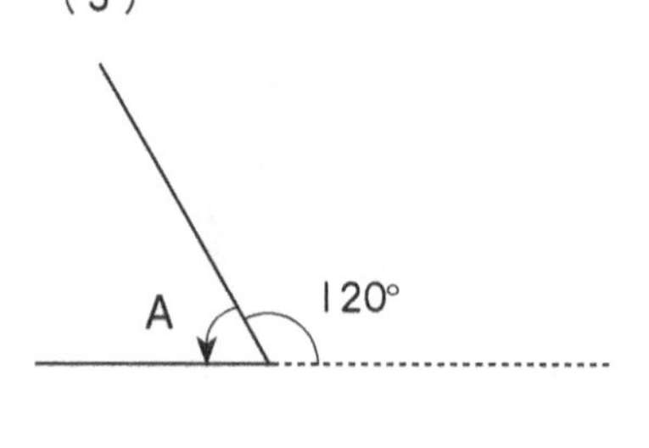

⟨Ans.⟩

(4)

A 40°

⟨Ans.⟩

(5)

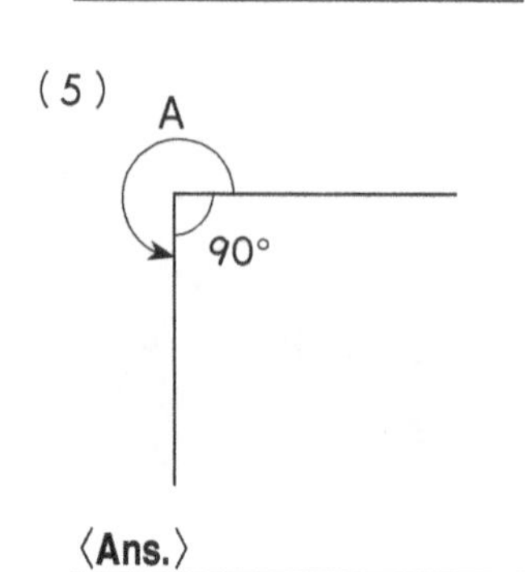

⟨Ans.⟩

(6)

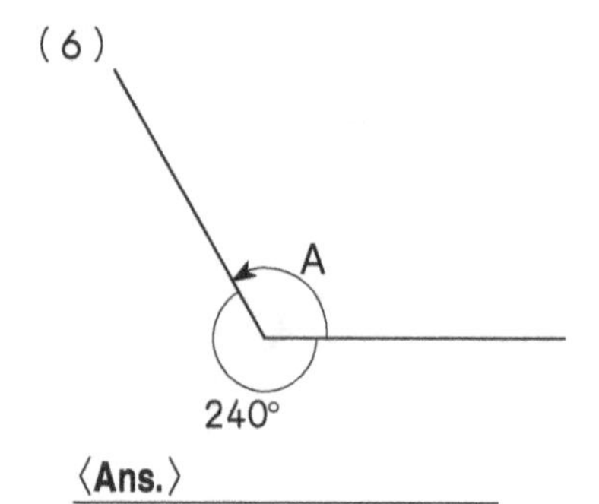

⟨Ans.⟩

Don't forget!

- The opposite angle at the point where two straight lines intersect is called a **vertical angle**.
- Vertical angles are equal.

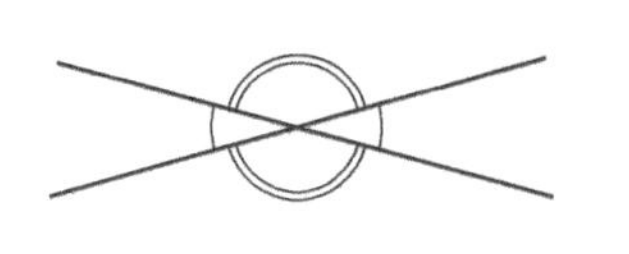

3 Answer the questions below.

6 points per question

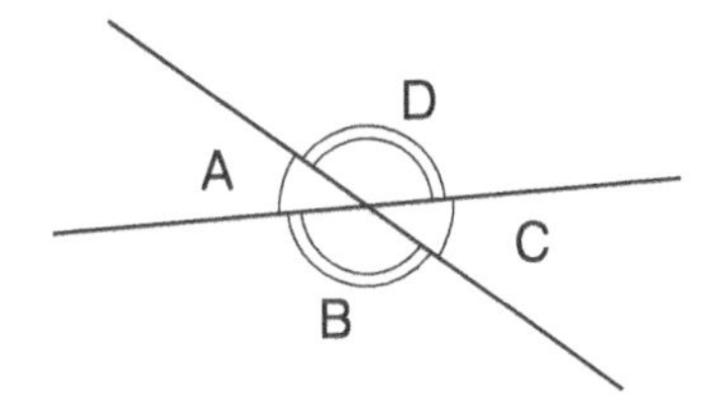

(1) Which angle is equal to angle A? 〈Ans.〉

(2) Which angle is equal to angle B? 〈Ans.〉

(3) What is the sum of angle B and C? 〈Ans.〉

4 Find the measure of angle A and B in each illustration below.

6 points per question

(1)

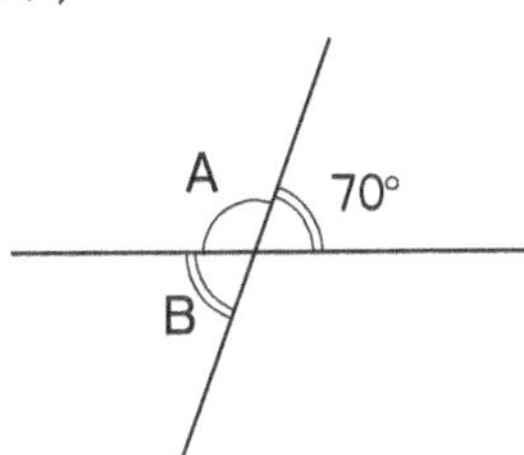

〈Ans.〉 A= B=

(2)

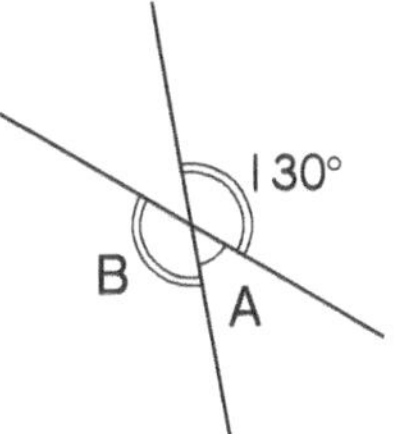

〈Ans.〉 A= B=

(3)

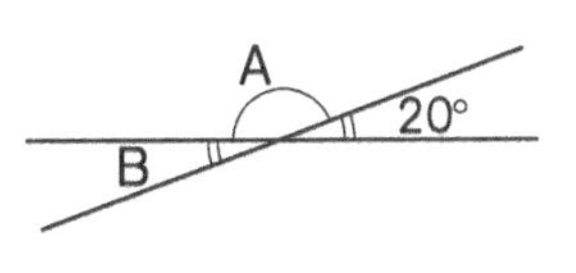

〈Ans.〉 A= B=

5 Find the measure of angle A, B, and C in each illustration below.

6 points per question

(1)

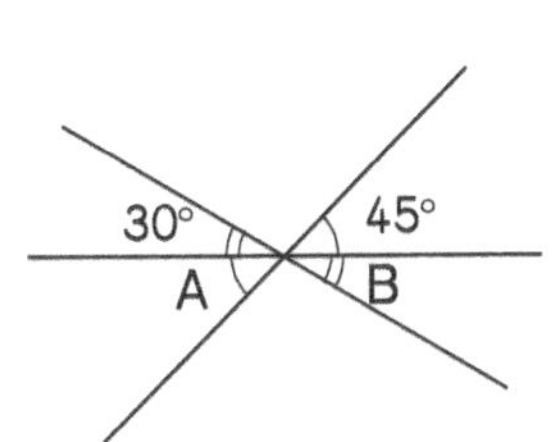

〈Ans.〉 A= B=

(2)

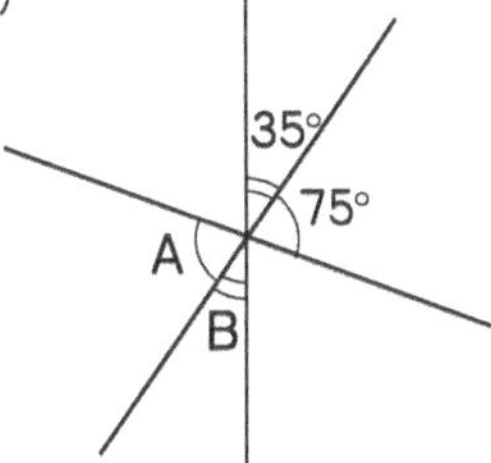

〈Ans.〉 A= B=

(3)

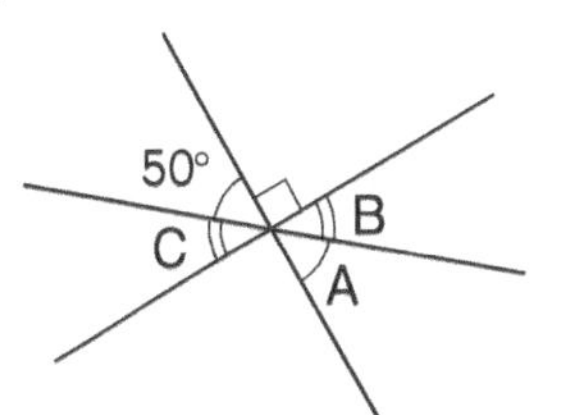

〈Ans.〉 A= B= C=

Angles Review 2

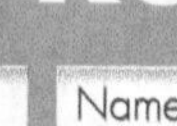

Date / /

Name

Level 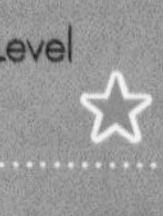

Score 

/100

■ The Answer Key is on page 88.

Don't forget!

●The sum of the interior angles in a triangle is 180°.

Angle A + Angle B + Angle C = 180°

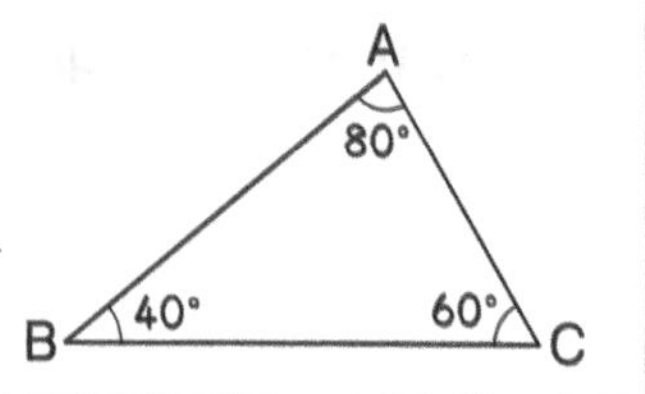

1 Find the measure of angle A in each triangle below.

5 points per question

(1)

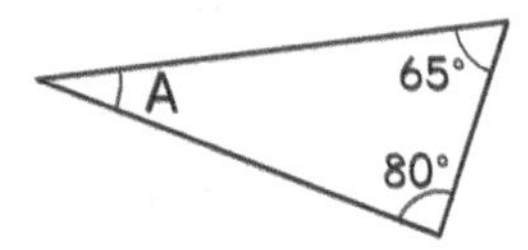

〈Ans.〉

(2)

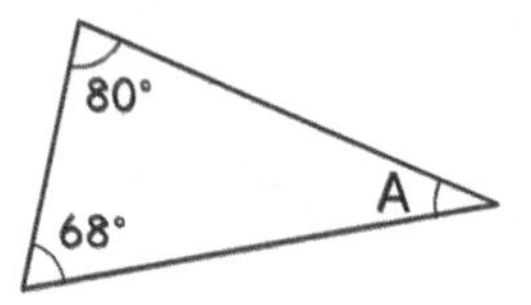

〈Ans.〉

(3)

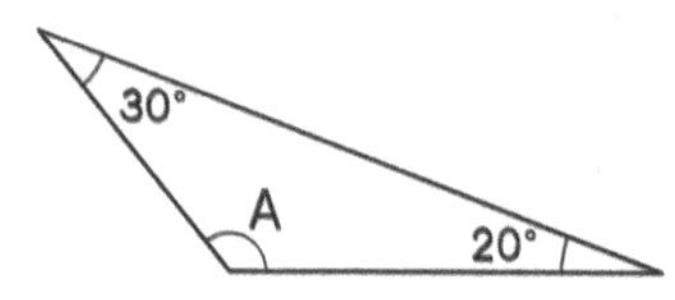

〈Ans.〉

(4)

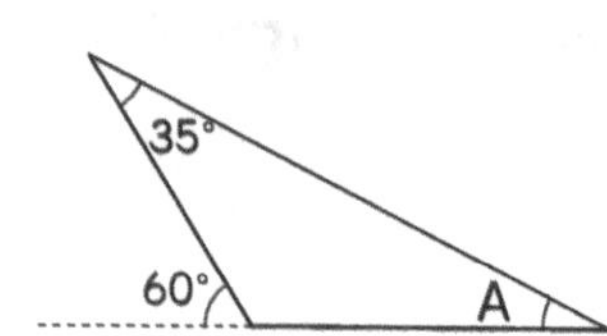

〈Ans.〉

(5)

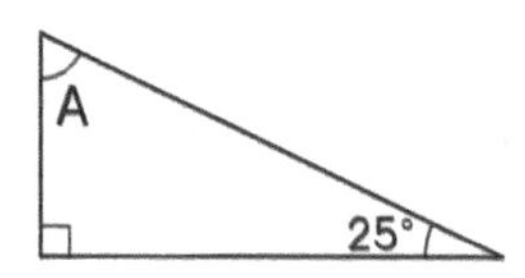

〈Ans.〉

(6)

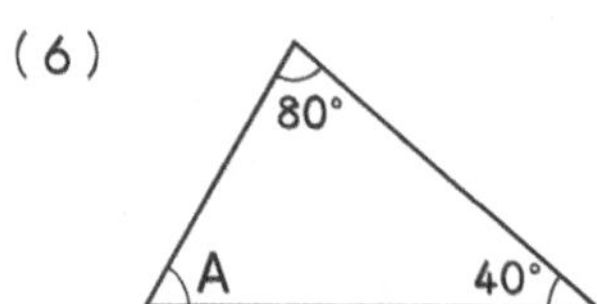

〈Ans.〉

(7)

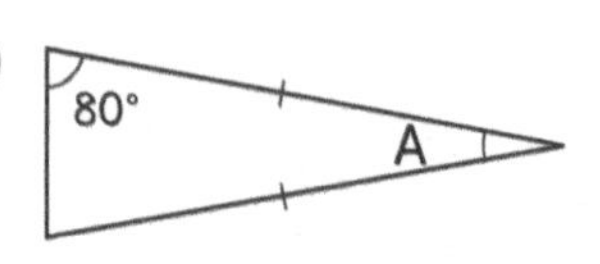

〈Ans.〉

(8)

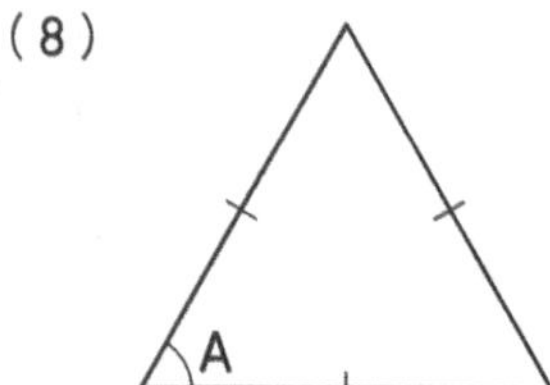

〈Ans.〉

2 Find the measure of angle A in each triangle below.

5 points per question

(1)

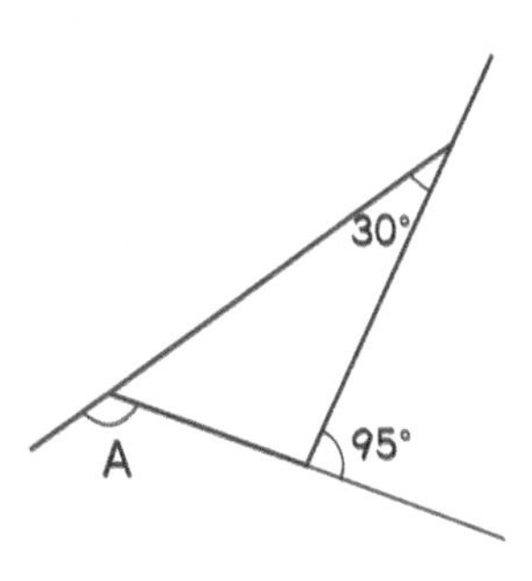

⟨Ans.⟩ ______

(2)

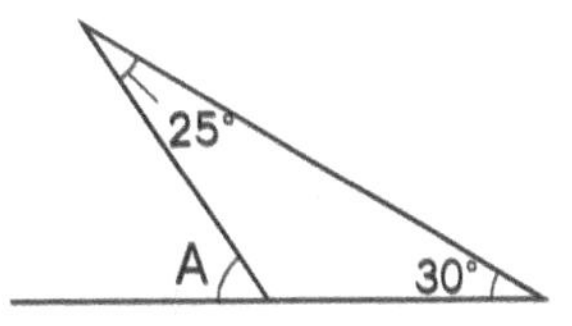

(3)

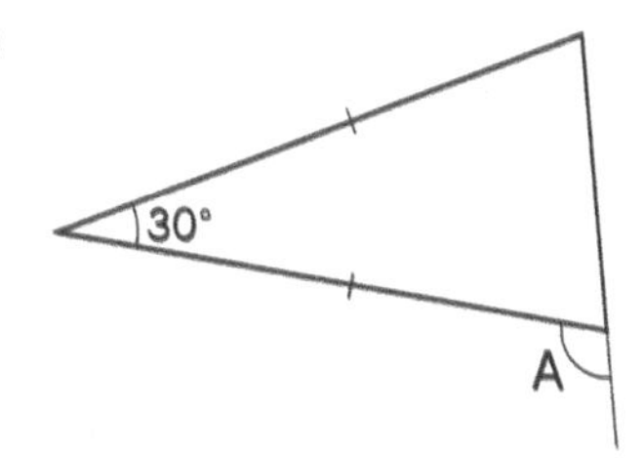

⟨Ans.⟩ ______

(4)

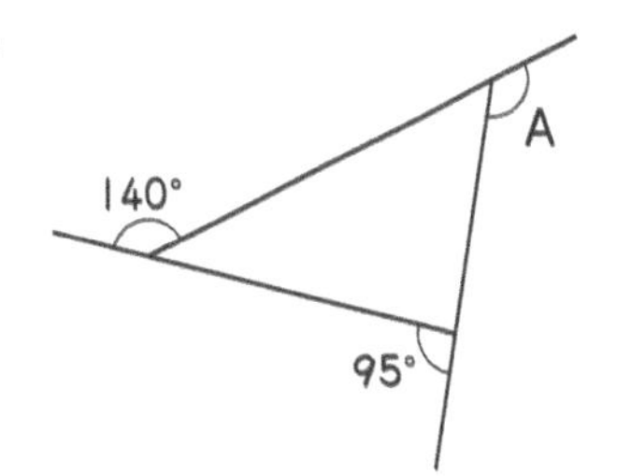

⟨Ans.⟩ ______

3 Find the measure of angle A and B in each drawing below.

10 points per question

(1)

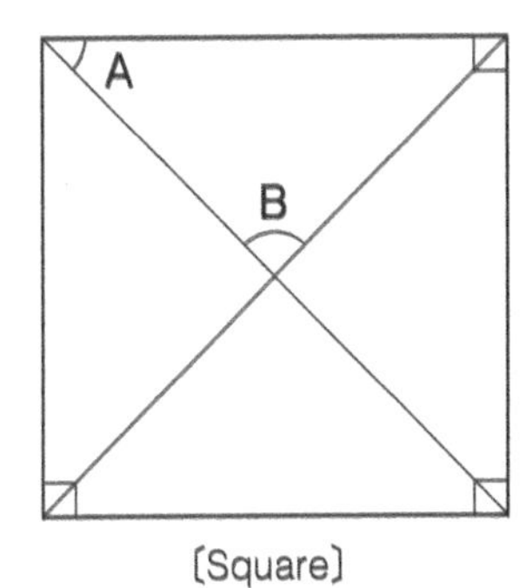

[Square]

⟨Ans.⟩ A ______

B ______

(2)

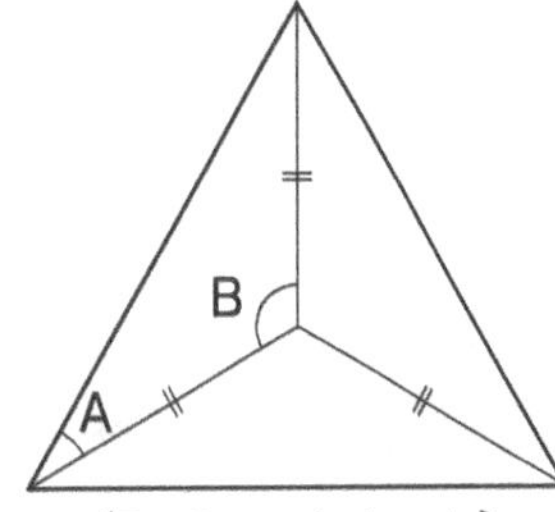

[Equilateral triangle]

⟨Ans.⟩ A ______

B ______

(3)

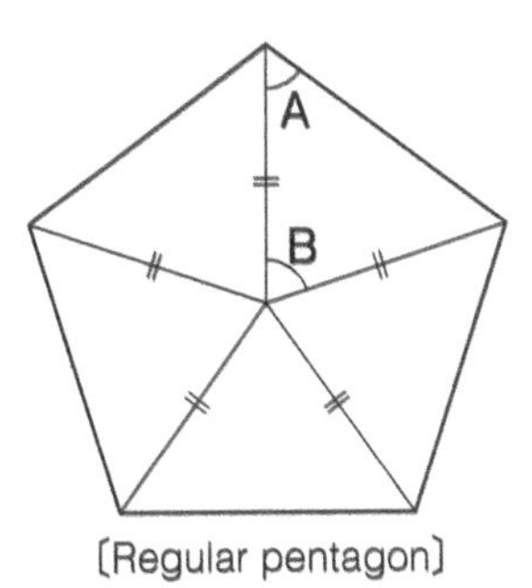

[Regular pentagon]

⟨Ans.⟩ A ______

B ______

(4)

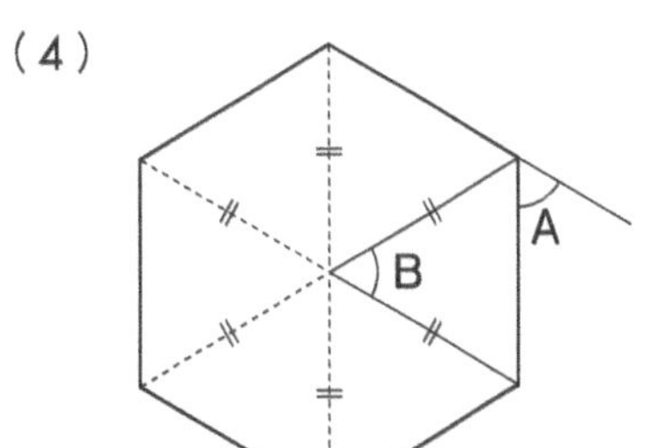

[Regular hexagon]

⟨Ans.⟩ A ______

B ______

6 Angles Review 3

Level 

Score 

Date / /

Name

100

■ The Answer Key is on page 88.

Don't forget!

●When a transversal intersects two lines, the angles at positions shown in the figure are **corresponding angles**.

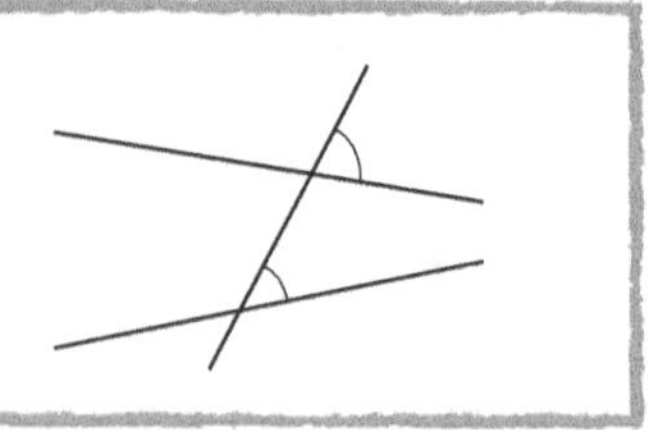

1 **Find the corresponding angle of the following angles shown in the figure to the right.**

6 points per question

(1) What is the corresponding angle of angle a? 〈Ans.〉

(2) What is the corresponding angle of angle b? 〈Ans.〉

(3) What is the corresponding angle of angle c? 〈Ans.〉

(4) What is the corresponding angle of angle d? 〈Ans.〉

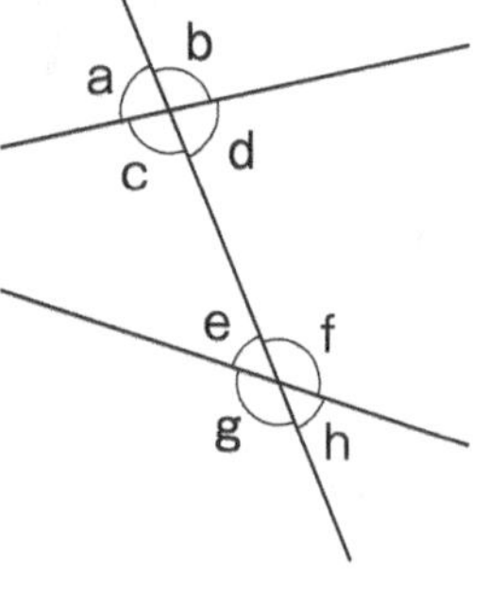

Don't forget!

●When the transversal intersects two lines, the angles at positions shown in the figure are **alternate angles**.

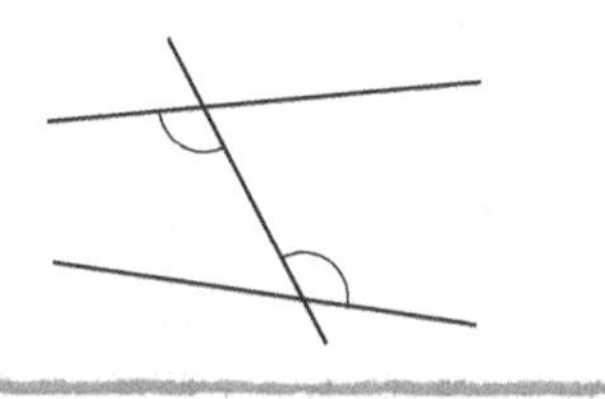

2 **Find the alternate angle of the following angles in the figure on the right.**

6 points per question

(1) What are the alternate angles of angle b? 〈Ans.〉 ,

(2) What are the alternate angles of angle c? 〈Ans.〉 ,

(3) What is the alternate angle of angle f? 〈Ans.〉

(4) What is the alternate angle of angle g? 〈Ans.〉

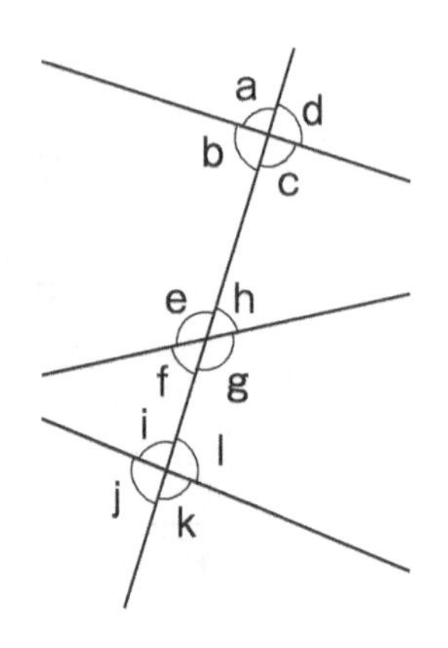

Don't forget!

- If two lines are perpendicular to the same straight line, they are **parallel**.
- Parallel lines are straight lines that do not intersect.
- When the transversal intersects two parallel lines, the corresponding angles are equal. Similarly, the alternate angles are equal.

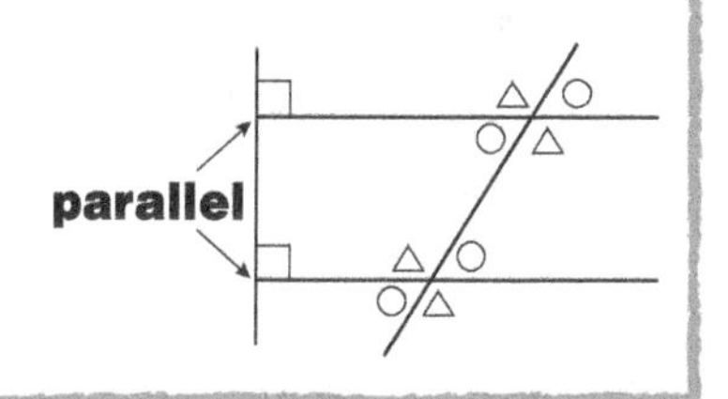

3 **Line A and line B are parallel. Find the following angles using the figure on the right.** 5 points per question

(1) Which angles are equal to angle a? 〈Ans.〉 , ,

(2) Which angles are equal to angle b? 〈Ans.〉 , ,

(3) Which angles are equal to angle c? 〈Ans.〉 , ,

(4) Which angles are equal to angle d? 〈Ans.〉 , ,

4 **Line A and line B are parallel. Find the following angles using the figure on the right.** 5 points per question

(1) What is the measure of angle a? 〈Ans.〉

(2) What is the measure of angle b? 〈Ans.〉

(3) What is the measure of angle c? 〈Ans.〉

(4) What is the measure of angle d? 〈Ans.〉

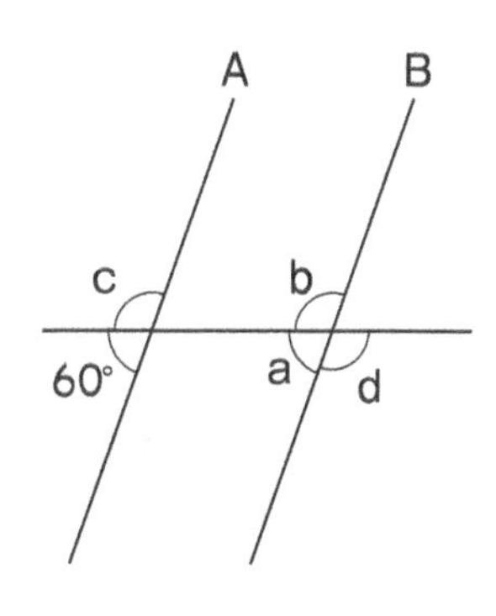

5 **Line A and line B are parallel and line C and line D are parallel. Find the following angles using the figure on the right.** 4 points per question

(1) What is the measure of angle a? 〈Ans.〉

(2) What is the measure of angle b? 〈Ans.〉

(3) What is the measure of angle c? 〈Ans.〉

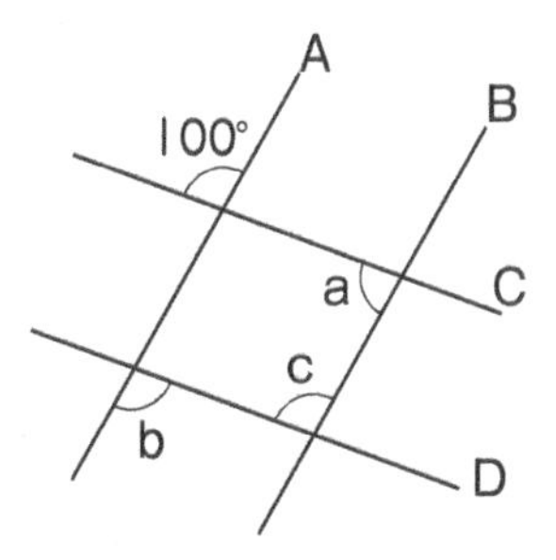

7 Drawing Lines Review

Level ☆

Date / /

Name

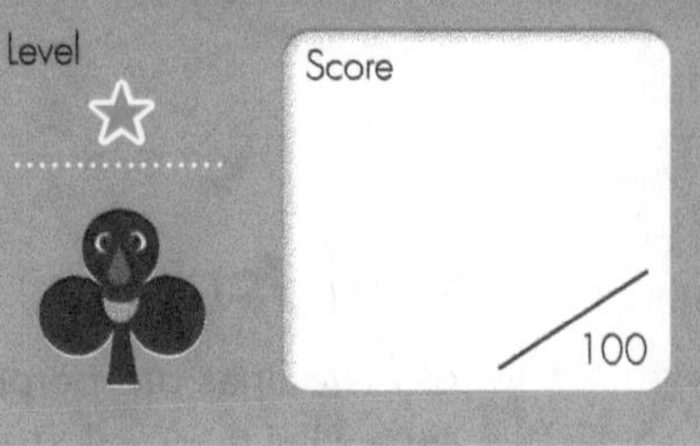

■ The Answer Key is on page 89.

1 **Using the example as a guide, use your triangular rulers to draw a line that is perpendicular to each line A below. The second line should go through point B.**

7 points per question

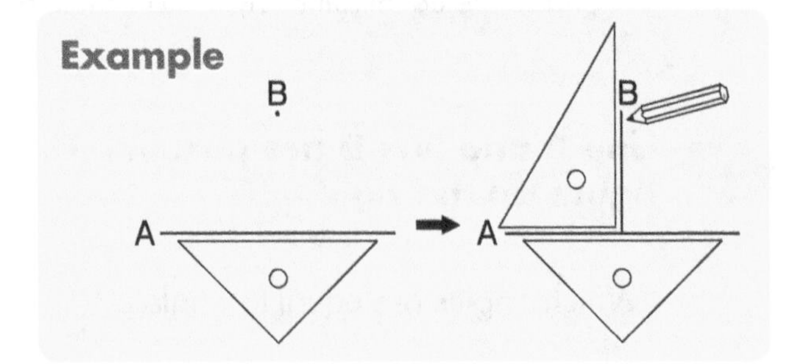

(1)

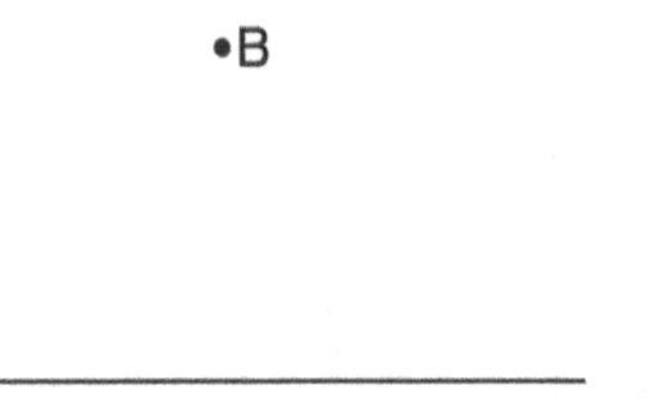

(2)

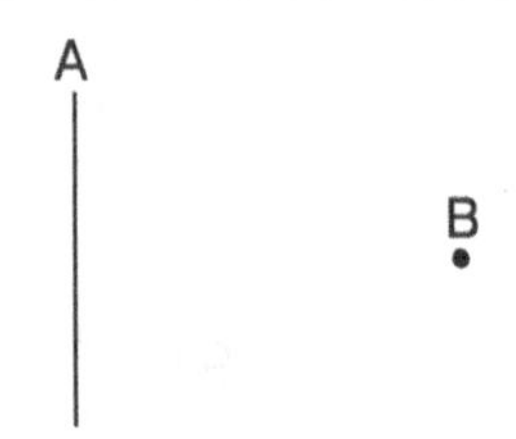

(3)

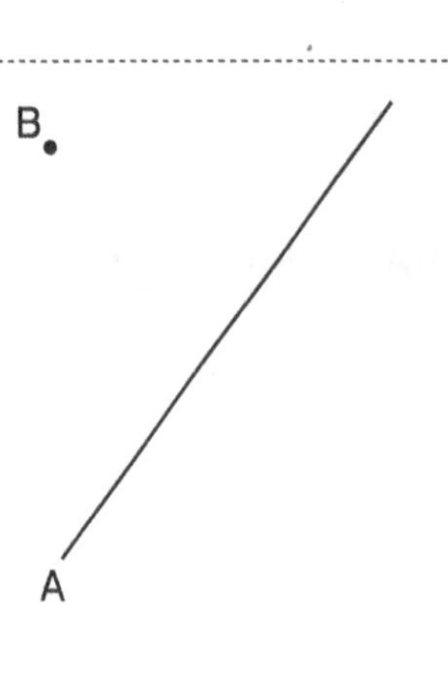

(4)

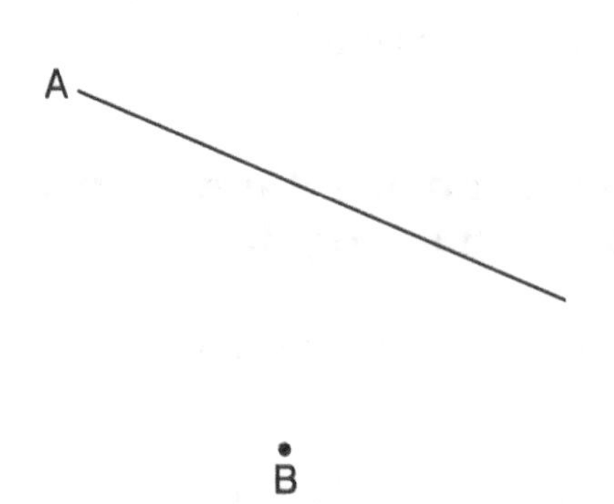

2 **Using the example as a guide, use your triangular rulers to draw a line that is parallel to each line A below. The second line should go through point B.**

8 points per question

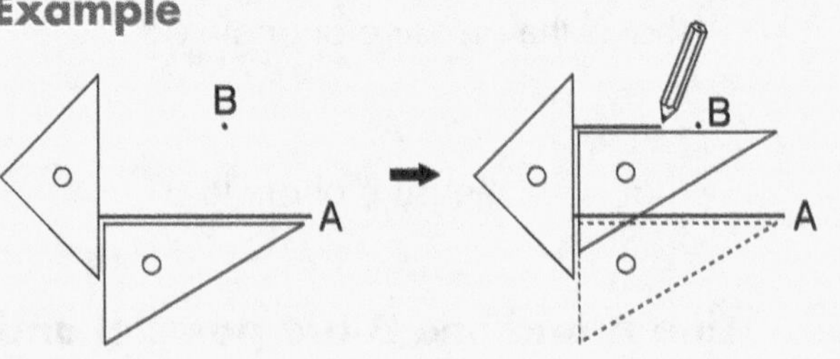

(1)

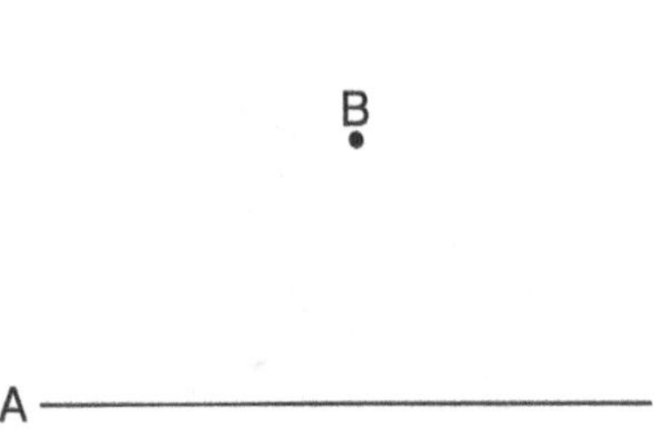

(2)

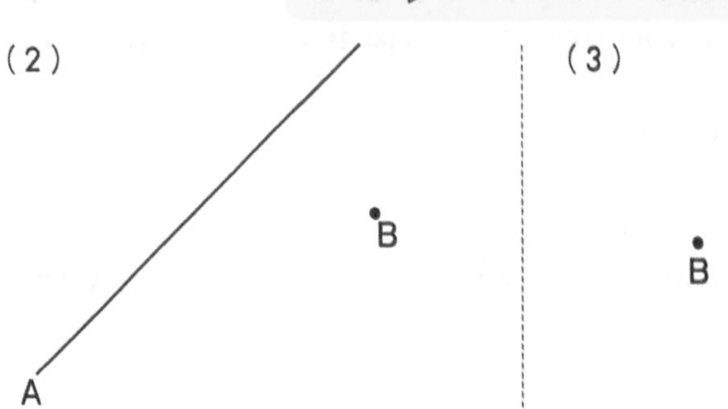

(3)

A

B

3 Draw the figures below.

8 points per question

(1) Draw two lines that are parallel to line A and are each 1.5 cm away from line A.

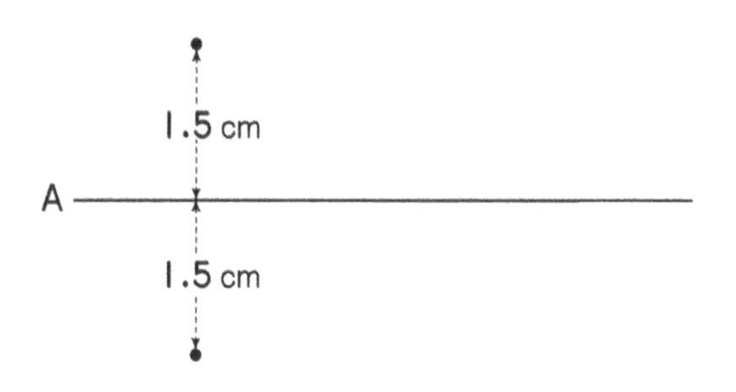

(2) Finish a quadrilateral that has two sets of parallel line segments. Draw a line that is parallel to line A and one that is parallel to line B.

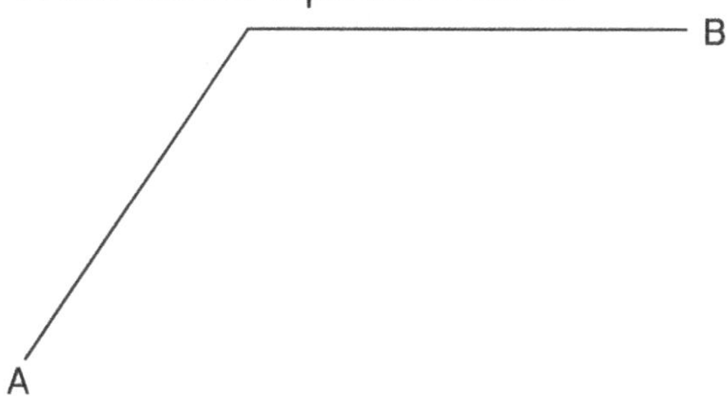

4 Use your triangular rulers to draw the shapes below.

8 points per question

(1) A square with 1 inch sides.

(2) A rectangle with sides of 3 cm and 4 cm.

5 Use your trianglular rulers to draw trapezoids below in the space provided.

8 points per question

(1)

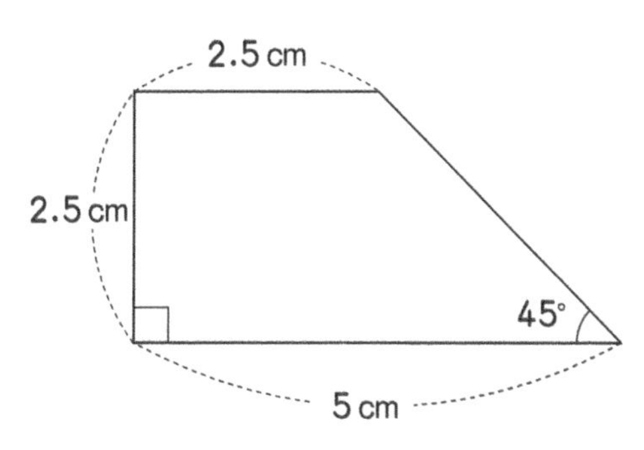

(2)

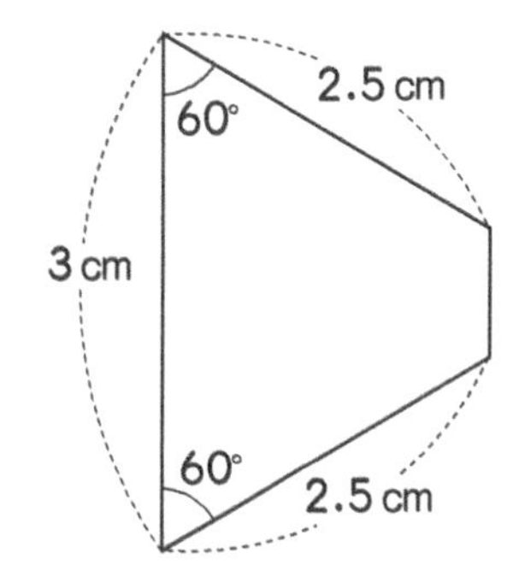

8 Drawing Quadrilaterals Review

Date / / Name

Level ☆ Score /100

■ The Answer Key is on page 89.

1 Draw each parallelogram below. Use your triangular rulers to draw the parallel lines as shown in the example on the right. 7 points per question

Example

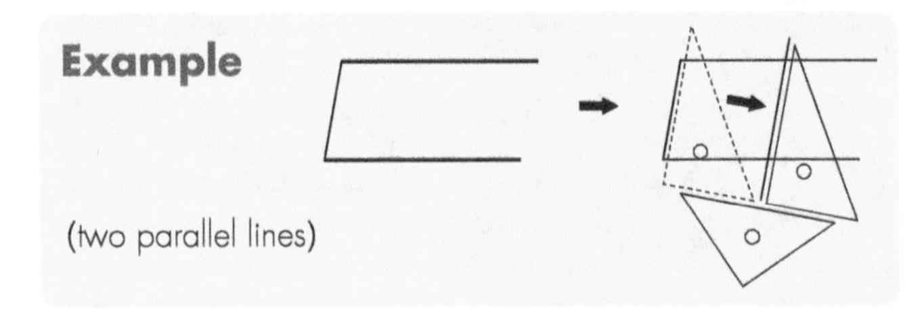

(two parallel lines)

(1) 〔Parallelogram〕

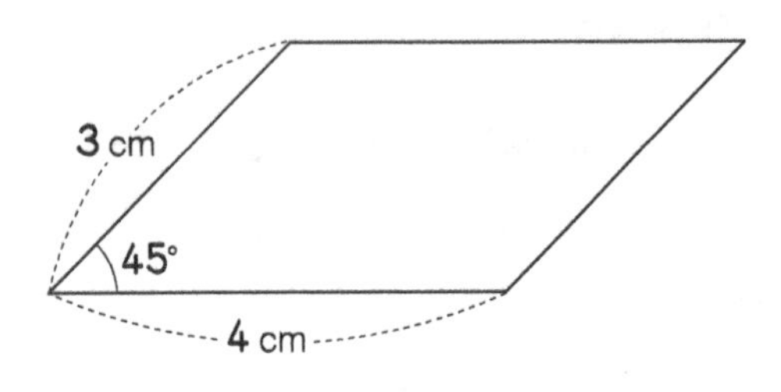

(2) 〔Rhombus〕

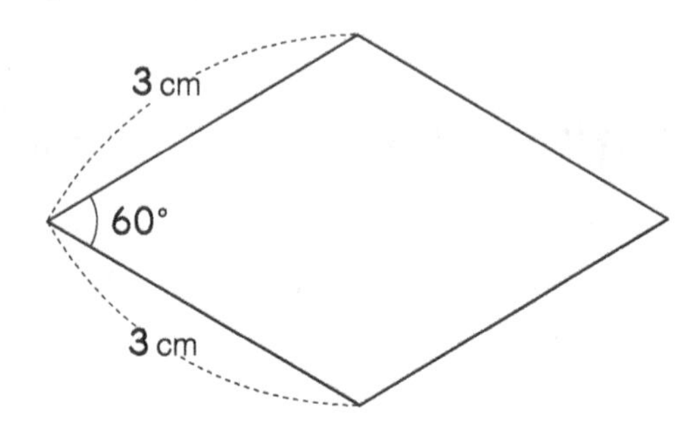

(3) 〔Rectangle〕

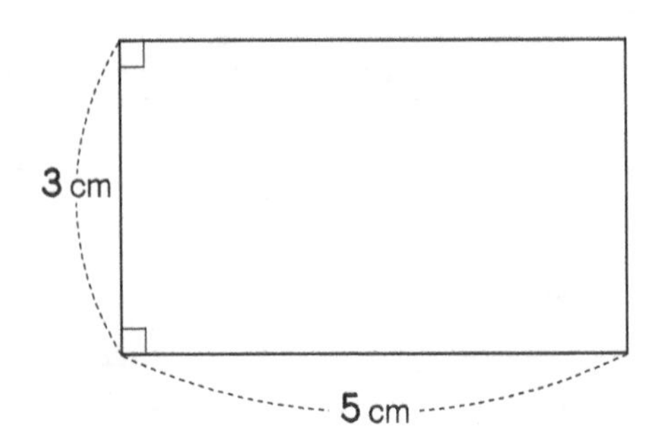

2 **Connect both sets of opposite vertices as shown in the example.** 8 points per question

Example

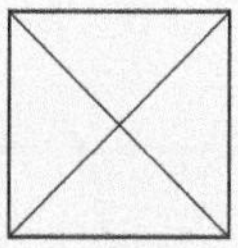

(1) (2)

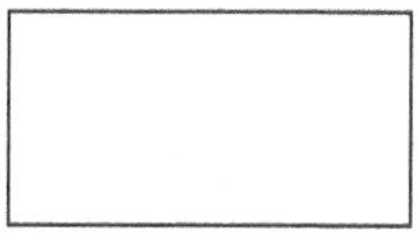

(3) 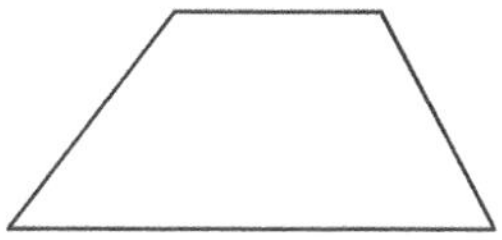 (4) 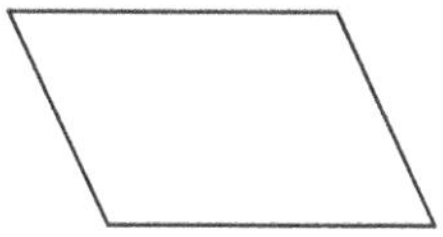 (5)

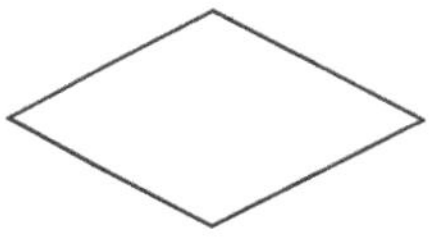

3 **The line segments connecting each pair of opposite vertices in a quadrilateral are called diagonals. Answer the questions below about the diagonals in the shapes pictured here.** 6 points per question

A 〔Square〕 B 〔Rectangle〕 C 〔Trapezoid〕 D 〔Parallelogram〕 E 〔Rhombus〕

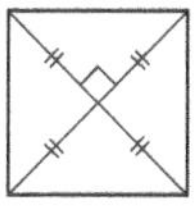

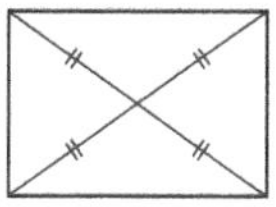

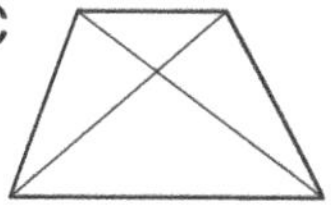

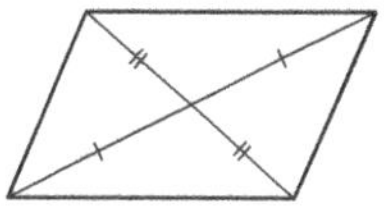

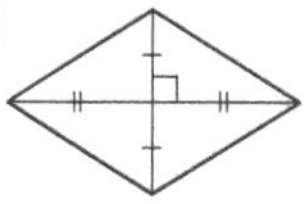

(1) Which quadrilaterals have two congruent diagonals?

〈Ans.〉 ,

(2) Which quadrilaterals have perpendicular diagonals?

〈Ans.〉 ,

(3) Which quadrilaterals have diagonals that intersect at the center of the shape?

〈Ans.〉 , , ,

4 **The diagonals for three different quadrilaterals are shown below. What kind of quadrilaterals would have these diagonals?** 7 points per question

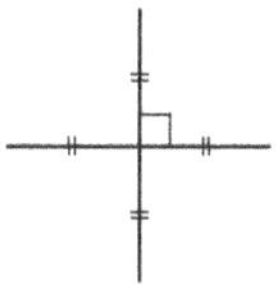

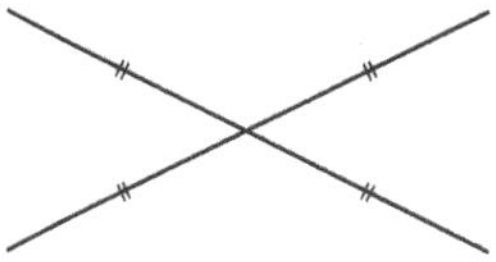

(1) (2) (3)

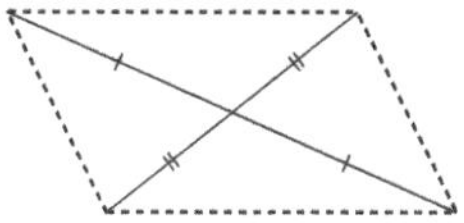

〈Ans.〉

〈Ans.〉

〈Ans.〉

Area Review

Date / /

Name

Level ★

■ The Answer Key is on page 89.

Don't forget!

- The area of a parallelogram is the base times the height.
- The area of a triangle is the base times the height divided by two.
- The base and height must be perpendicular to each other.

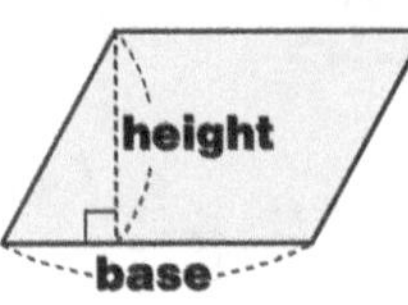

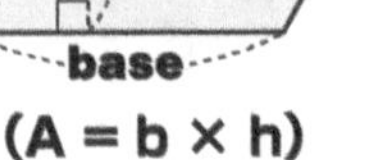

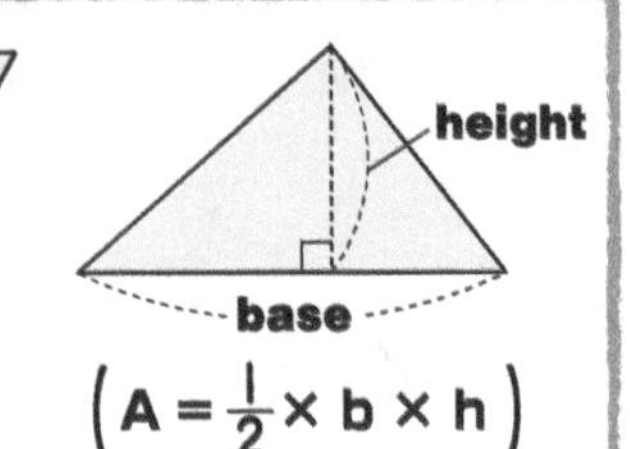

$\left(A = \frac{1}{2} \times b \times h\right)$

1 Find the area of each quadrilateral below.

6 points per question

(1)

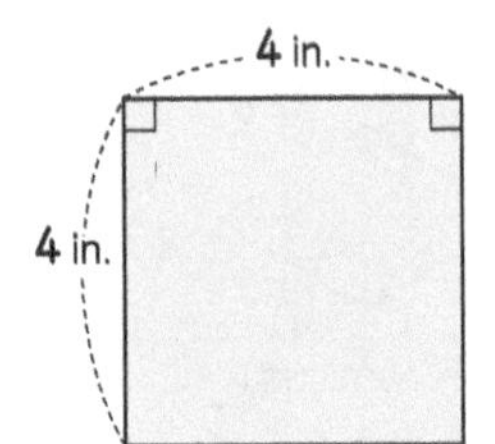

⟨Ans.⟩

(2)

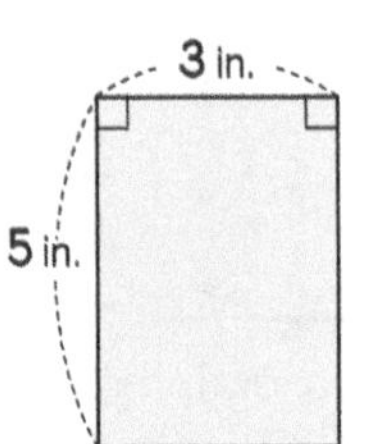

⟨Ans.⟩

(3)

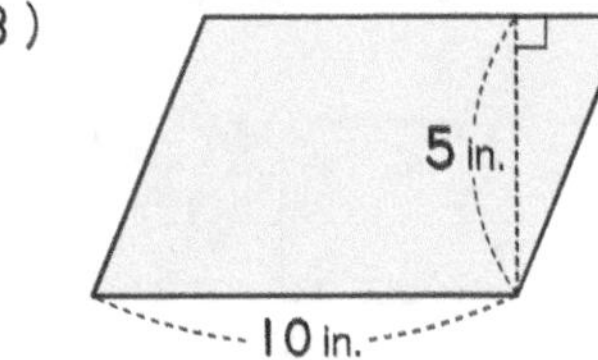

⟨Ans.⟩

(4)

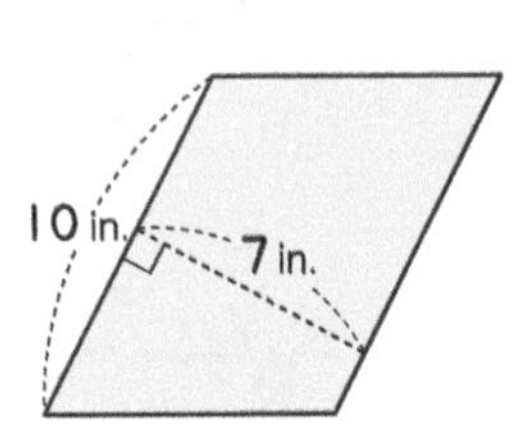

⟨Ans.⟩

(5)

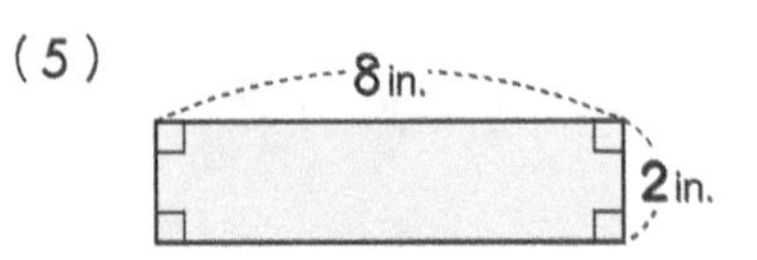

⟨Ans.⟩

(6)

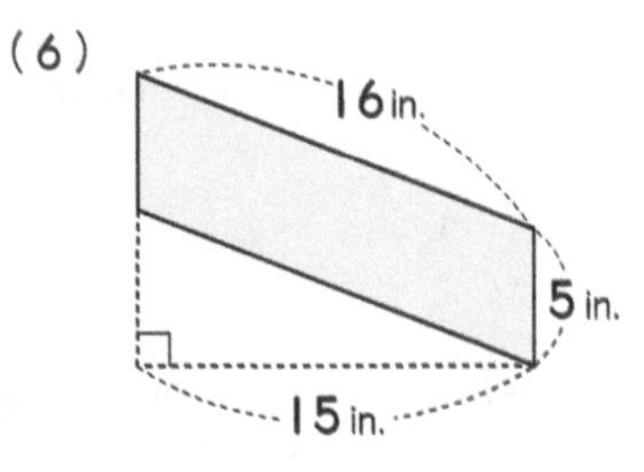

⟨Ans.⟩

2 Find the area of each triangle below.

6 points per question

(1)

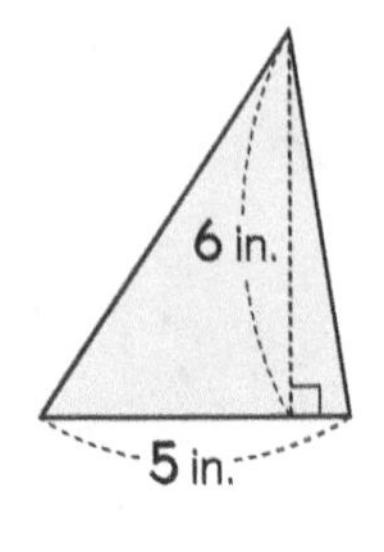

⟨Ans.⟩

(2)

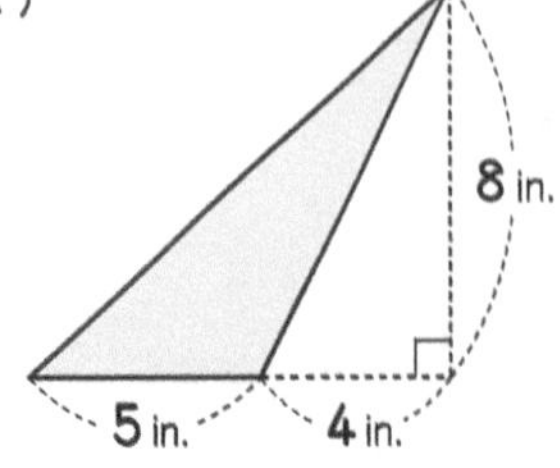

⟨Ans.⟩

(3)

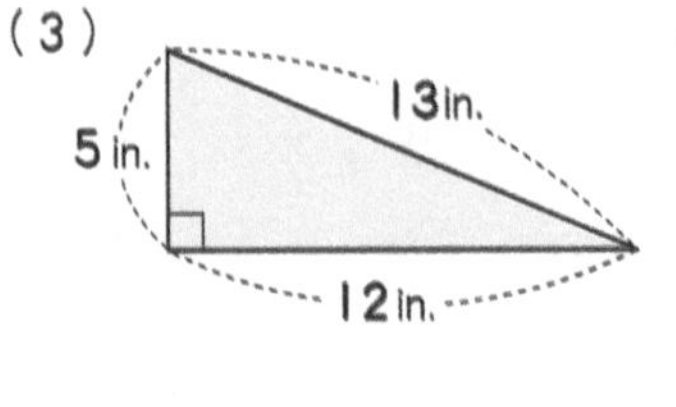

⟨Ans.⟩

(4)

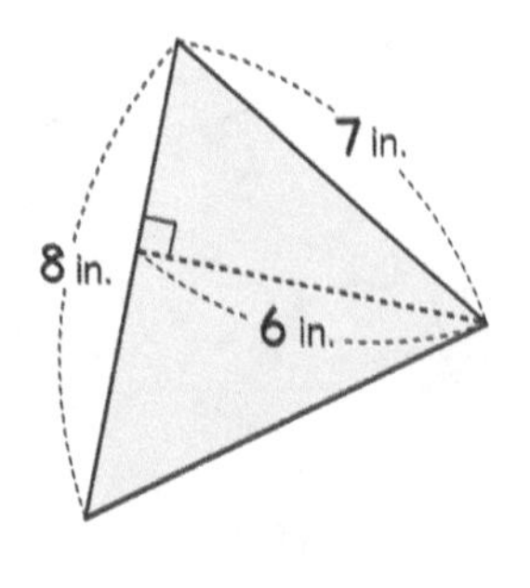

⟨Ans.⟩

3 Find the area of each quadrilateral below.

5 points per question

(1)

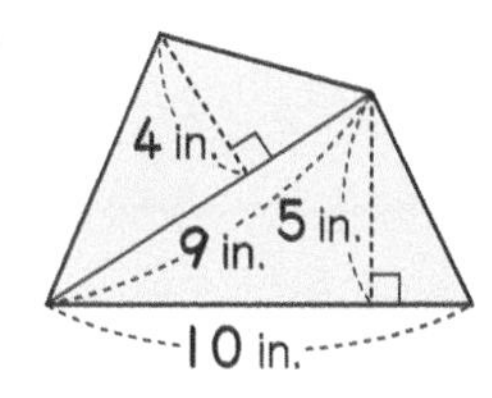

$\frac{1}{2}\times10\times5+\frac{1}{2}\times9\times4=$

⟨Ans.⟩

(2)

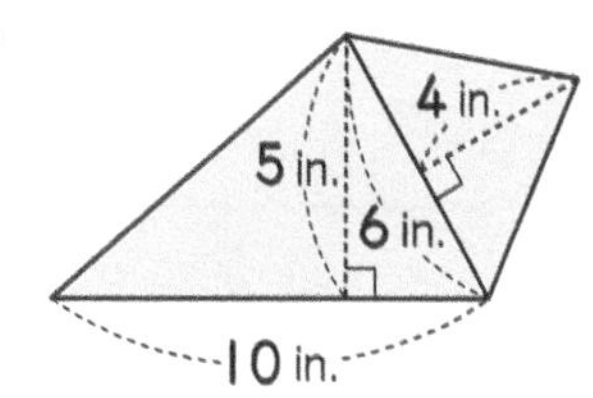

⟨Ans.⟩

(3)

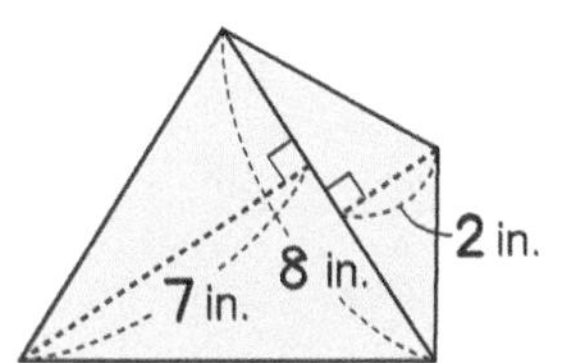

⟨Ans.⟩

(4)

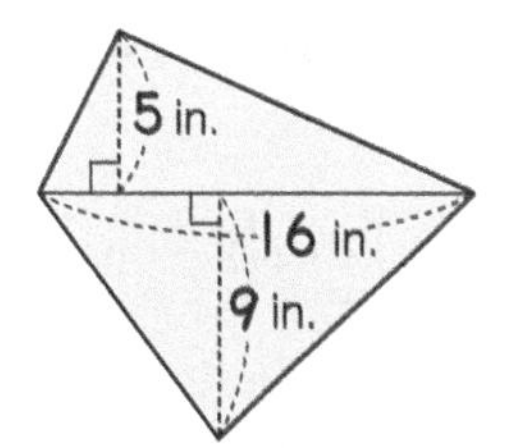

⟨Ans.⟩

4 Find the area of the shaded regions of the figures below.

5 points per question

(1)

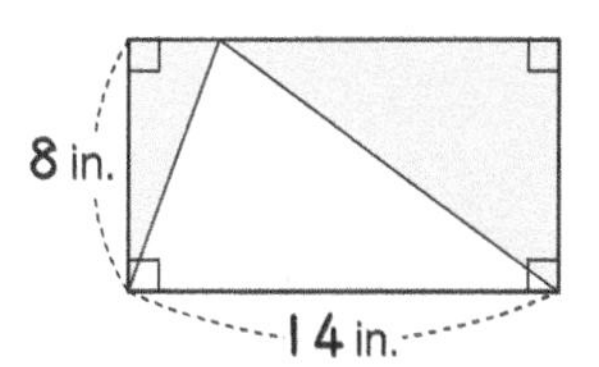

⟨Ans.⟩

(2)

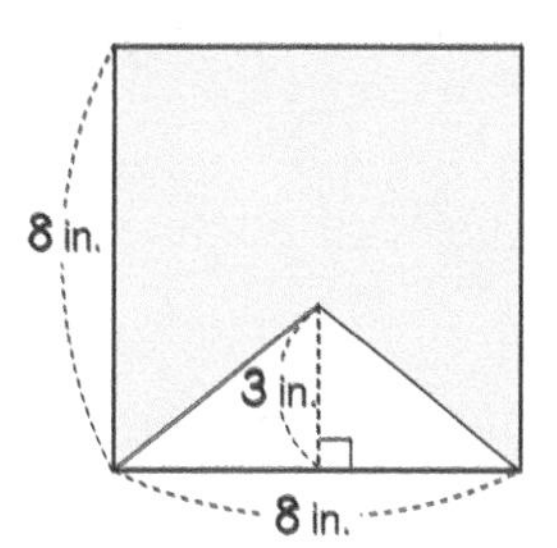

⟨Ans.⟩

(3)

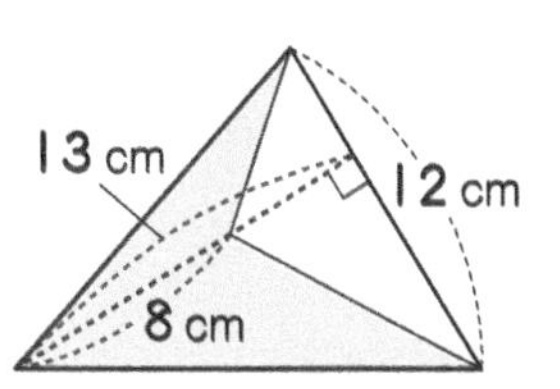

⟨Ans.⟩

(4)

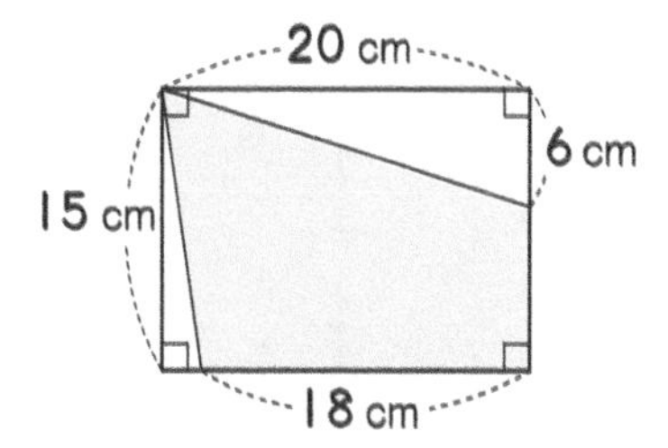

⟨Ans.⟩

10 Circles & Perimeter Review

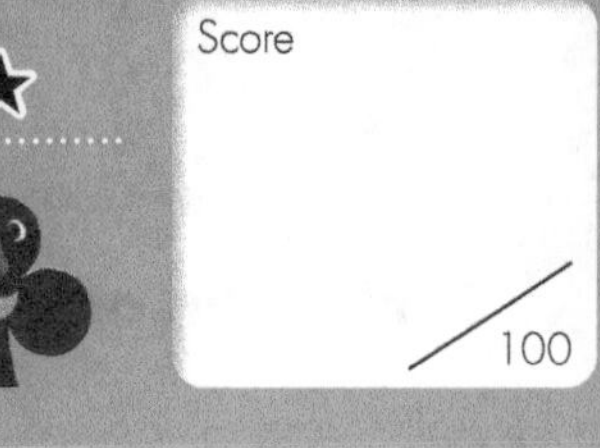

Level ★

Date / /

Name

■ The Answer Key is on page 89.

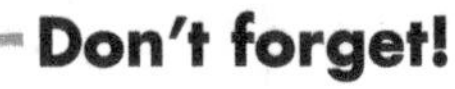

Don't forget!

Circle

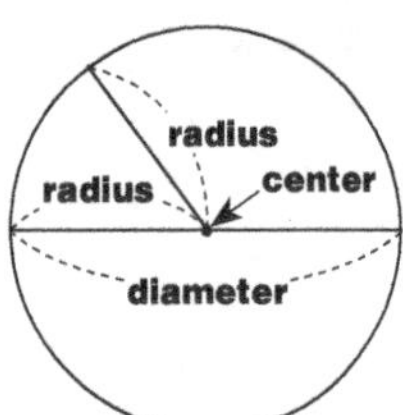

The length of a diameter is the radius times two.

(d = 2r)

1 **Two circles of the same size are inside a larger circle that has a diameter of 12 inches.**

5 points per question

(1) What is the radius of the big circle? 〈Ans.〉

(2) What is the diameter of each small circle? 〈Ans.〉

(3) What is the radius of each small circle? 〈Ans.〉

12 in.

2 **Two circles of the same size are inside a larger circle that has a radius of 8 inches.**

5 points per question

(1) What is the diameter of the big circle? 〈Ans.〉

(2) What is the diameter of each small circle? 〈Ans.〉

(3) What is the radius of each small circle? 〈Ans.〉

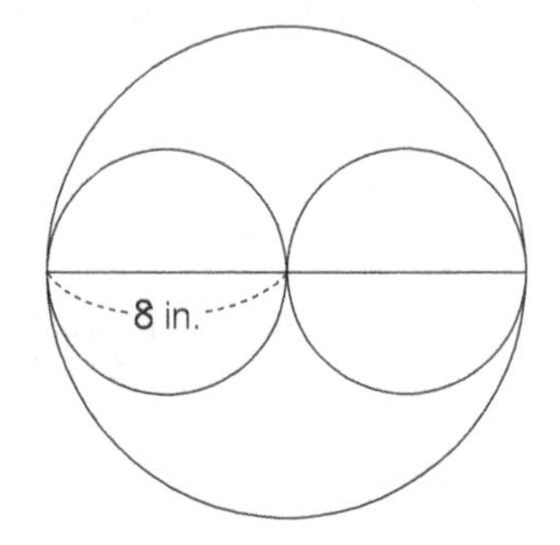

3 **How long is a side of each square below?**

5 points per question

(1)

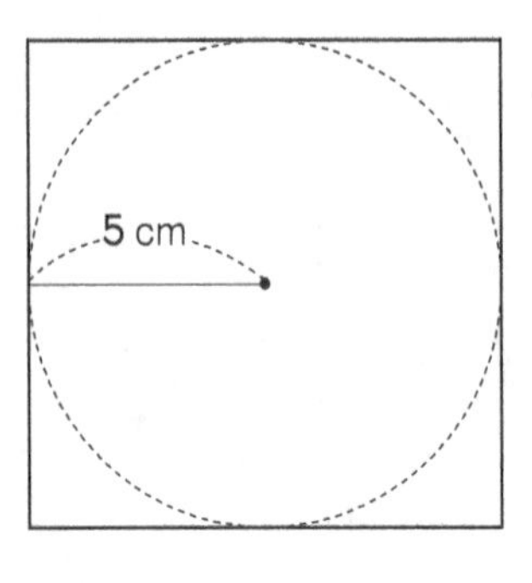

(2)

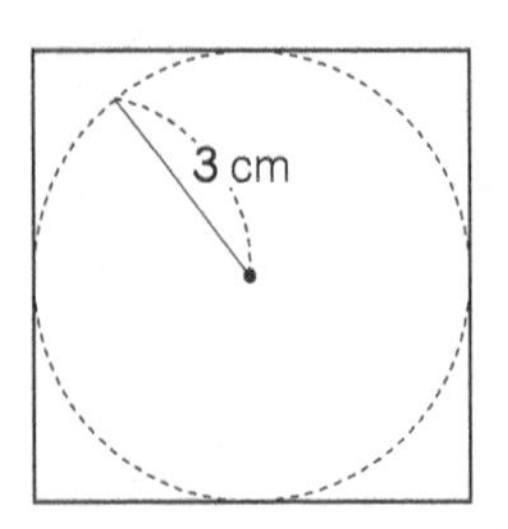

(3)

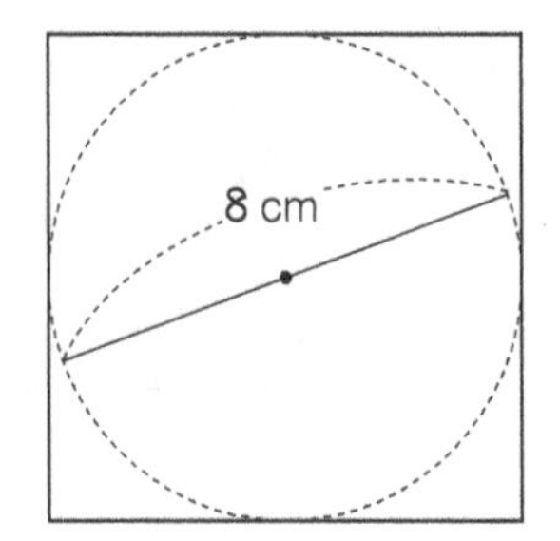

〈Ans.〉 〈Ans.〉 〈Ans.〉

4 Find the perimeter of the following quadrilaterals.

7 points per question

(1) [Rectangle]

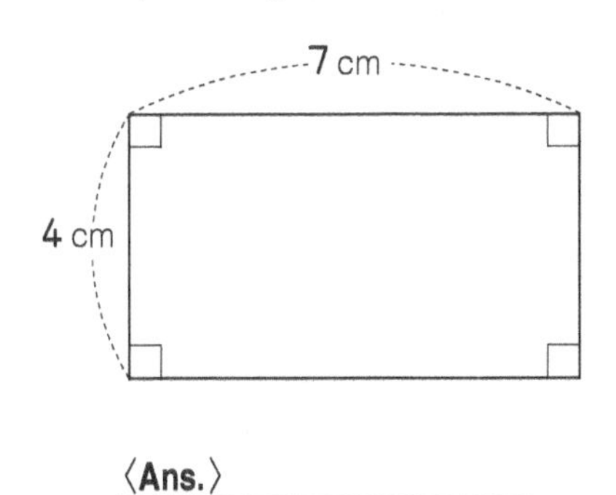

〈Ans.〉 ____________

(2) [Square]

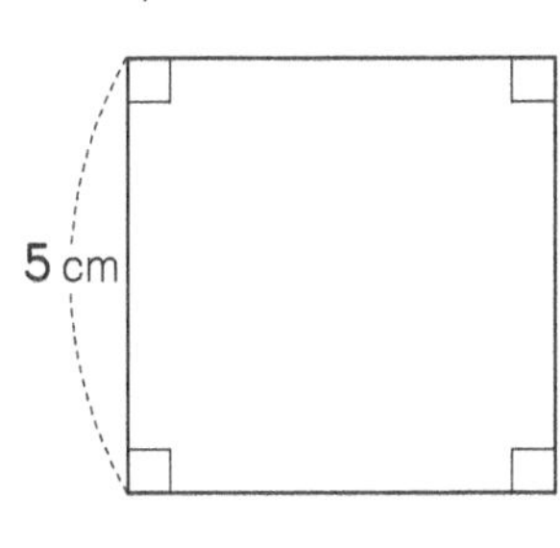

〈Ans.〉 ____________

(3) [Rectangle]

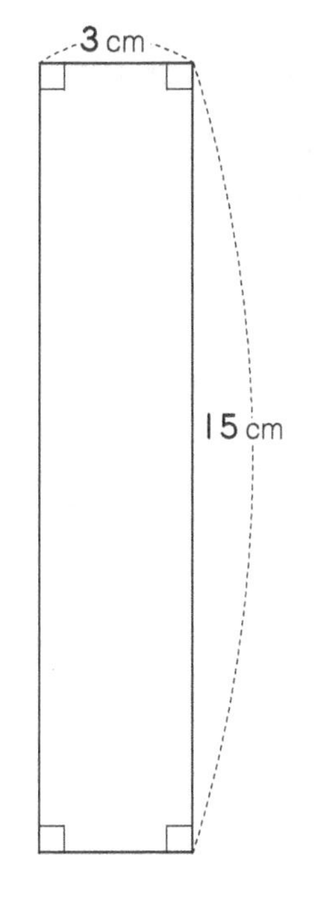

〈Ans.〉 ____________

(4) [Parallelogram]

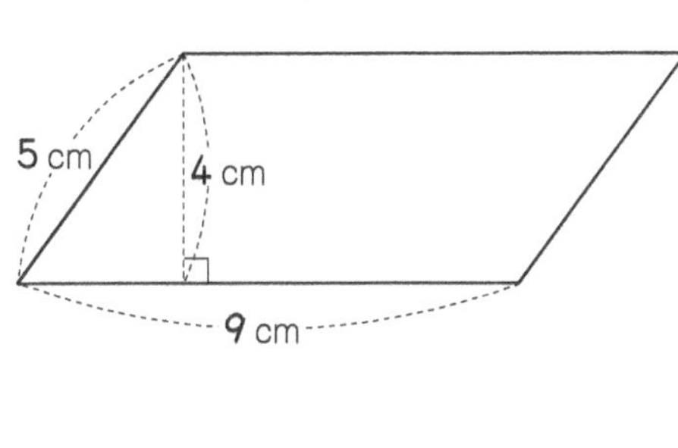

〈Ans.〉 ____________

(5) [Rhombus]

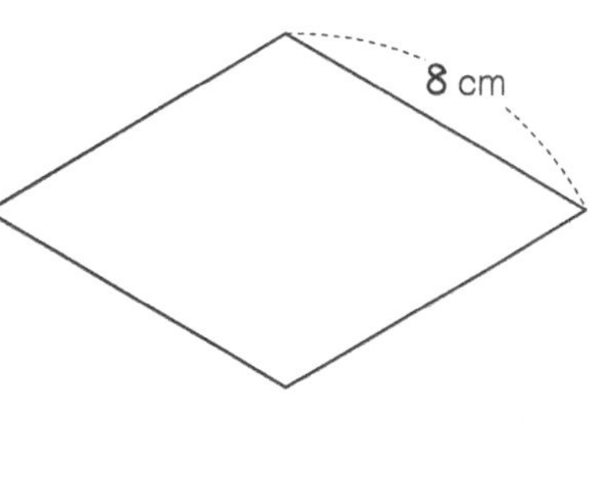

〈Ans.〉 ____________

5 Find the perimeter of the following figures.

5 points per question

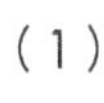

(1)

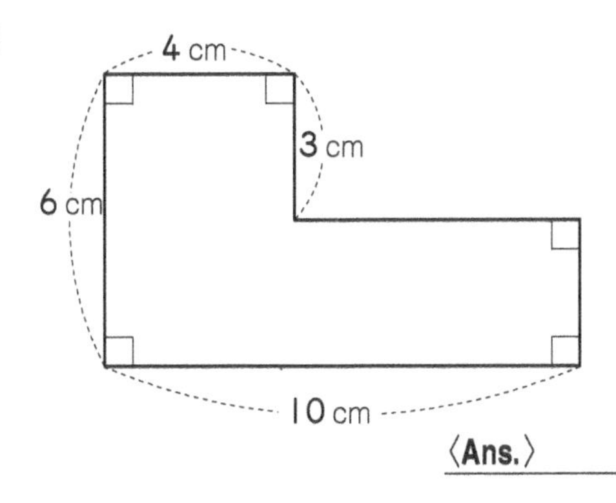

〈Ans.〉 ____________

(2)

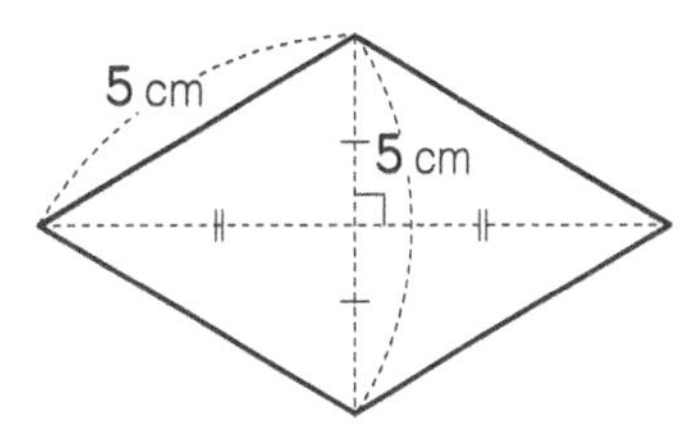

〈Ans.〉 ____________

(3)

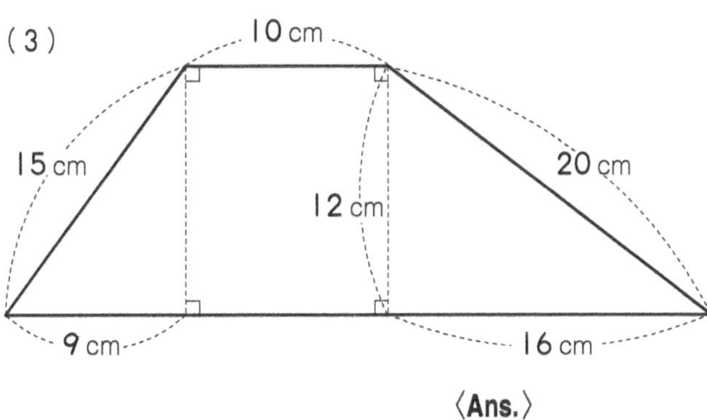

〈Ans.〉 ____________

(4)

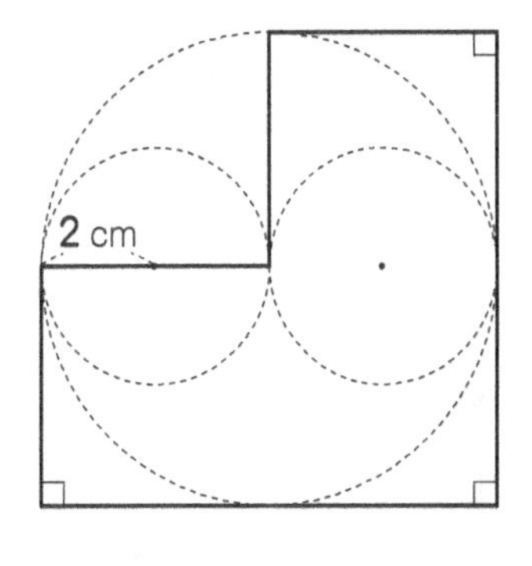

〈Ans.〉 ____________

Volume & Net Review

Date / / Name

Level ★

■ The Answer Key is on page 89.

Don't forget!

- The volume of a rectangular prism is length times width times height.
- The volume of a cube is edge times edge times edge.

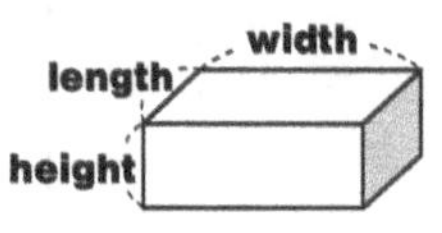

(V = l × w × h)

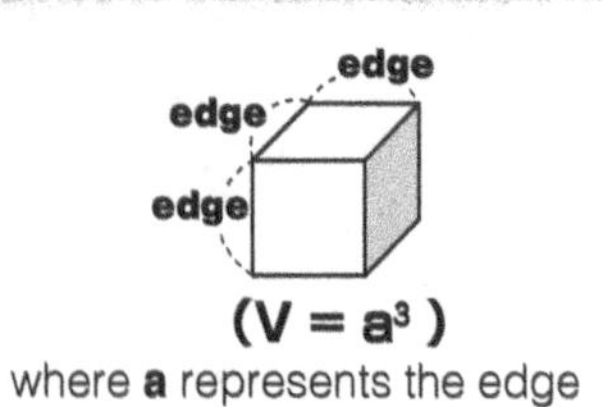

(V = a^3)
where **a** represents the edge

1 Find the volume of the rectangular prisms below. Be careful of the unit.

7 points per question

(1)

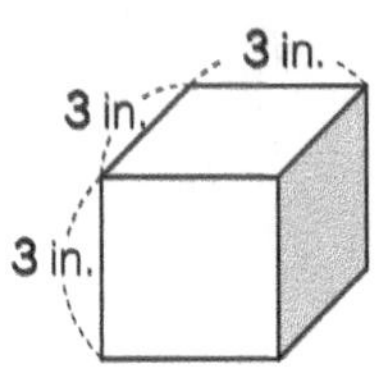

〈Ans.〉 ____________

(2)

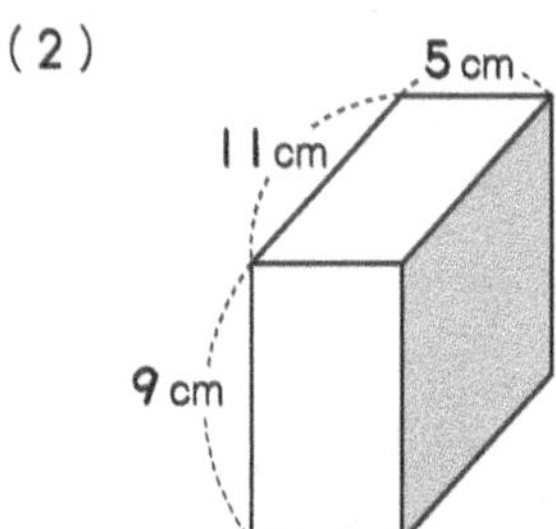

〈Ans.〉 ____________

(3)

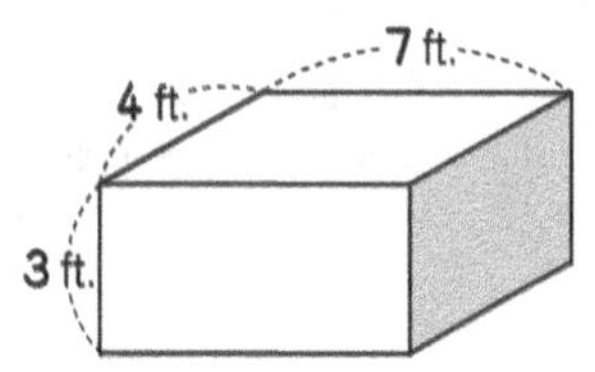

〈Ans.〉 ____________

(4)

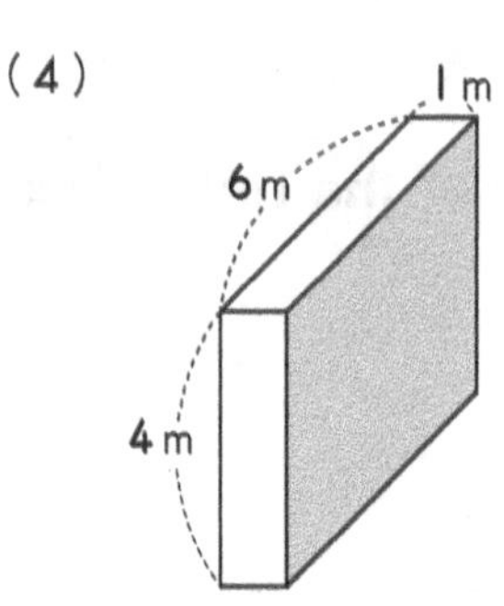

〈Ans.〉 ____________

(5)

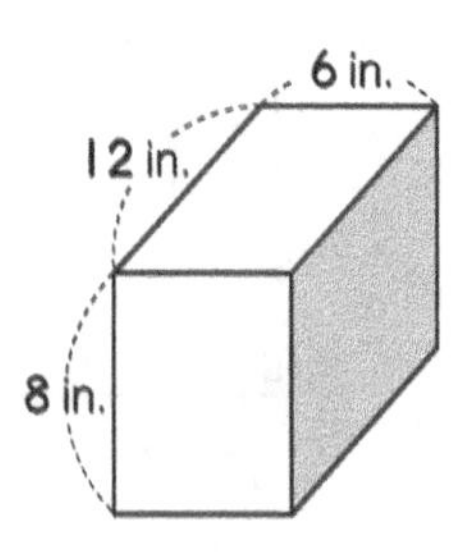

〈Ans.〉 ____________

(6)

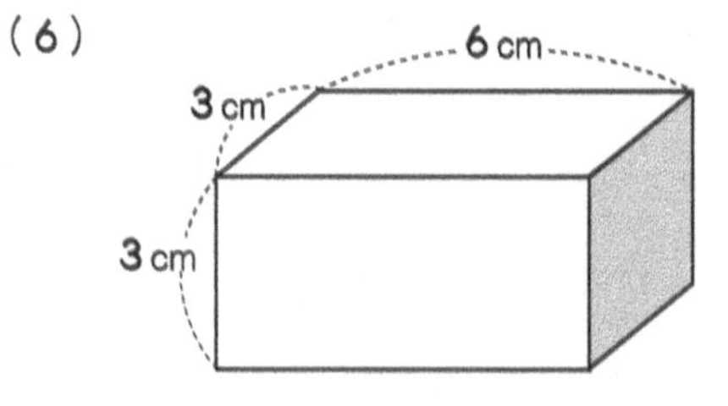

〈Ans.〉 ____________

(7)

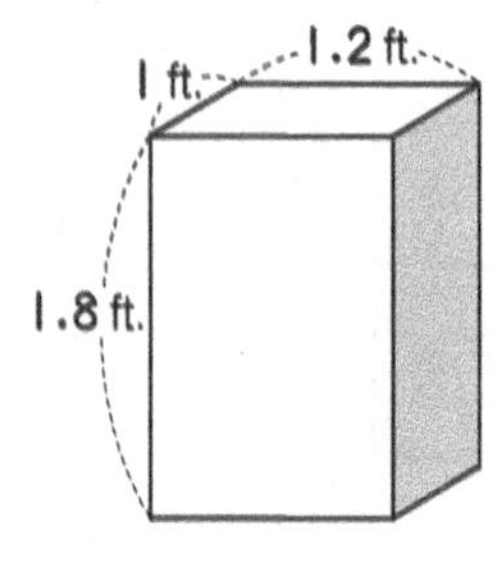

〈Ans.〉 ____________

(8)

2 cm
2 cm
2 cm

〈Ans.〉 ____________

2 Find the net of the rectangular prisms below.

7 points per question

(1)

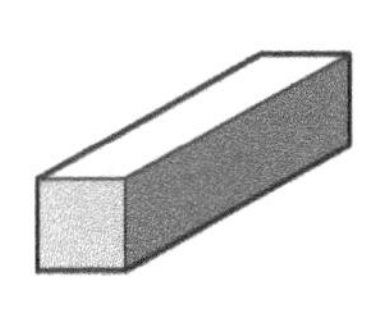

⟨Ans.⟩

(2)

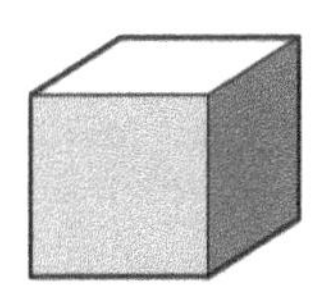

⟨Ans.⟩

(3)

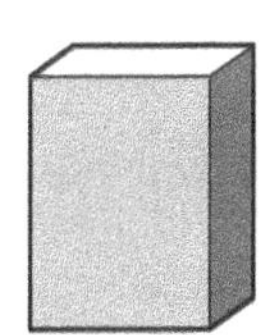

⟨Ans.⟩

a

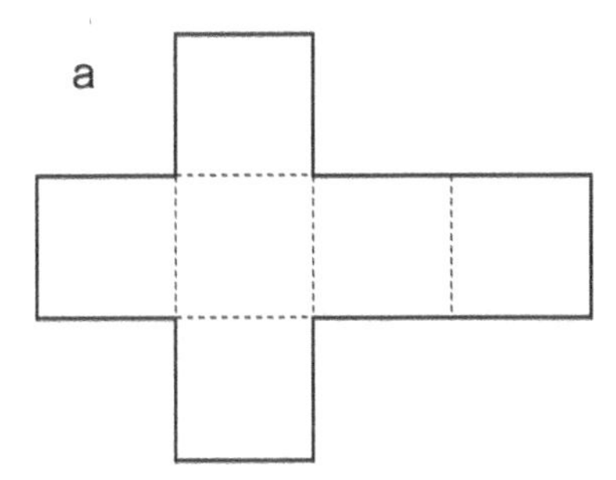

b

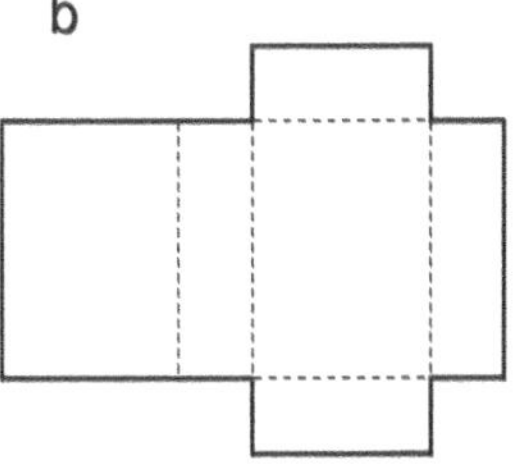

c

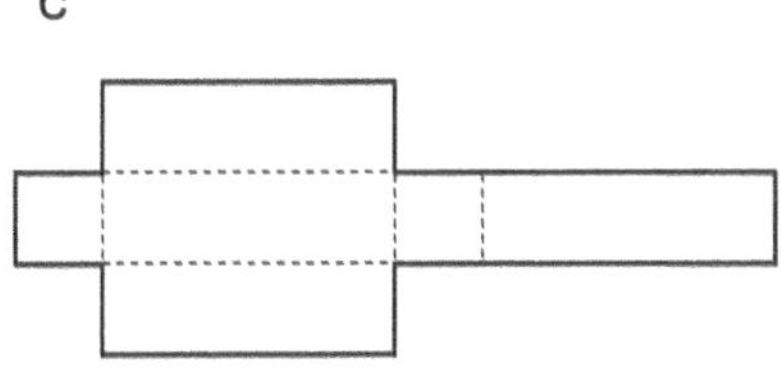

3 Answer the questions about the net on the right.

7 points per question

(1) Which face is parallel to A?

⟨Ans.⟩

(2) Which face is parallel to C?

⟨Ans.⟩

A

B C D E

F

4 Which nets cannot build a cube?

9 points for completion

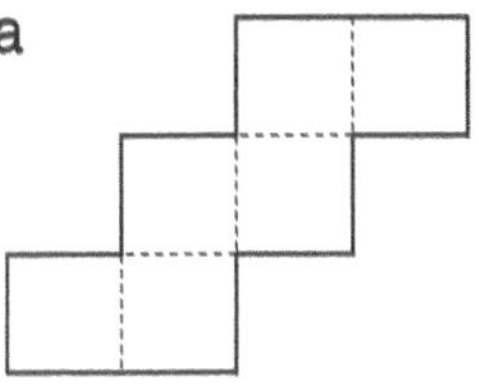

b

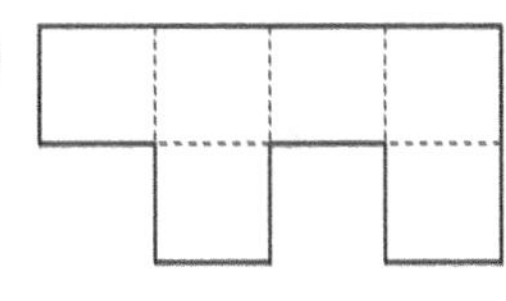

c

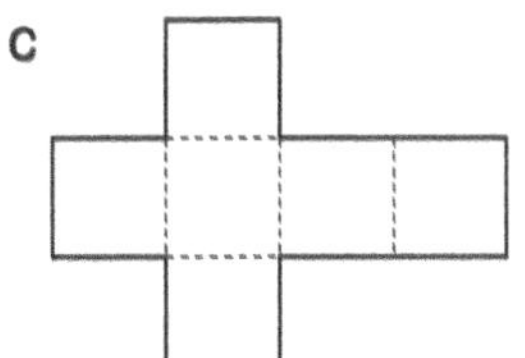

d

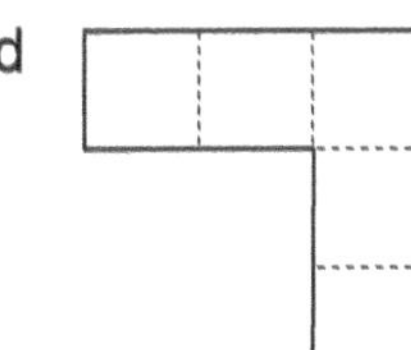

⟨Ans.⟩ ,

12 Congruent Figures Review

Level ★

Date / /　Name

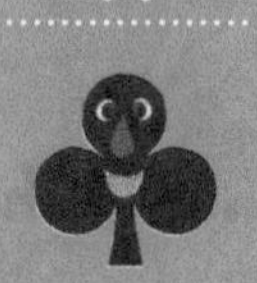

Score /100

■ The Answer Key is on page 90.

Don't forget!

● Two plane figures are called **congruent** when they have the same size and shape.
● Congruent figures can exactly overlap each other when moving or turning over.

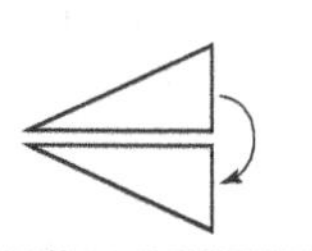

1 **Which are the pairs of congruent shapes below?** 10 points for completion

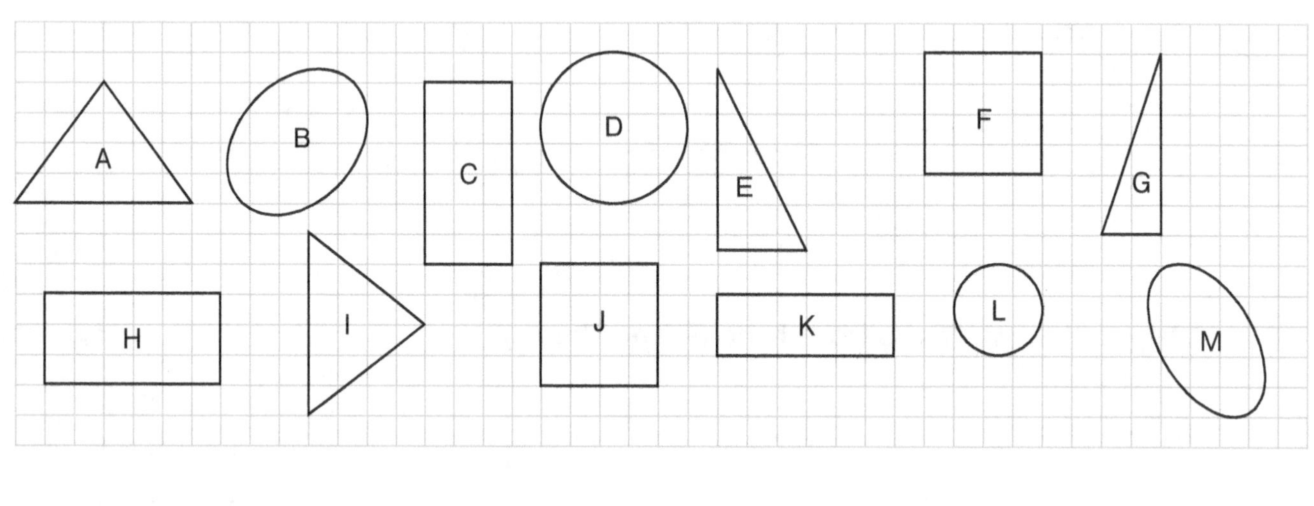

〈**Ans.**〉 (　　) and (　　), (　　) and (　　), (　　) and (　　)

2 **Which figures are the same size and shape as triangle A?** 15 points for completion

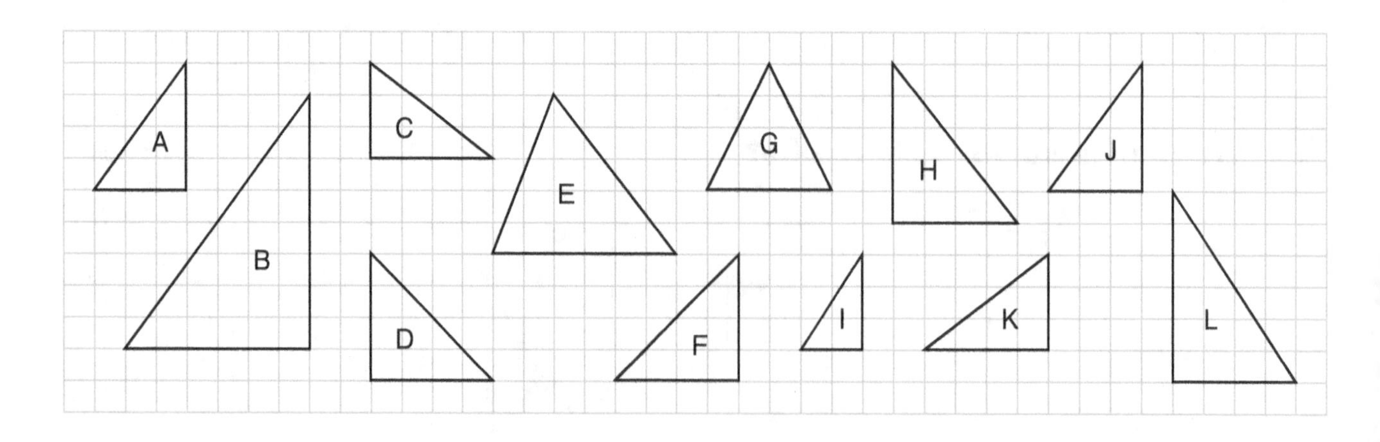

〈**Ans.**〉 (　　), (　　), (　　)

Don't forget!

● In congruent figures, the vertices, overlapping sides, and angles that are the same are called **corresponding vertices, sides,** and **angles.**

3 Use the figures on the right to answer the questions below.

5 points per question

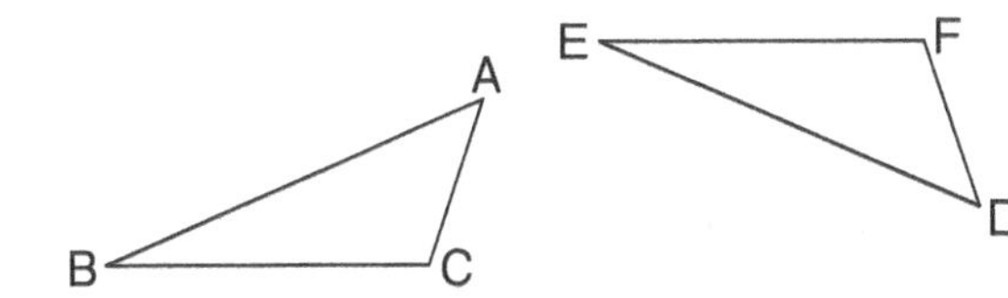

(1) What is the corresponding vertex to A?

⟨Ans.⟩

(2) What is the corresponding vertex to B?

⟨Ans.⟩

(3) What is the corresponding vertex to C?

⟨Ans.⟩

(4) What is the corresponding side to AB?

⟨Ans.⟩

(5) What is the corresponding side to BC?

⟨Ans.⟩

(6) What is the corresponding side to CA?

⟨Ans.⟩

(7) What is the corresponding angle to A?

⟨Ans.⟩

(8) What is the corresponding angle to B?

⟨Ans.⟩

(9) What is the corresponding angle to C?

⟨Ans.⟩

4 Use the congruent quadrilaterals on the right to answer the questions below.

5 points per question

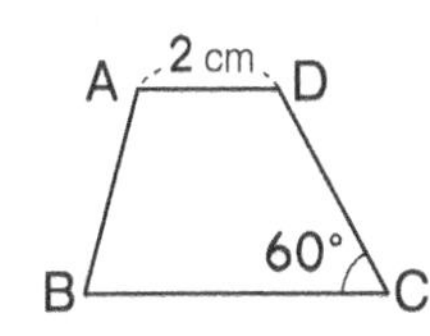

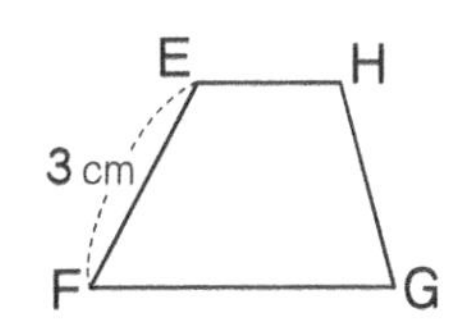

(1) What is the corresponding vertex to G?

⟨Ans.⟩

(2) What is the corresponding side to BC?

⟨Ans.⟩

(3) What is the corresponding angle to D?

⟨Ans.⟩

(4) How long is side EH?

⟨Ans.⟩

(5) How long is side CD?

⟨Ans.⟩

(6) Find the measure of angle F.

⟨Ans.⟩

13 Scale Drawing Review

Level 

Date / /

Name

Score 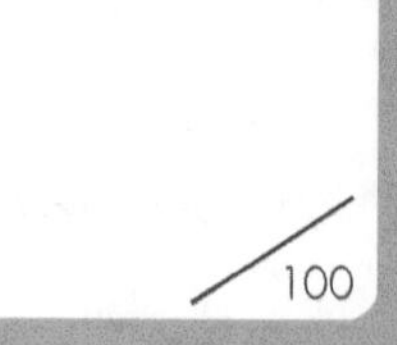

/100

■ The Answer Key is on page 90.

Don't forget!

- Figures with the same shape and different size are called **similar** figures.
- When similar figures are enlarged or reduced to become the same size, they are called **congruent** figures.

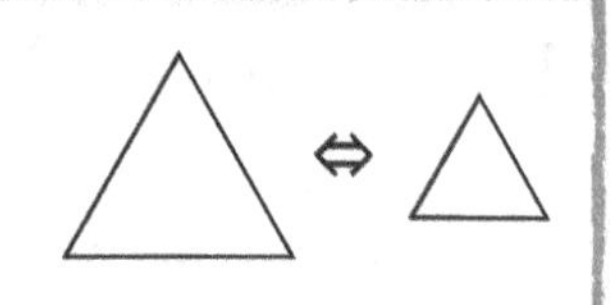

1 **Find the enlarged shape, similar to A, from the figures below.** 20 points for completion

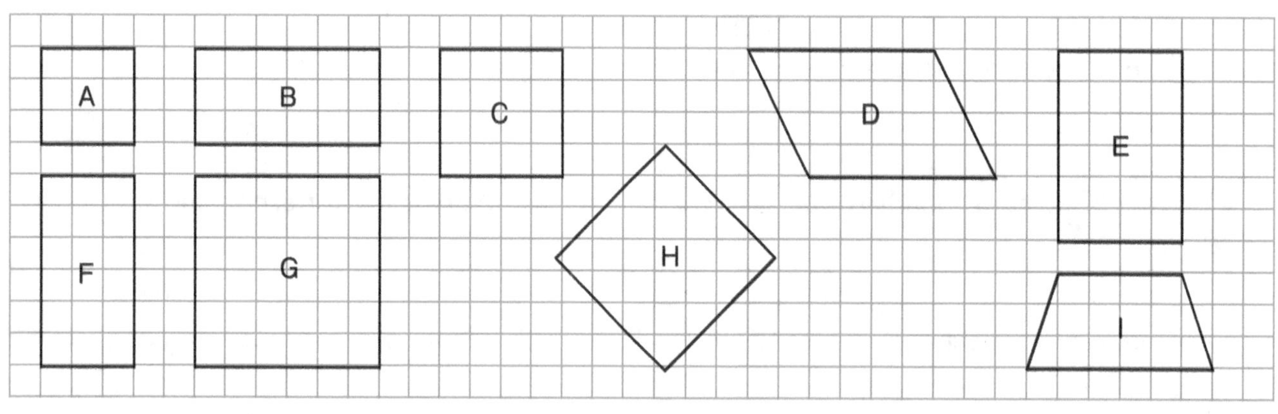

[Use your triangle rulers to make sure.]

⟨Ans.⟩ (   ), (   ), (   )

2 **Find the reduced shape, similar to A, from the figures below.** 20 points for completion

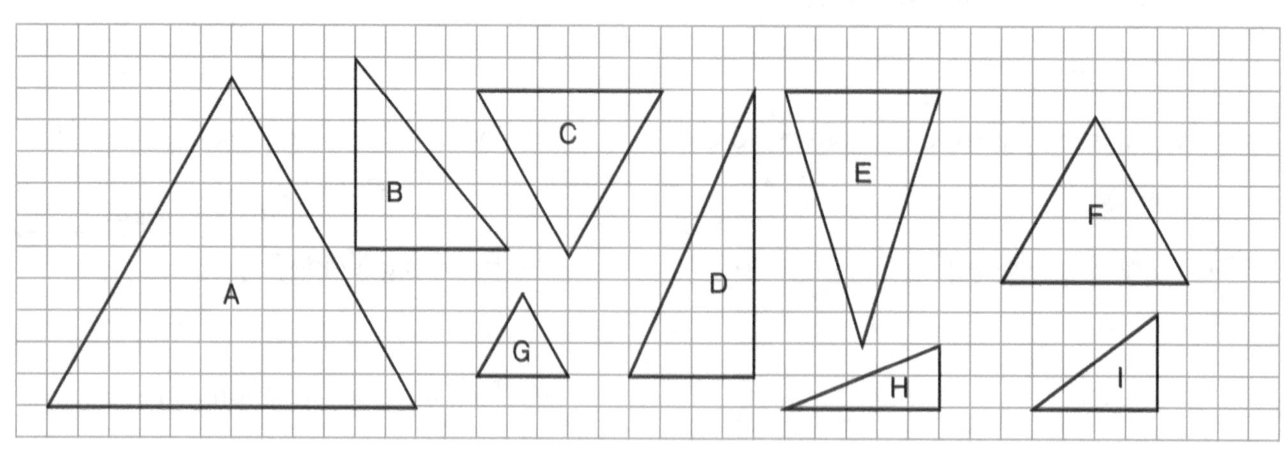

[Use your triangle rulers to make sure.]

⟨Ans.⟩ (   ), (   ), (   )

Don't forget!

● In similar figures, the corresponding vertices and their angles are equal in size, but the lengths of the corresponding sides are not equal.

3 **Triangle DEF is an enlarged drawing of triangle ABC. Answer the questions about the two triangles below.**

15 points per question

(1) What is the corresponding vertex for each vertex?

Vertex A ⟨Ans.⟩ ______

Vertex B ⟨Ans.⟩ ______

Vertex C ⟨Ans.⟩ ______

(2) What is the corresponding side for each side?

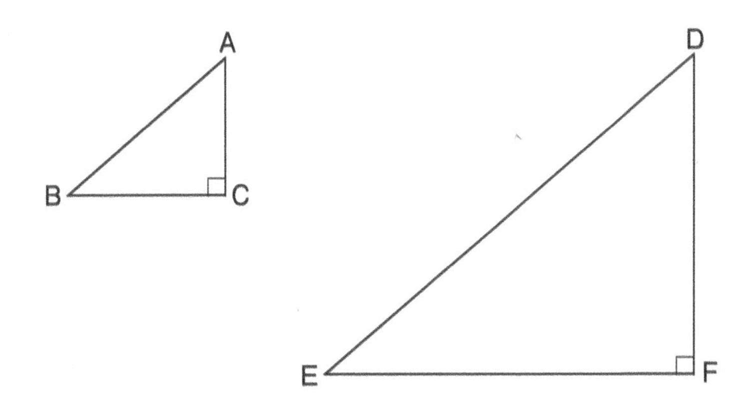

Side AB ⟨Ans.⟩ ______

Side BC ⟨Ans.⟩ ______

Side CA ⟨Ans.⟩ ______

4 **Quadrilateral EFGH is a reduced drawing of quadrilateral ABCD. Answer the questions about the two quadrilaterals below.**

15 points per question

(1) What is the corresponding vertex for each vertex?

Vertex A ⟨Ans.⟩ ______

Vertex B ⟨Ans.⟩ ______

Vertex C ⟨Ans.⟩ ______

(2) What is the corresponding side for each side?

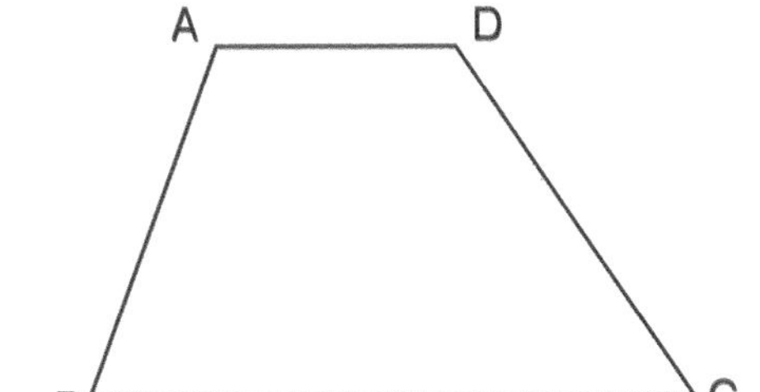

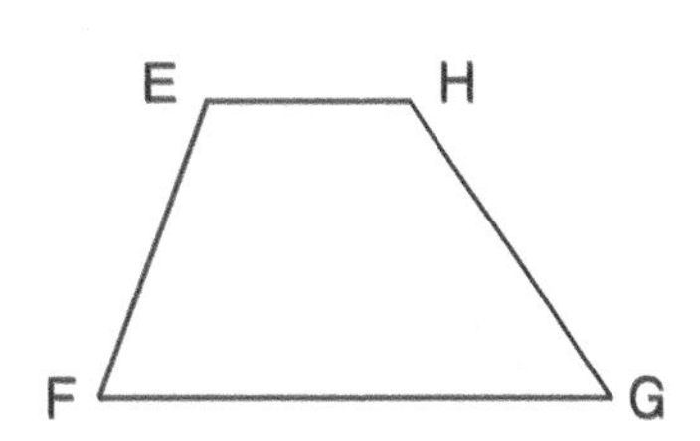

Side AB ⟨Ans.⟩ ______

Side CD ⟨Ans.⟩ ______

Side DA ⟨Ans.⟩ ______

14 Line Symmetry Review

Level ★

Date / /

Name

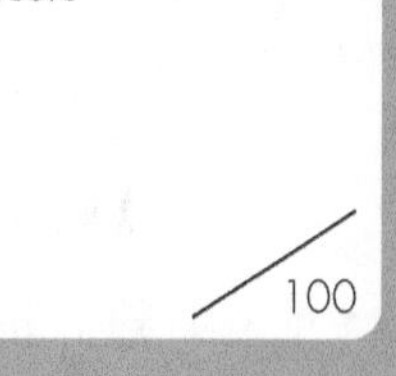

Score /100

■ The Answer Key is on page 90.

Don't forget!

●If you can fold a shape in half so that both halves are overlapped perfectly, that shape has **line symmetry**. That means that the form of the shape is the same on both sides of one axis.

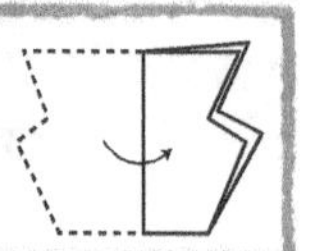

1 Which shapes below have line symmetry?

20 points for completion

A 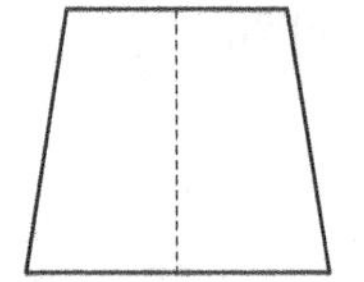

B 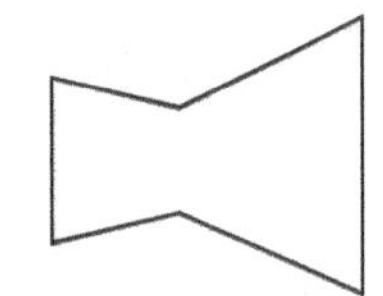

C 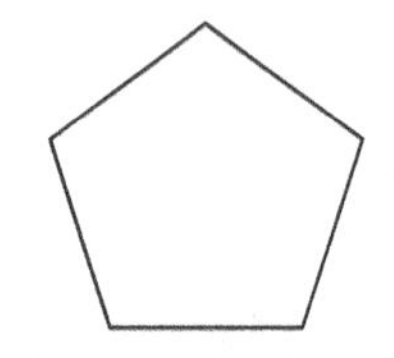

D 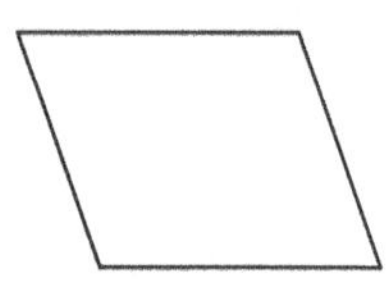

E 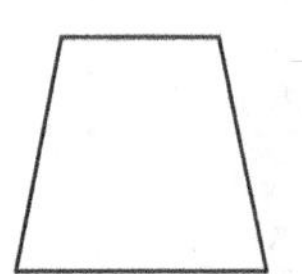

F 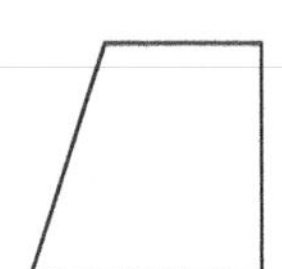

G 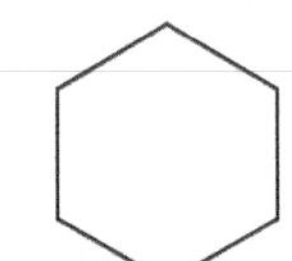

H 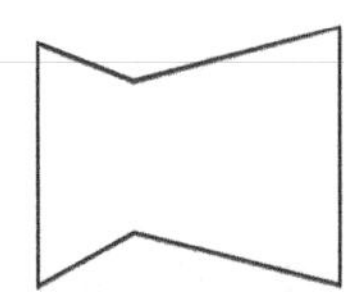

I 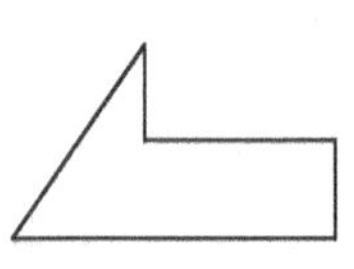

J 

〈Ans.〉 , , , , , ,

2 Which shapes below have line symmetry?

20 points for completion

A 〔Rectangle〕

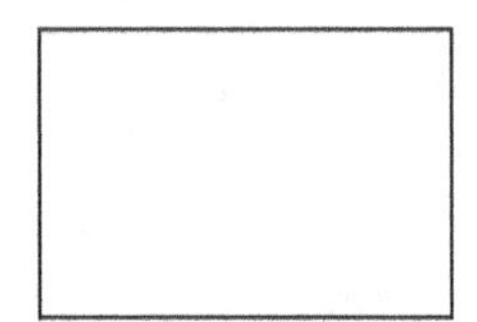

B 〔Parallelogram〕

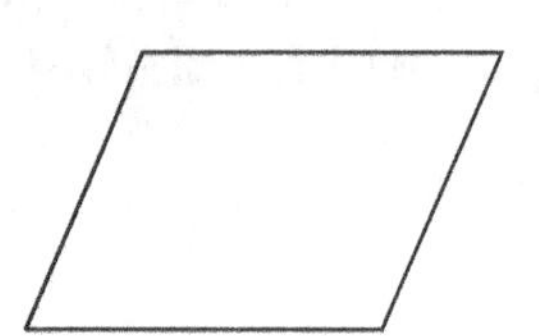

C 〔Rhombus〕

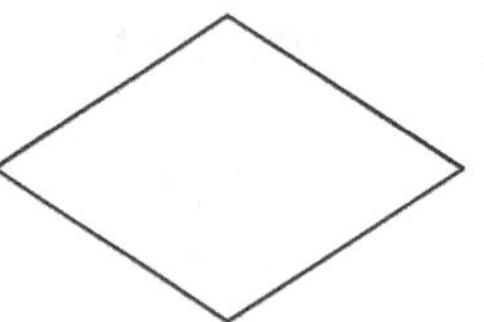

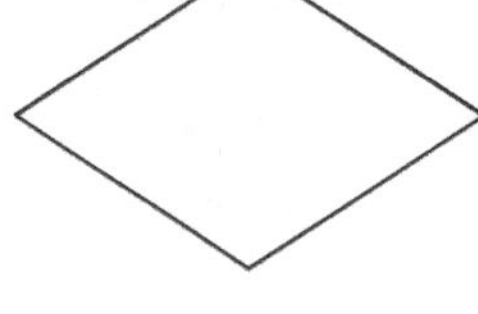

D 〔Square〕

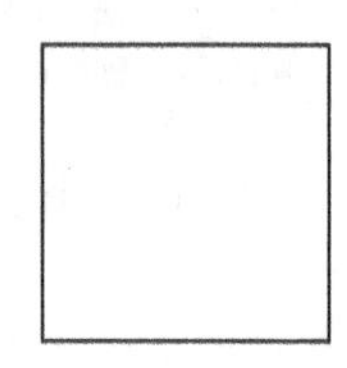

E 〔Isosceles triangle〕

F 〔Equilateral triangle〕

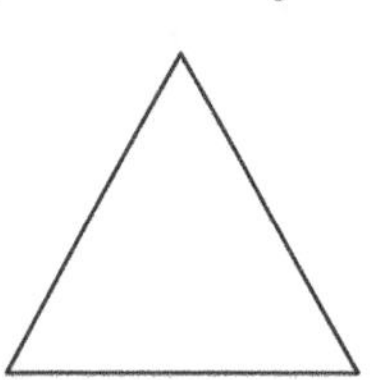

G 〔Right triangle〕

H 〔Isosceles right triangle〕

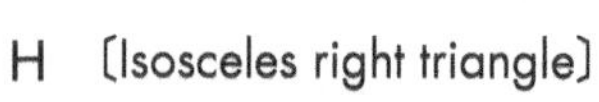

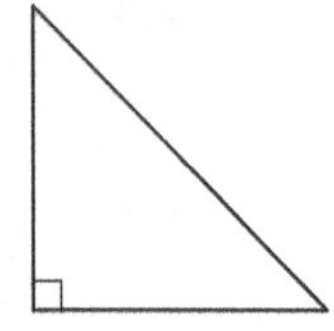

I 〔Regular pentagon〕

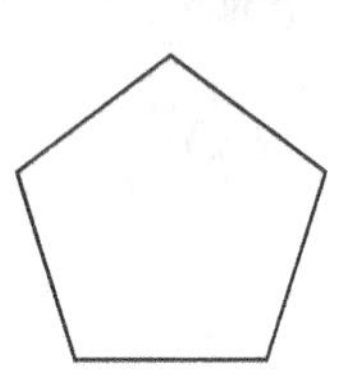

J 〔Circle〕

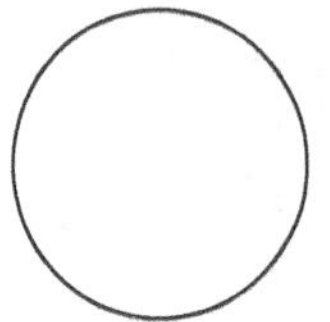

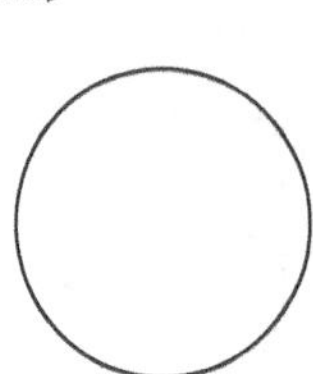

〈Ans.〉 , , , , , , ,

Don't forget!

●The line along which you could fold a shape into two halves is called the **line of symmetry**.

3 **Each of the shapes below has line symmetry. Where could you fold each shape in half exactly? Draw the folding line on each shape.** 5 points per question

(1)

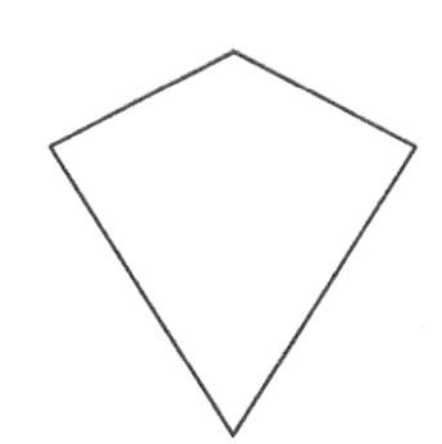

(2)

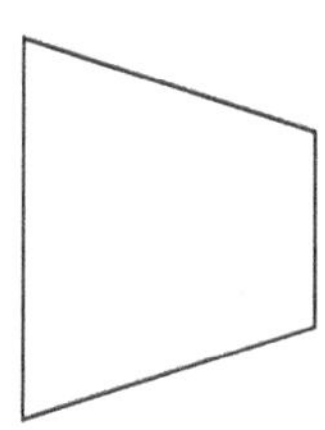

(3)

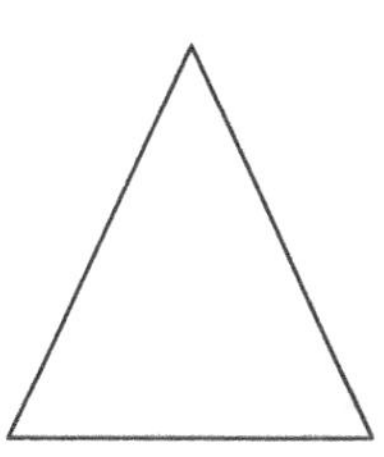

(4)

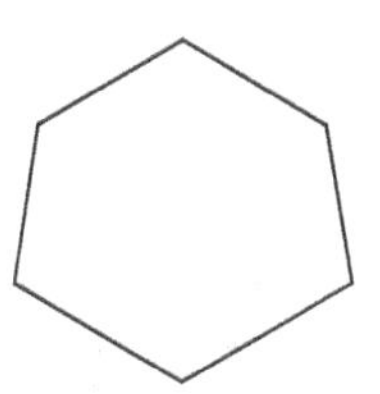

4 **Each of the shapes below has line symmetry. How many lines of symmetry (axes of symmetry) does each shape have?** 5 points per question

(1)

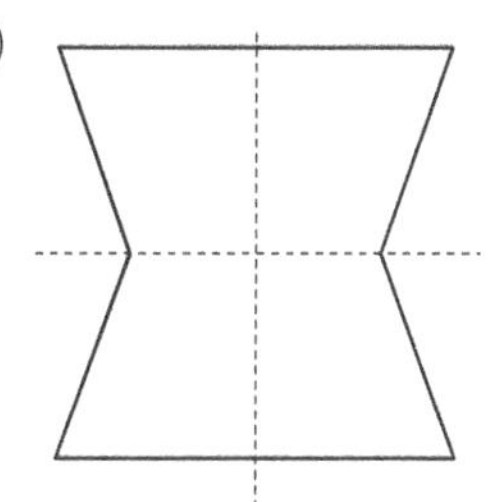

⟨**Ans.**⟩

(2)

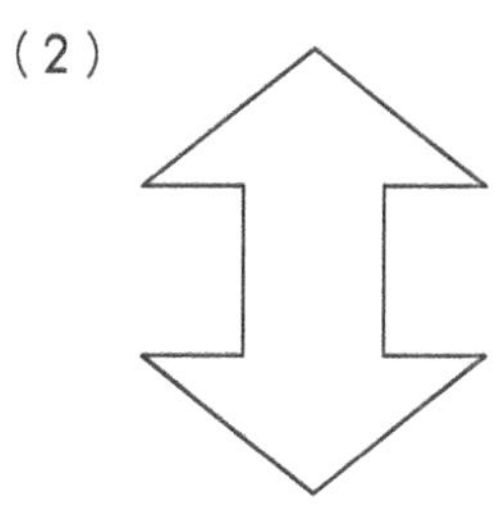

⟨**Ans.**⟩

(3)

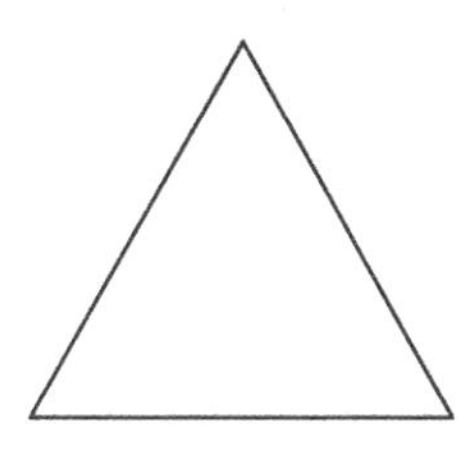

⟨**Ans.**⟩

(4)

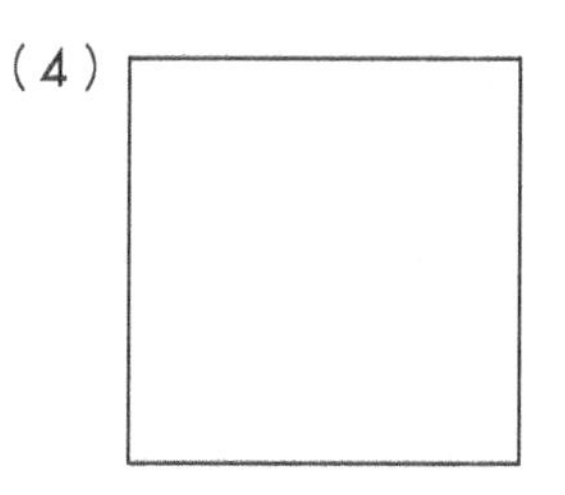

⟨**Ans.**⟩

(5)

⟨**Ans.**⟩

(6)

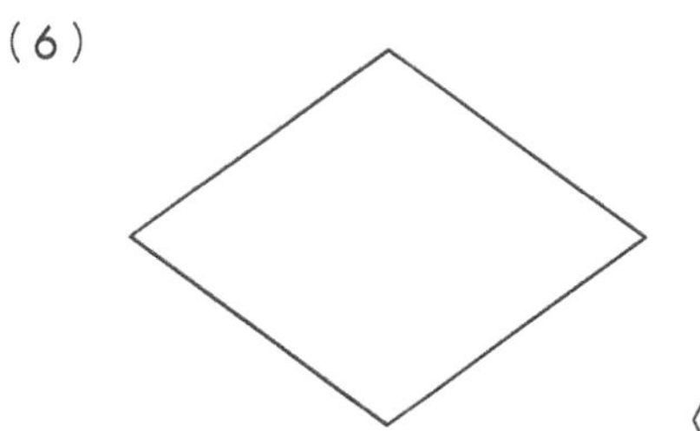

⟨**Ans.**⟩

(7)

⟨**Ans.**⟩

(8)

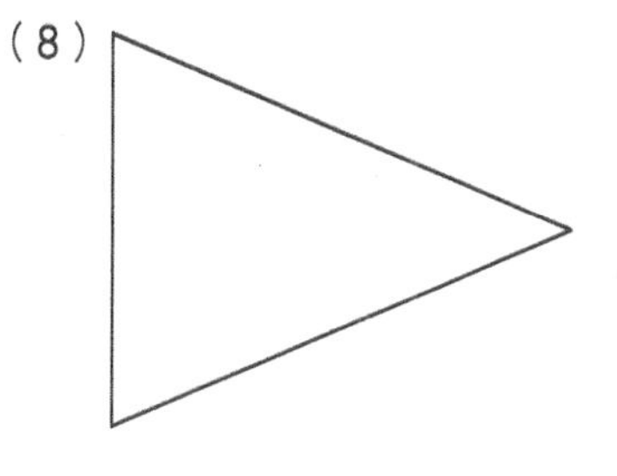

⟨**Ans.**⟩

Rotational Symmetry Review

Level ★

Date / /

Name

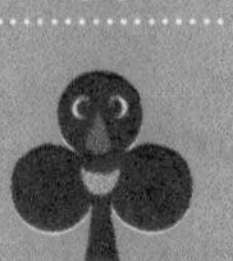

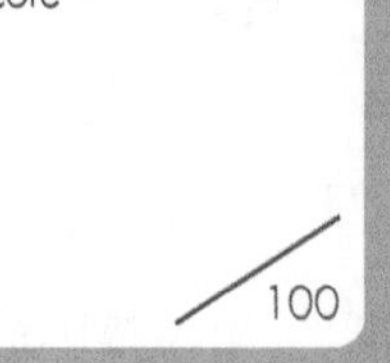

Score /100

■ The Answer Key is on page 90.

Don't forget!

●If you rotate a shape 180° around the point O, and the rotated figure overlaps the original figure exactly, then that shape has **rotational symmetry**.

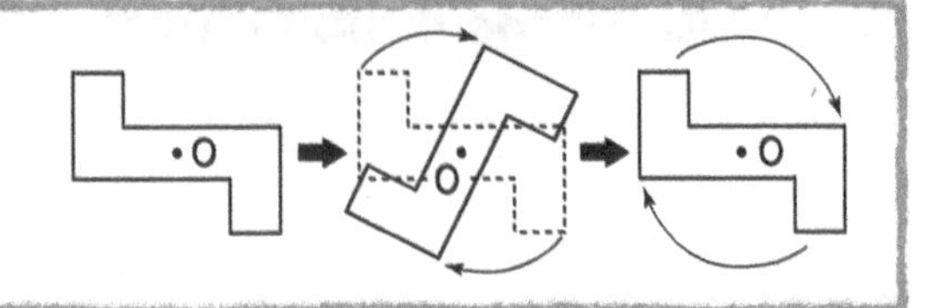

1 **Rotate each of the following shapes 180°. Which shapes will overlap the original shape exactly?** 15 points for completion

A

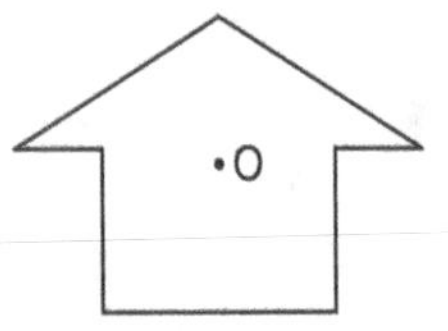

B

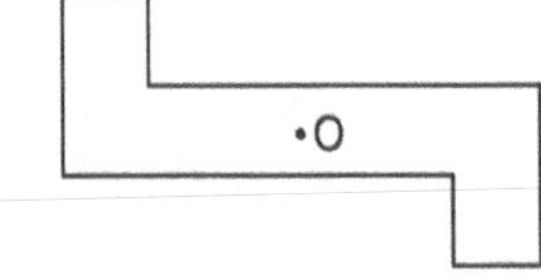

C

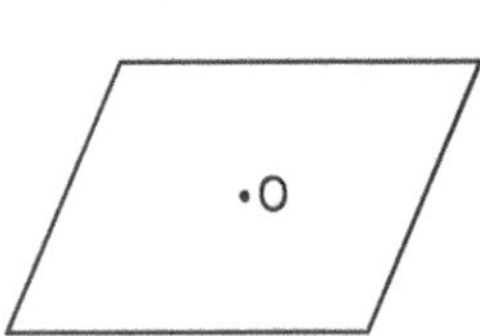

D

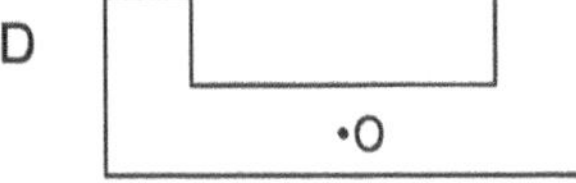

〈**Ans.**〉 ,

2 **Which shapes below have rotational symmetry?** 15 points for completion

A

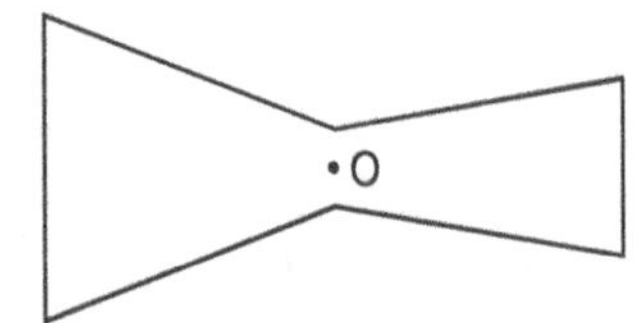

B

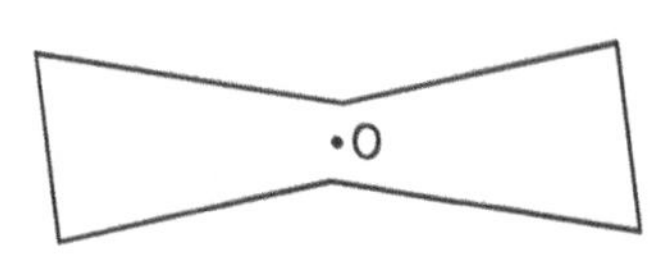

C

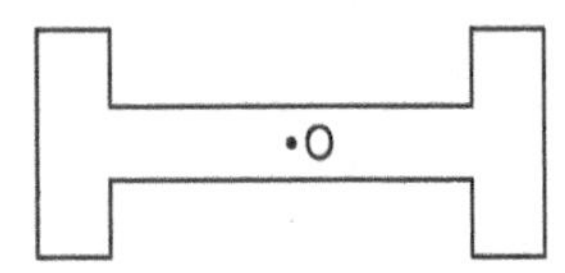

D

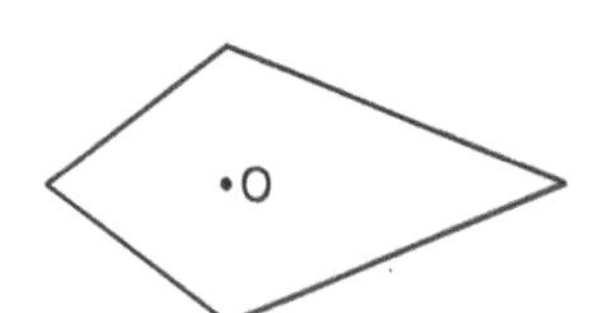

E

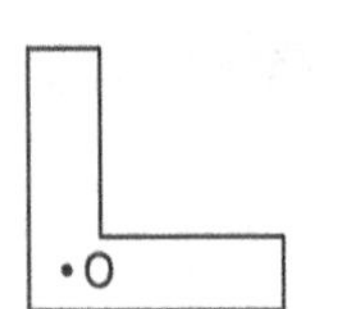

F

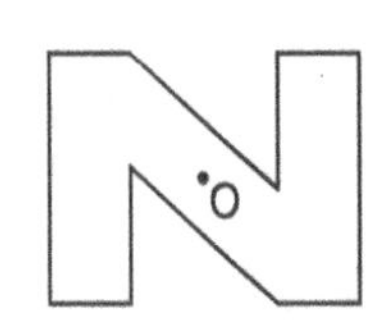

G

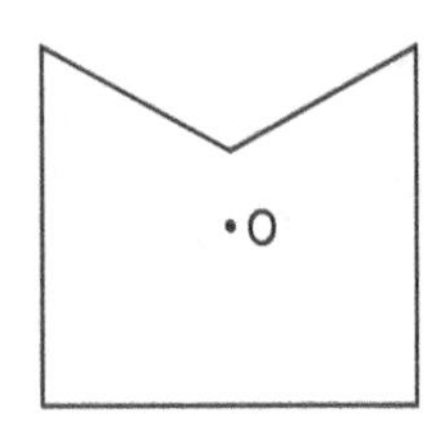

H

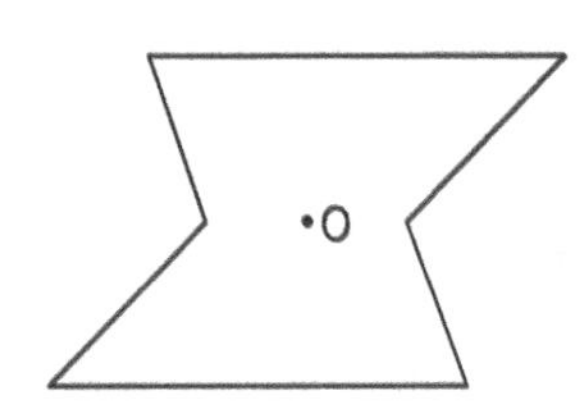

〈**Ans.**〉 , , ,

Don't forget!

- The point around which you can rotate a shape with rotational symmetry is called the **center of symmetry**.
- If you connect corresponding points in a figure with rotational symmetry, those lines will pass through the center of symmetry.
- The length from the center of symmetry to each corresponding point is equal.

3 **When a figure with rotational symmetry is rotated 180° around the center of symmetry, the overlapping vertices, sides, and angles are called corresponding vertices, sides, and angles. Answer the questions below about the rotationally symmetric figure pictured here.** 8 points per question

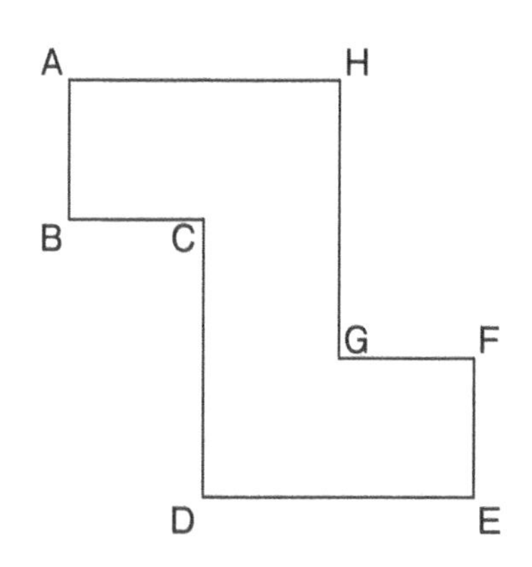

(1) What is the corresponding vertex to A?

〈Ans.〉

(2) What is the corresponding vertex to B?

〈Ans.〉

(3) What is the corresponding vertex to G?

〈Ans.〉

(4) What is the corresponding side to AH?

〈Ans.〉

(5) What is the corresponding side to CD?

〈Ans.〉

4 **The figure pictured below has rotational symmetry. Answer the questions below.** 15 points for completion

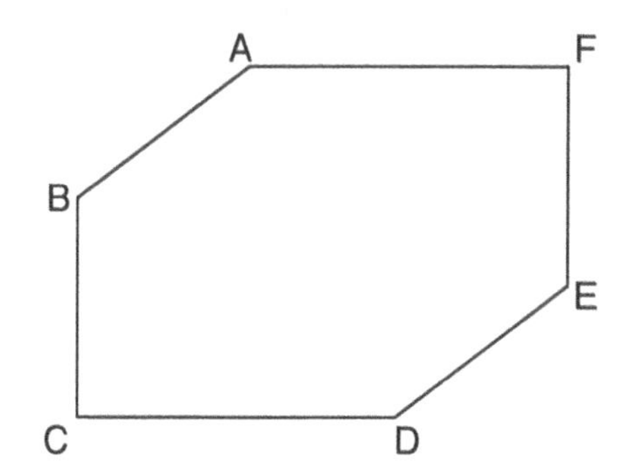

(1) Draw a line connecting point A and point D.

(2) Draw a line connecting point B and point E.

(3) Draw a line connecting point C and point F.

(4) Plot the center of symmetry, O of this figure.

5 **Plot the center of symmetry, O of the figure below.** 15 points

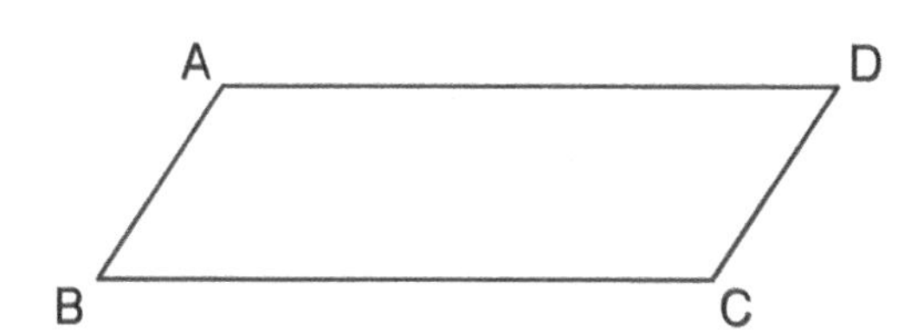

16 Line Symmetry & Rotational Symmetry Review

Level ★

Date / / Name

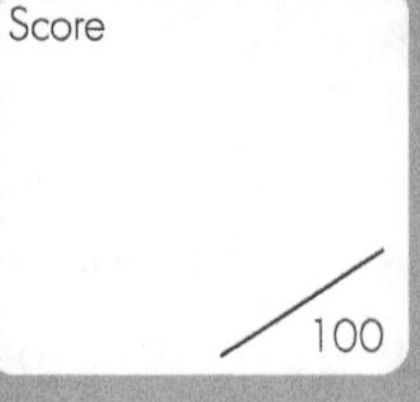

■ The Answer Key is on page 90.

Don't forget!

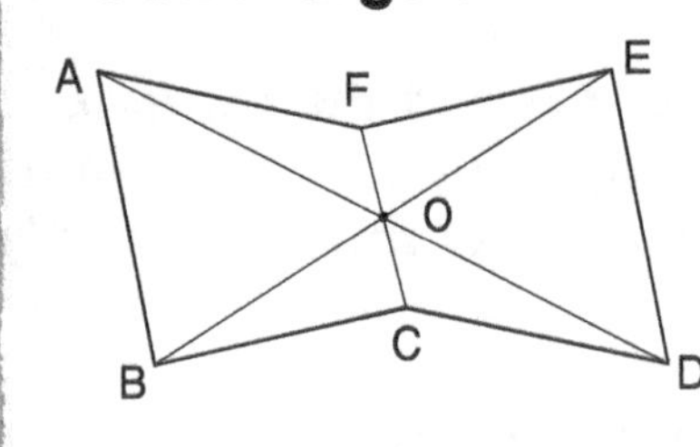

- In a figure with rotational symmetry, lines connecting all the corresponding vertices all cross through the center of symmetry.
- In the figure on the left, lines AD, BE, and CF all cross through the center of symmetry, O.
- If you connect two corresponding points of a figure with rotational symmetry and divide the figure, the opposing figures are congruent.

1 **The figure pictured here has rotational symmetry. Answer the questions below.** 6 points per question

(1) Plot the center of symmetry, O of this figure.

(2) What is the corresponding vertex to A? 〈Ans.〉

(3) What is the corresponding vertex to B? 〈Ans.〉

(4) What is the corresponding side to EF? 〈Ans.〉

(5) What is the corresponding side to CD? 〈Ans.〉

(6) What is the corresponding angle to C? 〈Ans.〉

(7) What is the same size as OF? 〈Ans.〉

(8) What is the same size as OB? 〈Ans.〉

2 **The figure pictured here has rotational symmetry. Answer the questions below.** 4 points per question

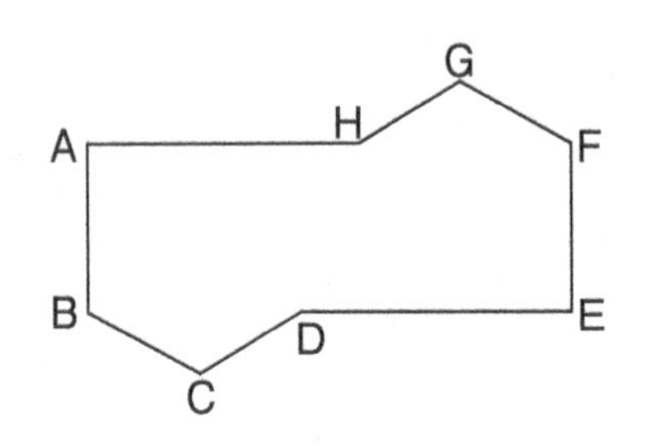

(1) Plot the center of symmetry, O of this figure.

(2) What is the corresponding side to AH? 〈Ans.〉

(3) Which triangle is congruent to triangle AHO? 〈Ans.〉

3 **Which figures below have line symmetry or rotational symmetry?** 10 points per question

A 〔Rectangle〕 B 〔Parallelogram〕 C 〔Rhombus〕 D 〔Square〕

E 〔Isosceles triangle〕 F 〔Equilateral triangle〕 G 〔Right triangle〕 H 〔Isosceles right triangle〕

I 〔Trapezoid〕 J 〔Regular pentagon〕 K 〔Regular hexagon〕 L 〔Circle〕

(1) Line symmetry 〈Ans.〉 ___ , ___ , ___ , ___ , ___ , ___ , ___ , ___ , ___ , ___

(2) Rotational symmetry 〈Ans.〉 ___ , ___ , ___ , ___ , ___ , ___

4 **Which letters below have line symmetry and rotational symmetry?** 10 points per question

A B F H I K M

N P S T U X Z

(1) Line symmetry 〈Ans.〉 ___ , ___ , ___ , ___ , ___ , ___ , ___ , ___

(2) Rotational symmetry 〈Ans.〉 ___ , ___ , ___ , ___ , ___ , ___

Circle 1 (Circumference)

Level ★★

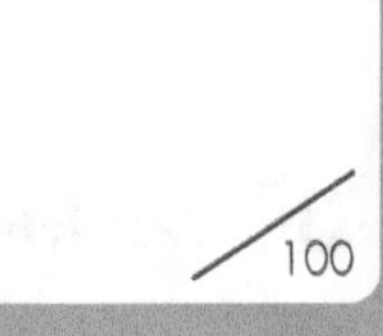

Date / / Name

■ The Answer Key is on page 90.

Don't forget!

- The perimeter of a circle is called the **circumference**.
- The length of the circumference is equal to the diameter times 3.14.

circumference = radius × 2 × 3.14

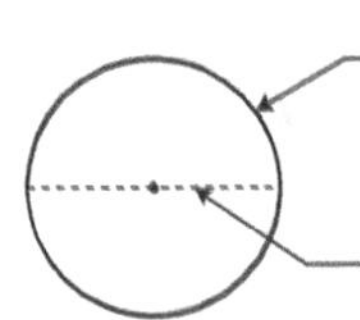

1 What is the circumference of each circle below?

5 points per question

(1)

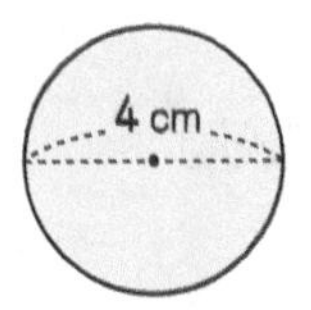

4 × 3.14 =

⟨Ans.⟩

(2)

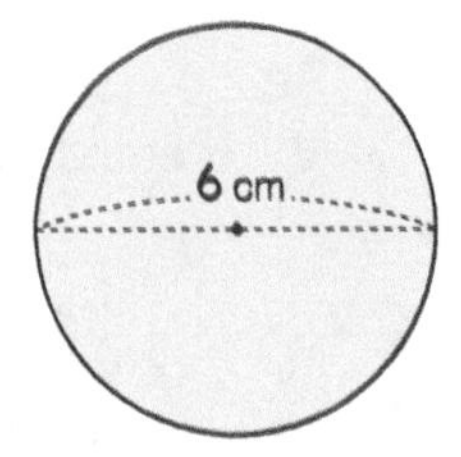

⟨Ans.⟩

(3)

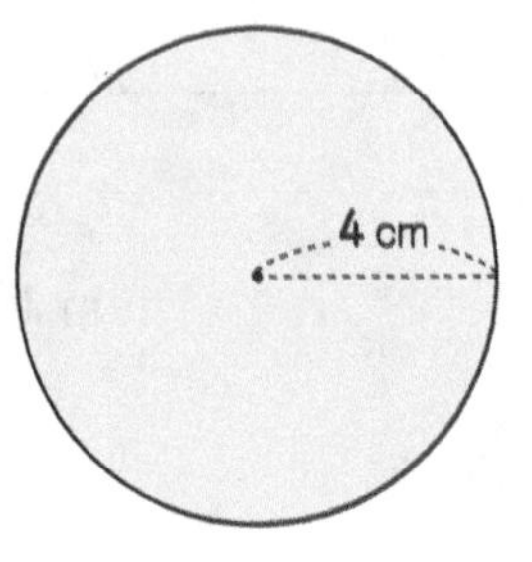

⟨Ans.⟩

(4)

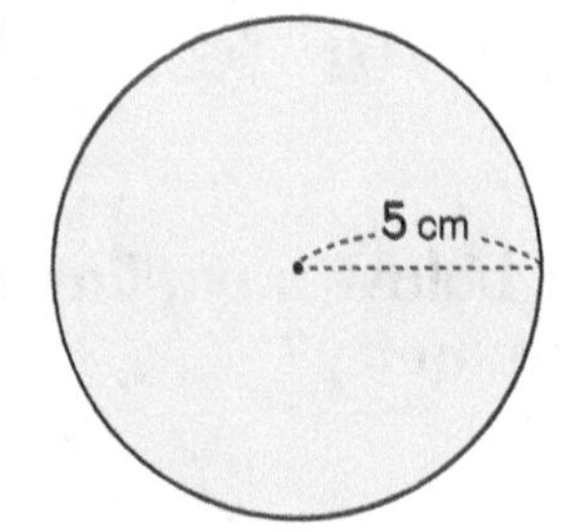

⟨Ans.⟩

(5)

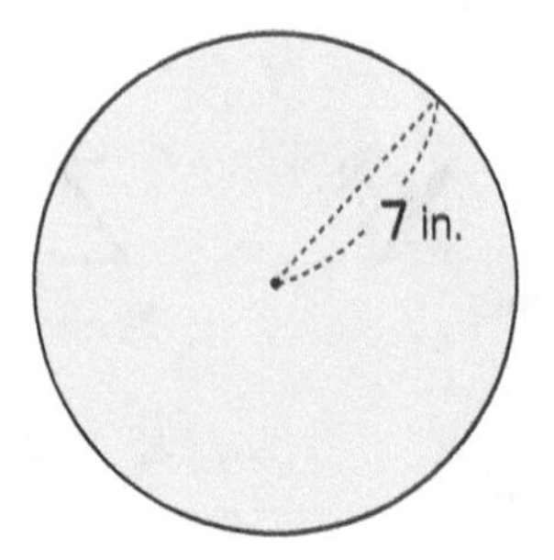

⟨Ans.⟩

(6)

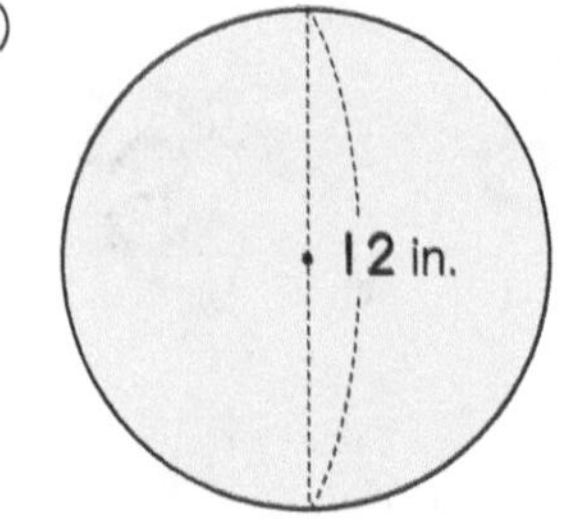

⟨Ans.⟩

2 What is the perimeter of each shape below?

7 points per question

(1)

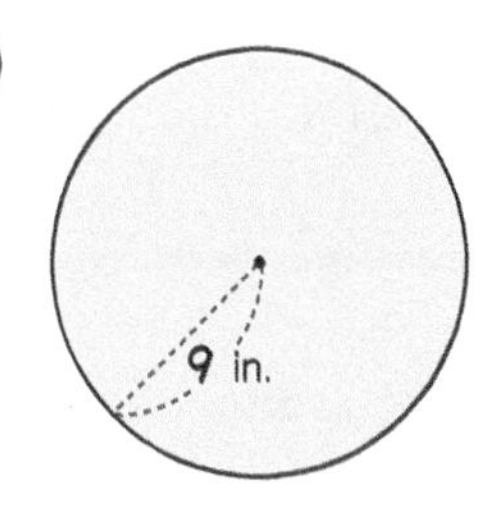

〈Ans.〉

(2)

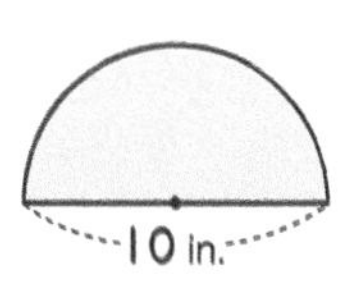

$10 \times 3.14 \times \frac{1}{2} + 10 =$

〈Ans.〉

(3)

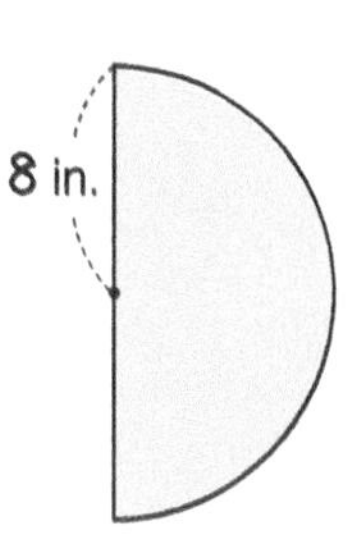

〈Ans.〉

(4)

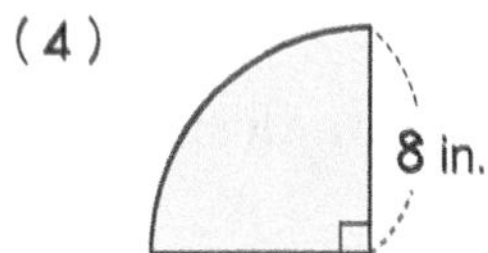

〈Ans.〉

(5)

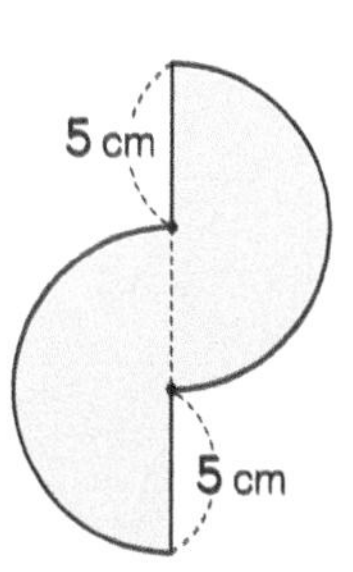

〈Ans.〉

(6)

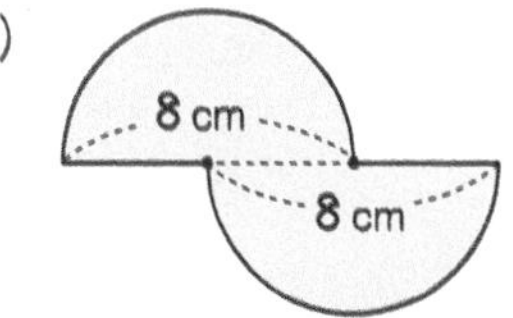

〈Ans.〉

(7)

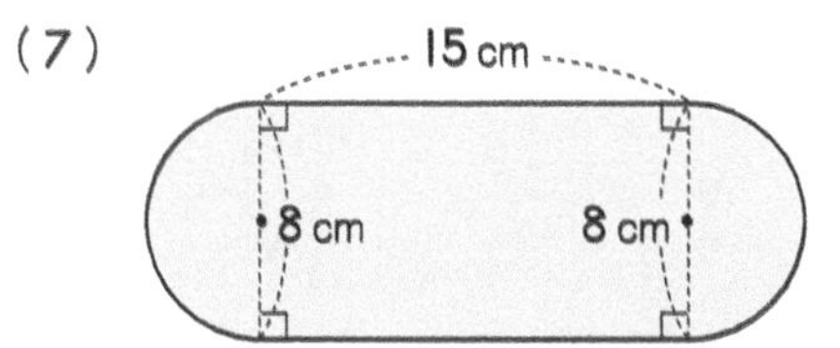

〈Ans.〉

(8)

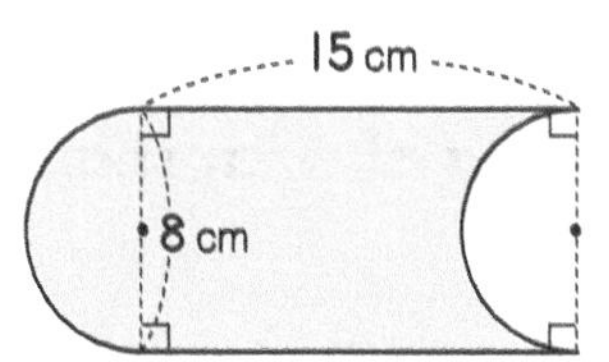

〈Ans.〉

(9)

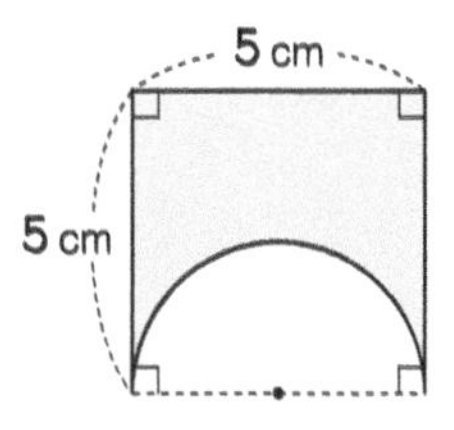

〈Ans.〉

(10)

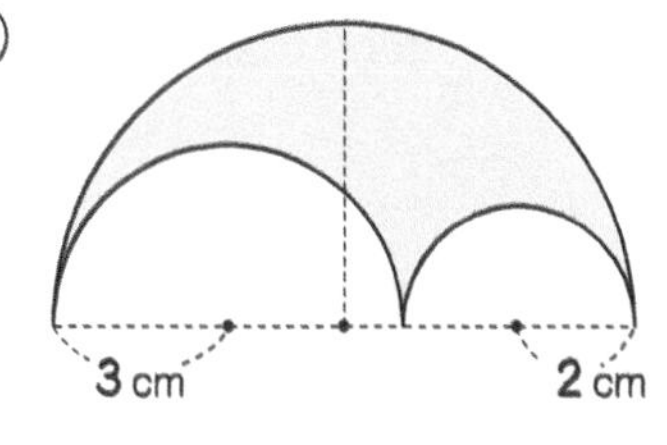

〈Ans.〉

18 Circle 2 (Area)

Level 

Date / / Name

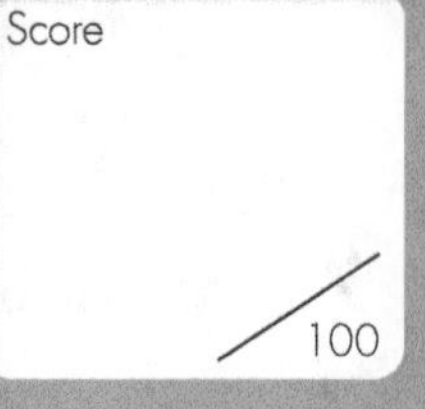

■ The Answer Key is on page 90.

Don't forget!

●The area of a circle is equal to the radius squared times 3.14.

area of circle = radius × radius × 3.14

= radius² × 3.14

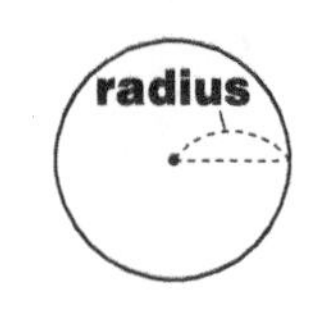

1 What is the area of each circle below?

4 points per question

(1)

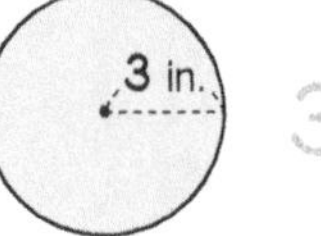

$3^2 \times 3.14 =$

〈Ans.〉

(2)

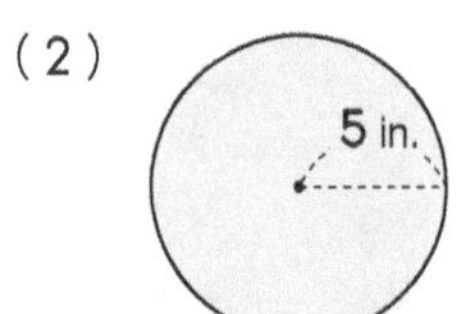

〈Ans.〉

(3)

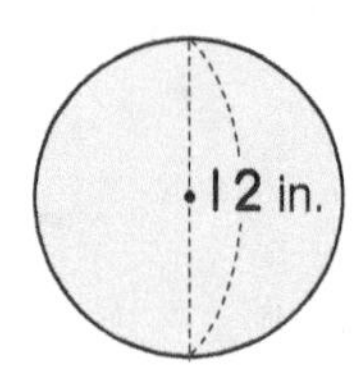

〈Ans.〉

(4)

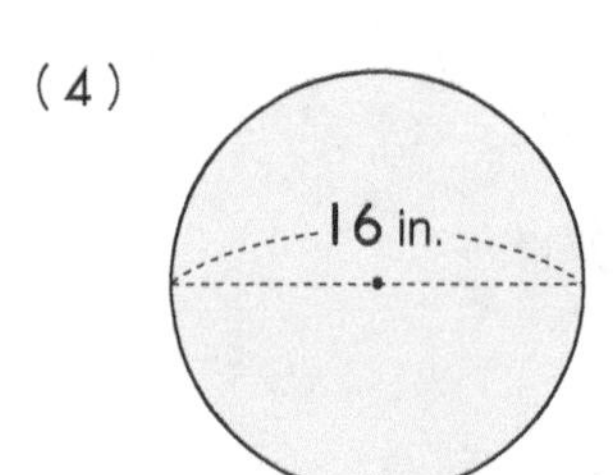

〈Ans.〉

2 What is the area of each shape below?

5 points per question

(1)

$4^2 \times 3.14 \times \frac{1}{2} =$

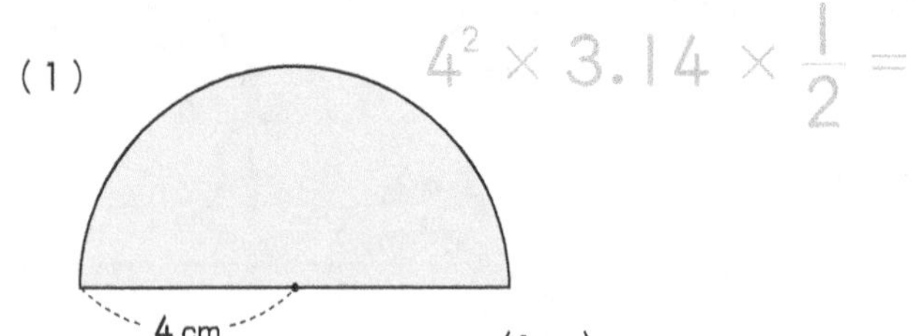

〈Ans.〉

(2)

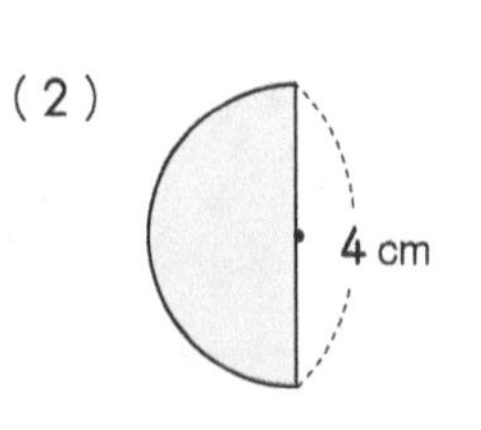

〈Ans.〉

(3)

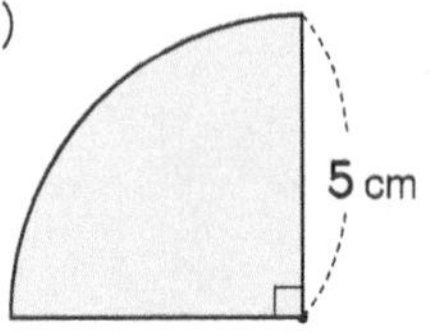

〈Ans.〉

(4)

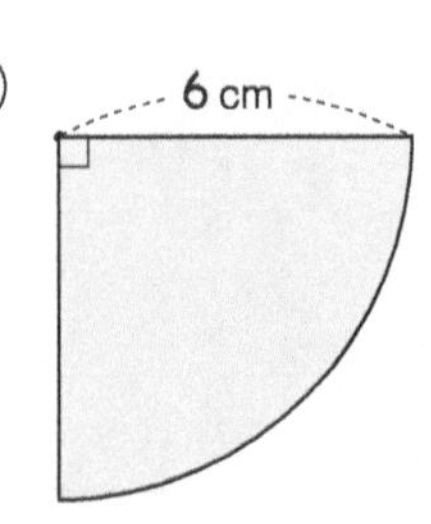

〈Ans.〉

3 What is the area of the shaded region in each shape below?

8 points per question

(1)

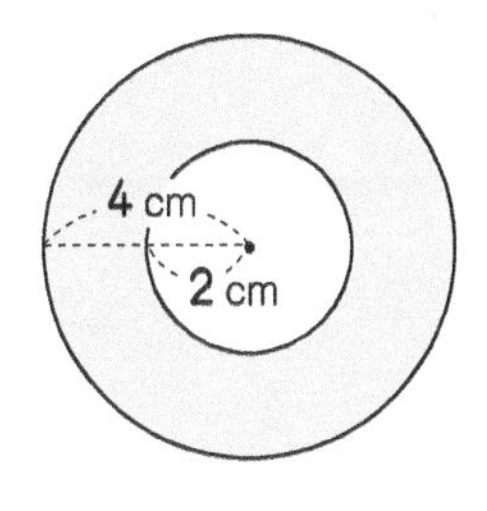

〈Ans.〉

(2)

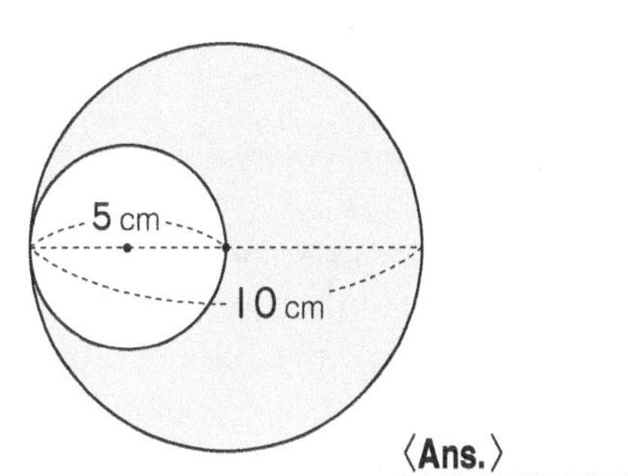

〈Ans.〉

(3)

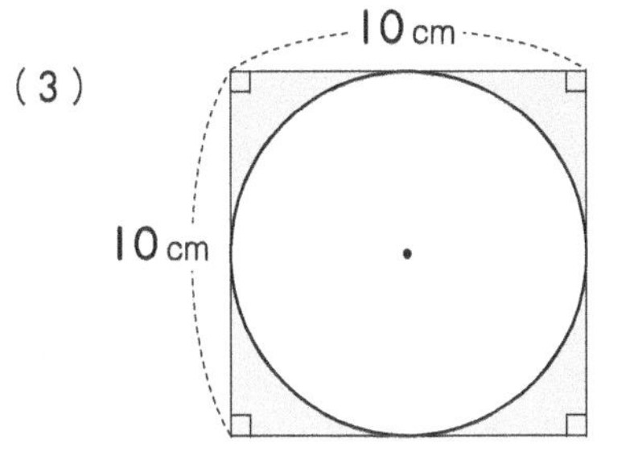

〈Ans.〉

(4)

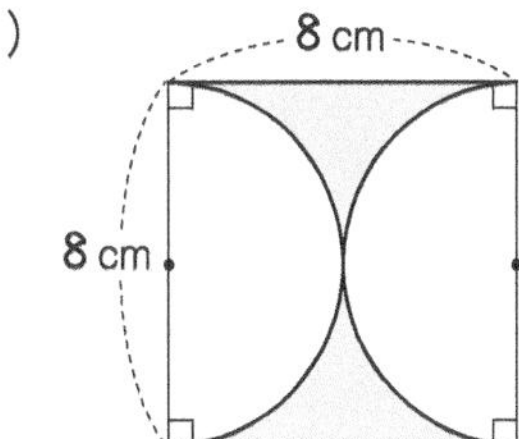

〈Ans.〉

(5)

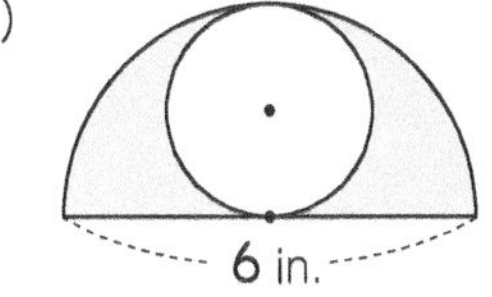

〈Ans.〉

(6)

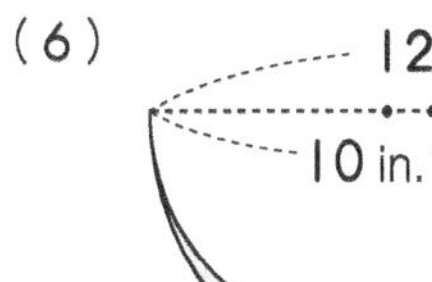

〈Ans.〉

(7)

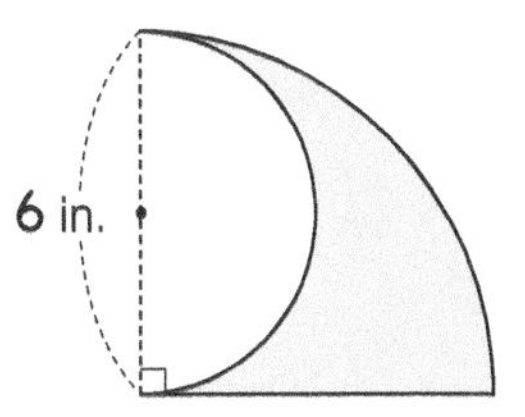

〈Ans.〉

(8)

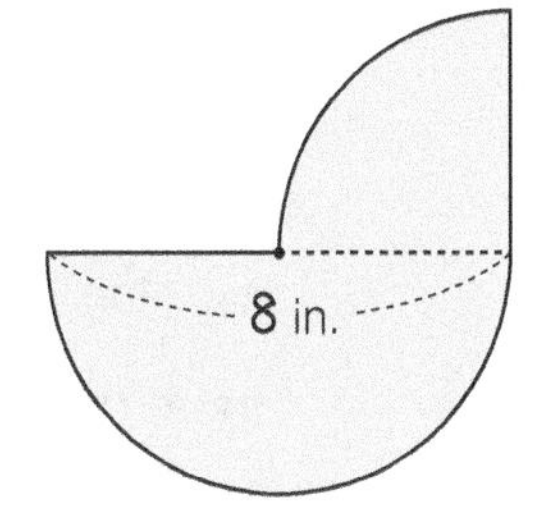

〈Ans.〉

19 Circle 3 (Arc & Sector)

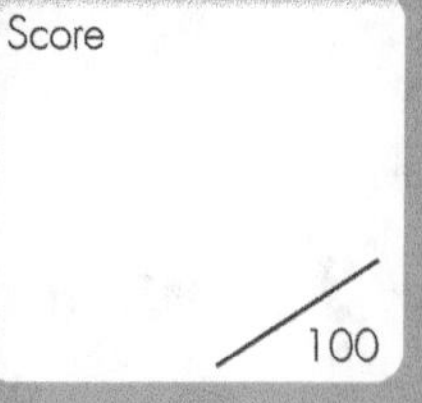

Level ★

Date / / Name

Score /100

■ The Answer Key is on page 91.

Don't forget!

- A part of the circumference is called an **arc**.
- The region bound by an arc and two radii of a circle is called a **sector**.
- The angle between two radii of a sector is called a **central angle**.
- The length of the arc is proportional to the measure of the central angle. For example, if the center angle is doubled, the arc length will be doubled.

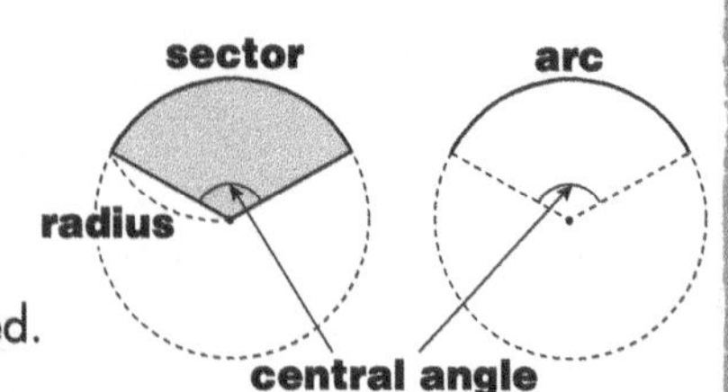

1 What is the length of each arc below?

10 points per question

(1)

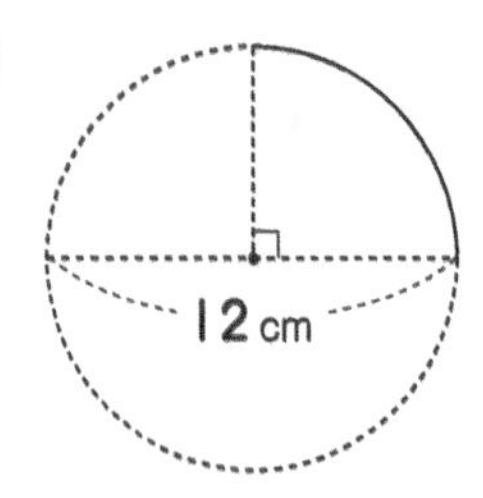

$\frac{\square}{4}$ of the circumference

⟨Ans.⟩

(2)

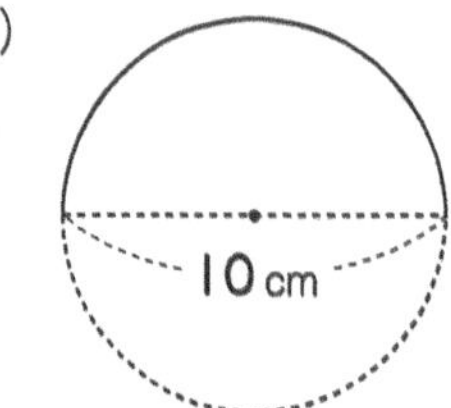

$\frac{\square}{2}$ of the circumference

⟨Ans.⟩

(3)

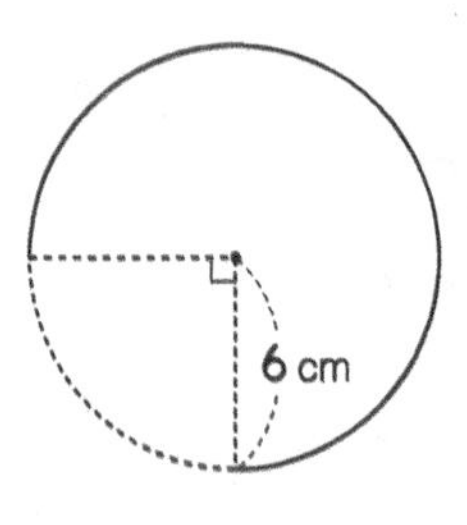

$\frac{\square}{4}$ of the circumference

⟨Ans.⟩

(4)

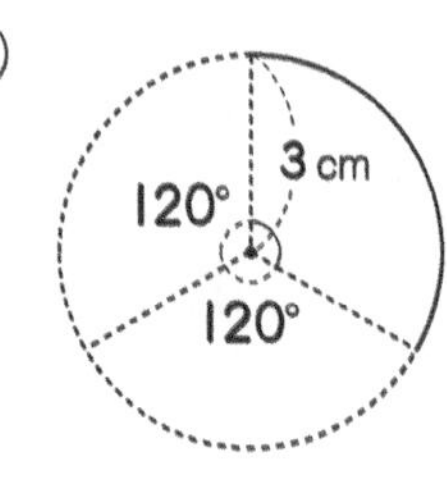

$\frac{1}{\square}$ of the circumference

⟨Ans.⟩

(5)

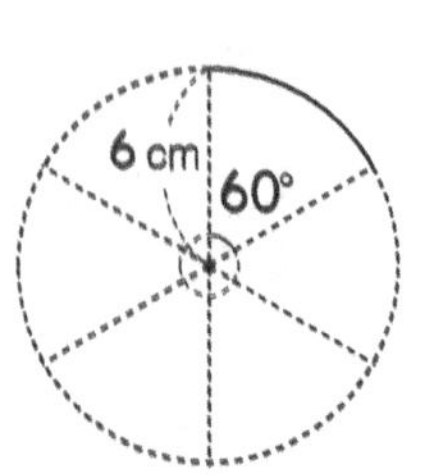

$\frac{1}{\square}$ of the circumference

⟨Ans.⟩

(6)

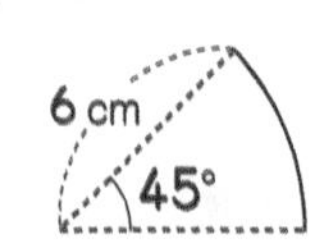

$\frac{1}{\square}$ of the circumference

⟨Ans.⟩

2 Which shapes are sectors below?

10 points for completion

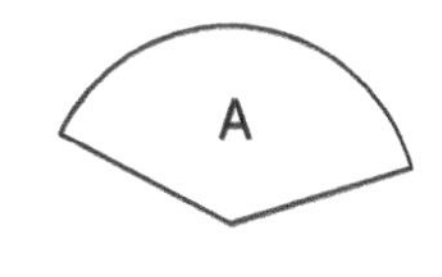

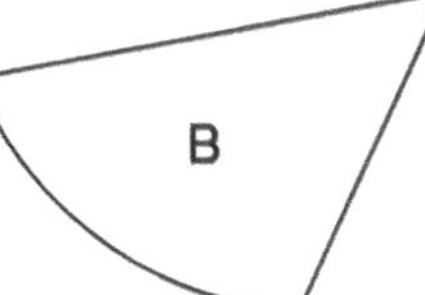

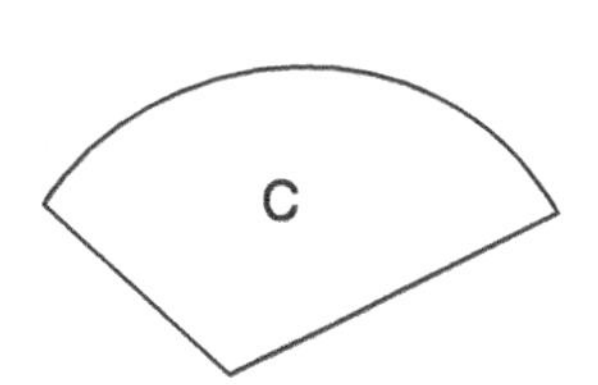

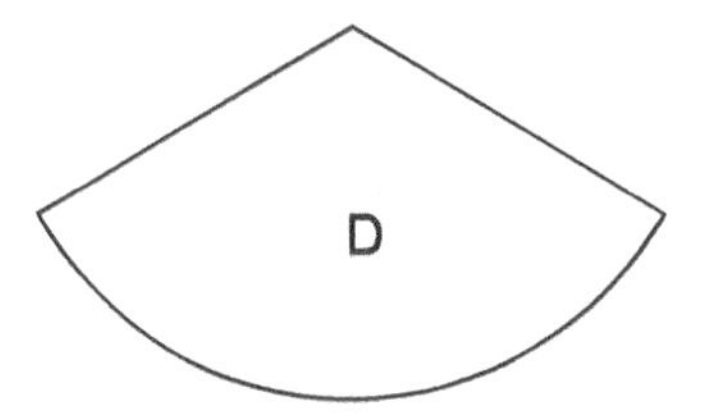

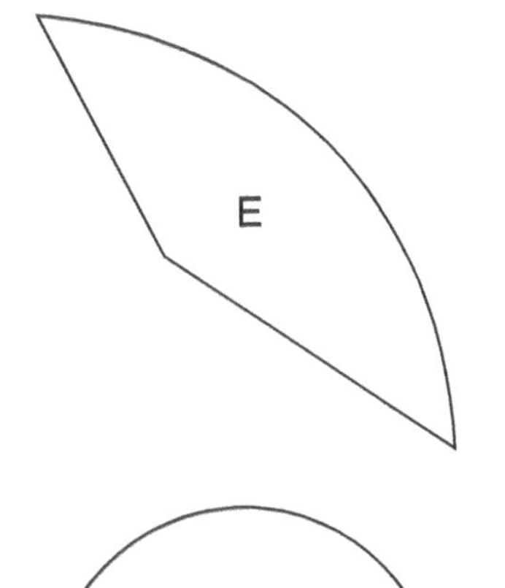

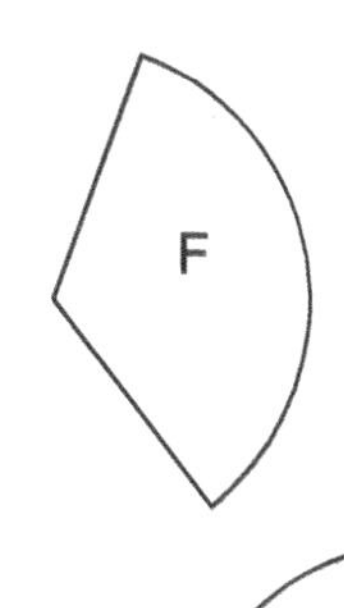

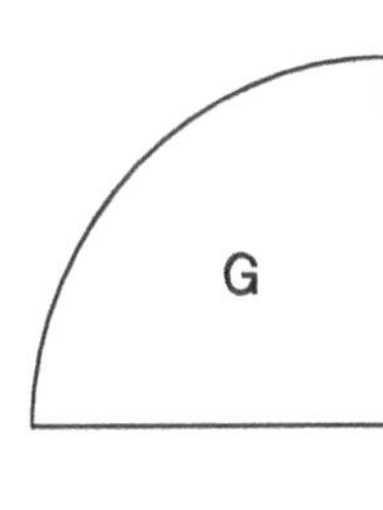

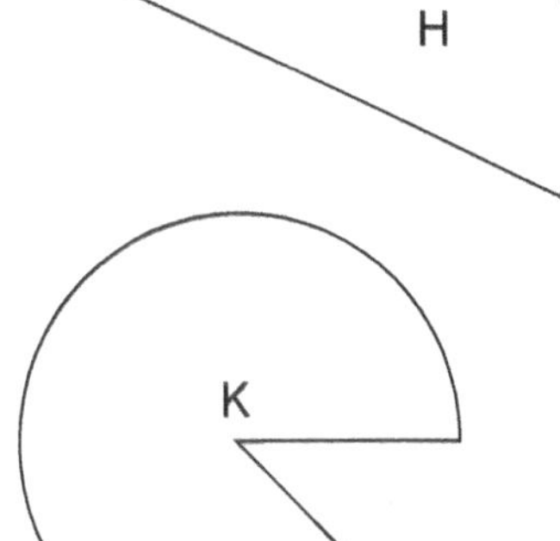

I

J

K

⟨Ans.⟩ , , , , ,

Don't forget!

The length of the curve in a sector
(The length of an arc)
= radius × 2 × 3.14 × $\frac{a}{360}$ (a is the central angle)

The perimeter of a sector
= radius × 2 × 3.14 × $\frac{a}{360}$ + 2 × radius

3 What is the perimeter of each sector below?

10 points per question

(1)

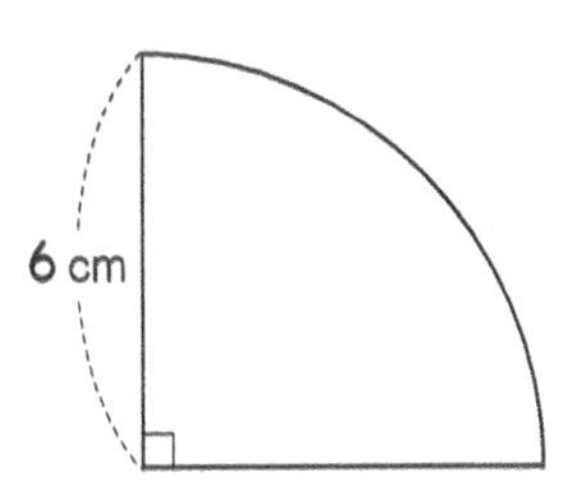

⟨Ans.⟩

(2)

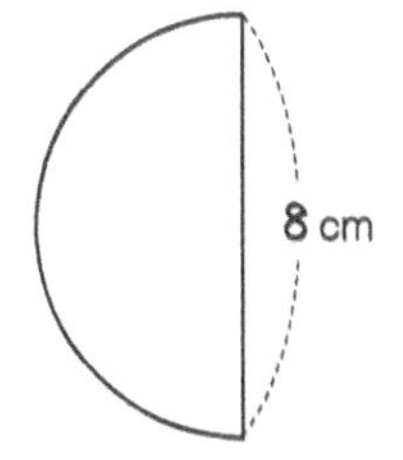

⟨Ans.⟩

(3)

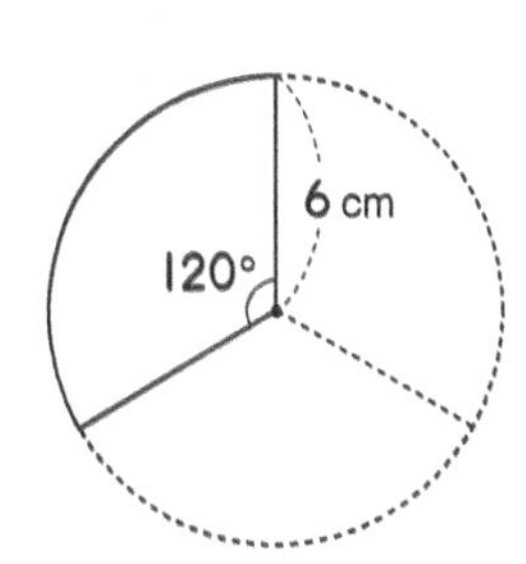

⟨Ans.⟩

20 Circle 4 (Area of Sector)

Level ★

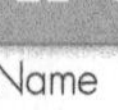

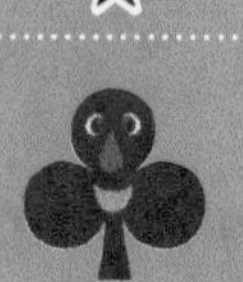

Date / / Name Score /100

■ The Answer Key is on page 91.

Don't forget!

The area of a sector = radius^2 × 3.14 × $\frac{a}{360}$ (a is the central angle of the sector)

1 What is the area of each sector below?

3 points per question

(1)

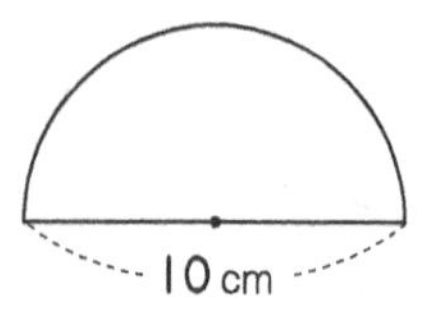

$\frac{1}{\boxed{2}}$ of the area of the circle

⟨Ans.⟩

(2)

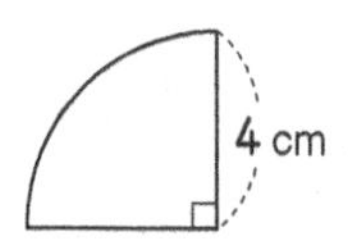

$\frac{1}{\square}$ of the area of the circle

⟨Ans.⟩

(3)

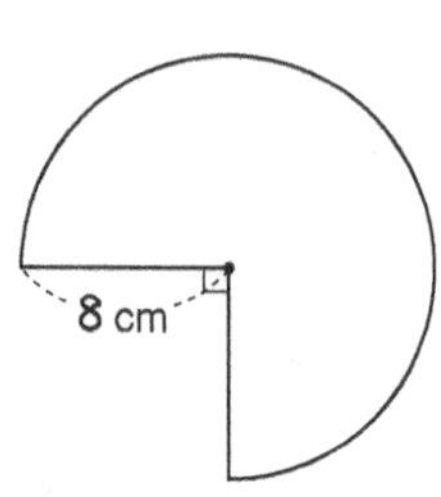

$\frac{3}{\square}$ of the area of the circle

⟨Ans.⟩

(4)

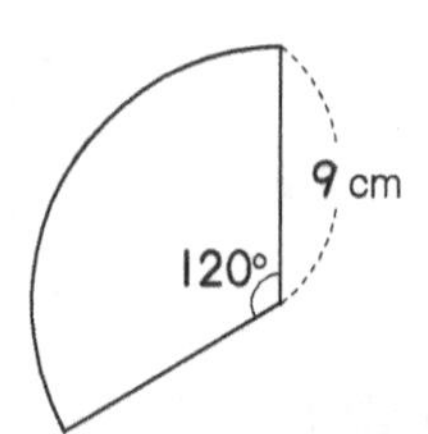

$\frac{120}{360} = \frac{1}{\square}$ of the area of the circle

⟨Ans.⟩

(5)

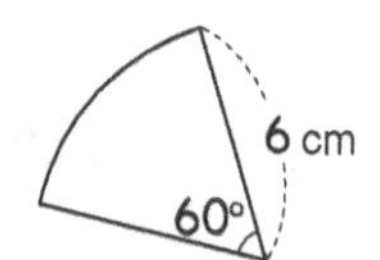

$\frac{60}{360} = \frac{1}{\square}$ of the area of the circle

⟨Ans.⟩

(6)

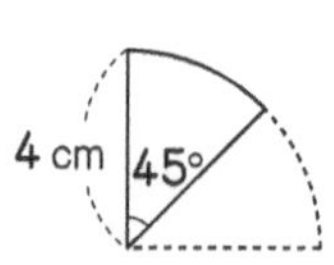

$\frac{\square}{360} = \frac{1}{\square}$ of the area of the circle

⟨Ans.⟩

2 What is the ratio of angles below at 360° to 1 in fractions?

4 points per question

(1) 90° ⟨Ans.⟩ $\frac{90}{360}$ =

(2) 240° ⟨Ans.⟩ $\frac{240}{360}$ =

(3) 270° ⟨Ans.⟩

(4) 135° ⟨Ans.⟩

(5) 30° ⟨Ans.⟩

(6) 210° ⟨Ans.⟩

(7) 15° ⟨Ans.⟩

(8) 75° ⟨Ans.⟩

3 What is the area of each sector below?

6 points per question

(1)

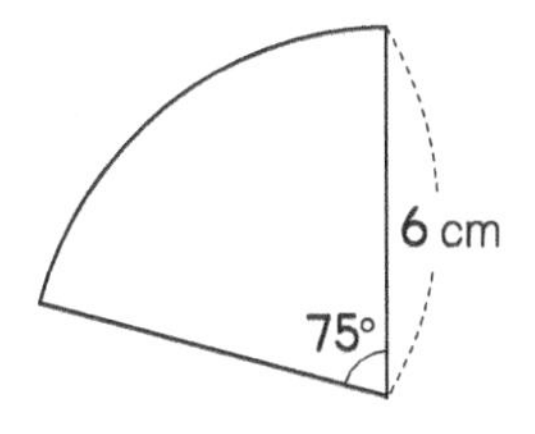

⟨Ans.⟩

(2)

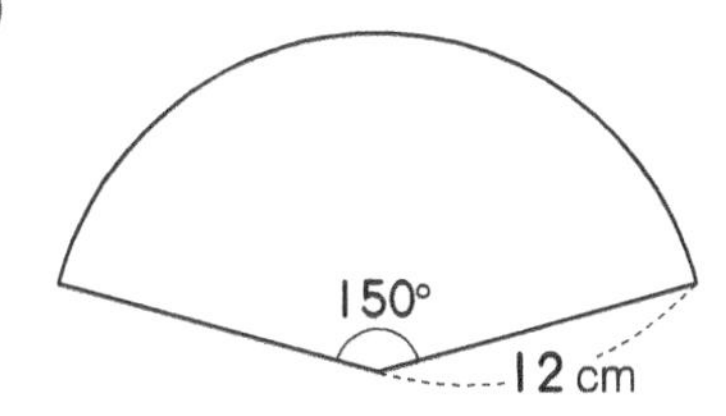

⟨Ans.⟩

(3)

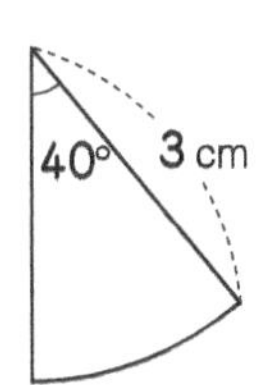

⟨Ans.⟩

(4) A sector with a radius of 8 cm and a central angle of 90°

⟨Ans.⟩

(5) A sector with a radius of 6 cm and a central angle of 120°

⟨Ans.⟩

Don't forget!

- The area of a sector is proportional to the measure of the central angle.
- For example, if the central angle is doubled, the area of the sector will be doubled.

4 What is the length of each arc below?

5 points per question

(1)

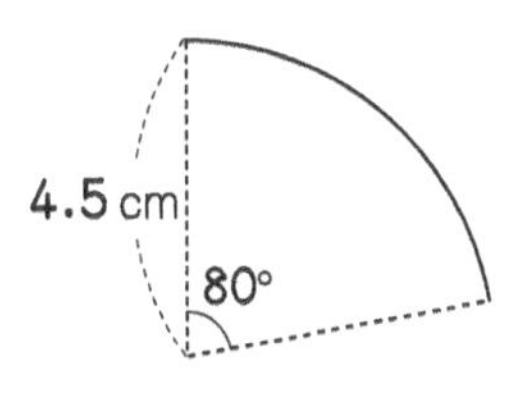

⟨Ans.⟩

(2)

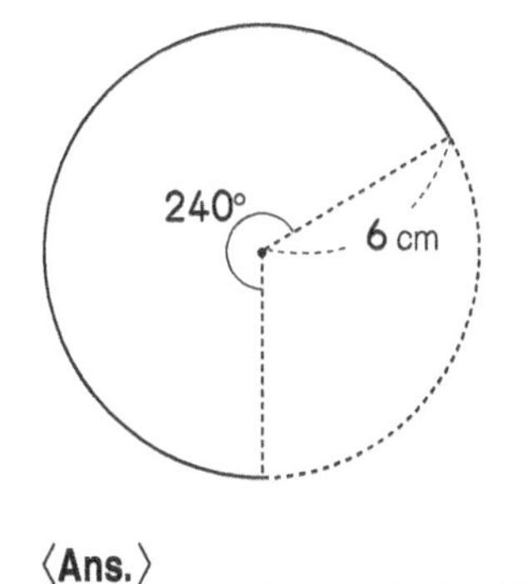

⟨Ans.⟩

(3) An arc with a radius of 9 cm and a central angle of 60°

⟨Ans.⟩

(4) An arc with a radius of 10 cm and a central angle of 36°

⟨Ans.⟩

21 Circle 5 (Use π for pi)

Level 

Date / /   Name

■ The Answer Key is on page 91.

Don't forget!

- The length of the circumference is about 3.14 times the diameter.
- The ratio of the circumference of a circle to the diameter is called **pi**.
- Pi is expressed as π, its magnitude is 3.1415926 ... and so on.

  Given the radius of a circle is represented with r, the length of the circumference (C) $\Rightarrow C = 2\pi r$

  the area of a circle (A) $\Rightarrow A = \pi r^2$

Find the length of each circumference below in terms of π. 8 points per question

Example

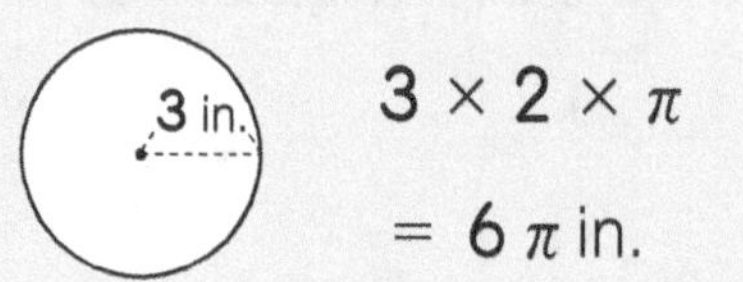

$3 \times 2 \times \pi$

$= 6\pi$ in.

(1)

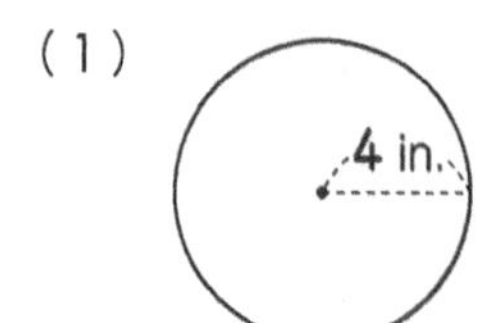

⟨Ans.⟩ ______

(2)

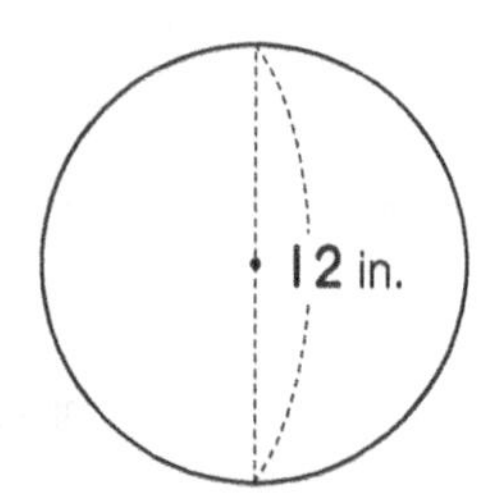

⟨Ans.⟩ ______

(3)

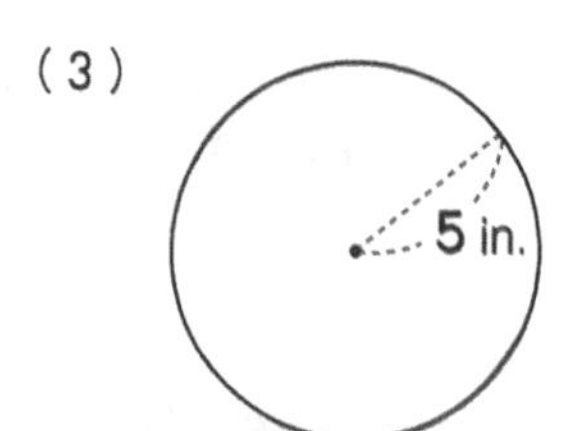

⟨Ans.⟩ ______

2 **Find the area of each circle below in terms of π.** 8 points per question

Example

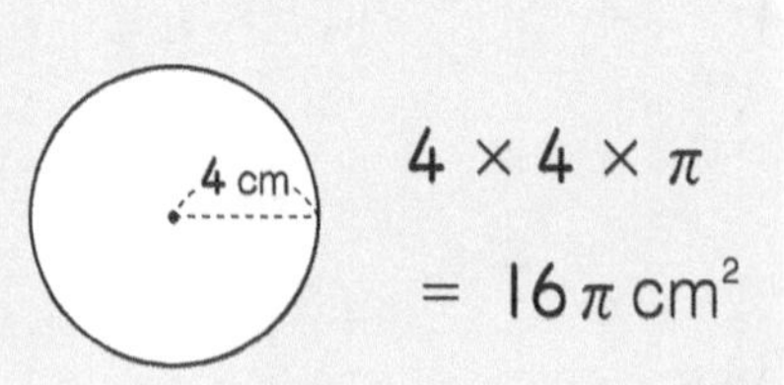

$4 \times 4 \times \pi$

$= 16\pi \text{ cm}^2$

(1)

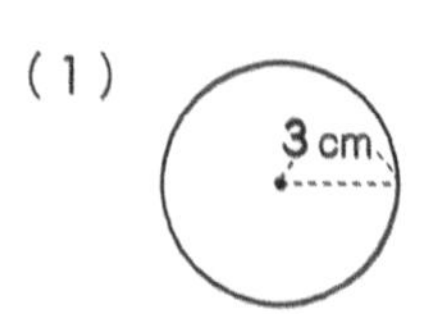

⟨Ans.⟩ ______

(2)

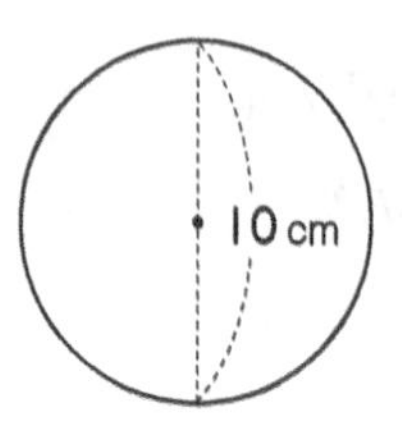

⟨Ans.⟩ ______

(3)

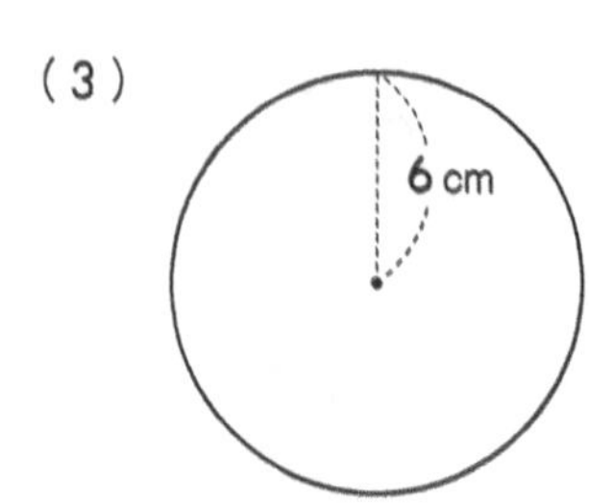

⟨Ans.⟩ ______

3 Find the length of each arc below in terms of π.

(1) 8 points (2)(3) 9 points per question

Example

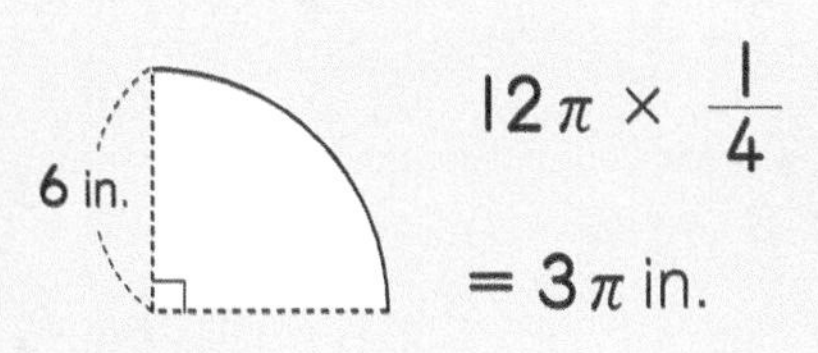

$12\pi \times \frac{1}{4}$

$= 3\pi$ in.

(1)

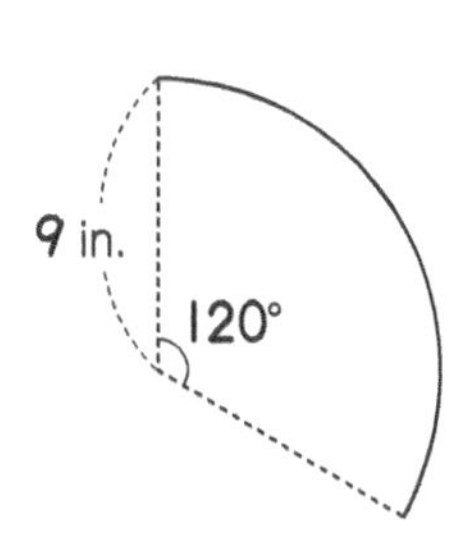

〈Ans.〉

(2)

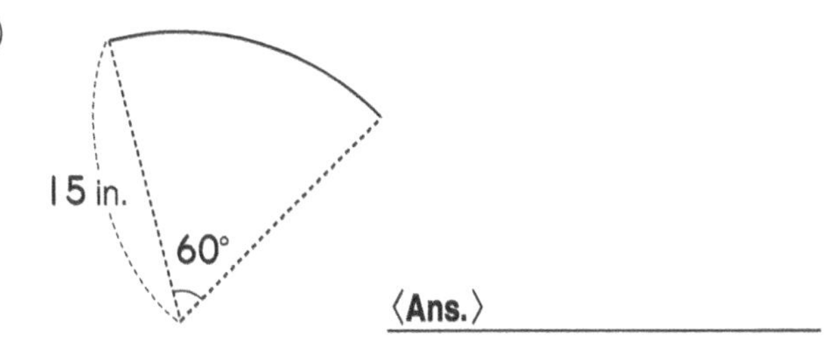

〈Ans.〉

(3)

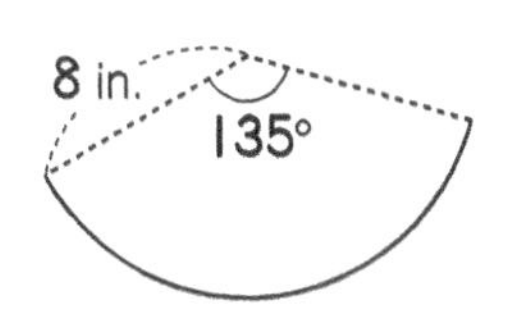

〈Ans.〉

4 Find the area of each sector below in terms of π.

(1) 8 points (2)(3) 9 points per question

Example

6 cm 30°

$6^2 \times \pi \times \frac{1}{12}$

$= 3\pi\ \text{cm}^2$

(1)

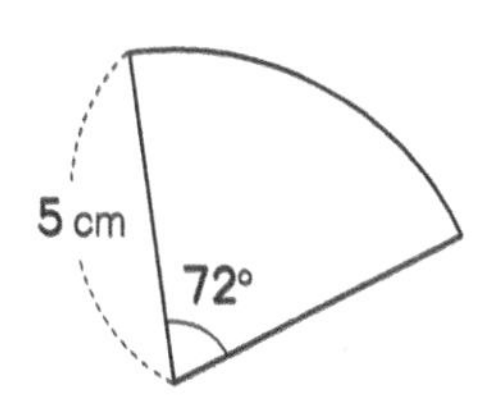

〈Ans.〉

(2)

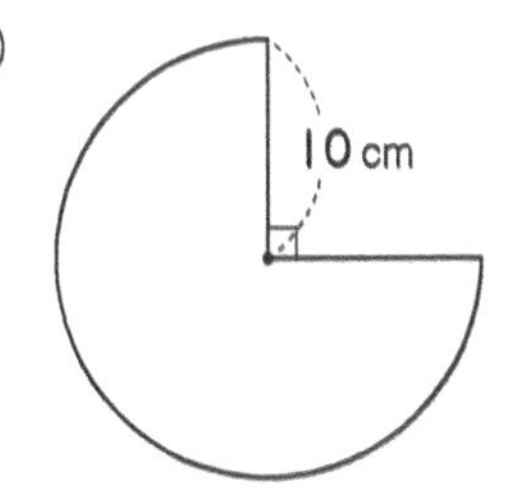

〈Ans.〉

(3)

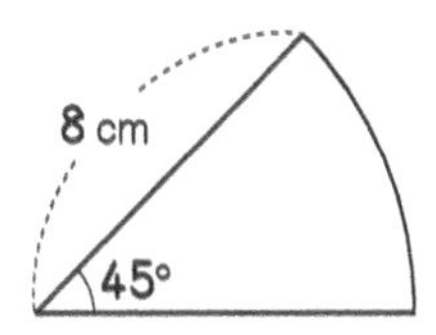

〈Ans.〉

22 Perimeter & Area of Shapes

Level ★★★

Date / /

Name

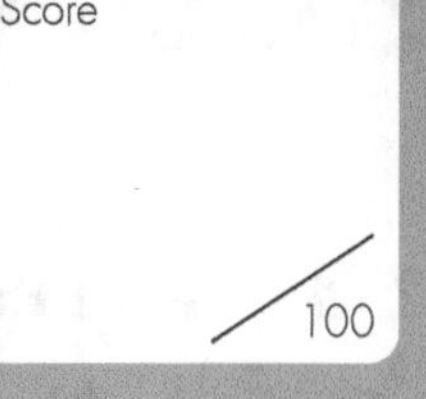

■ The Answer Key is on page 91.

1 What is the perimeter of each shape below?

3 points per question

(1) [Isosceles triangle]

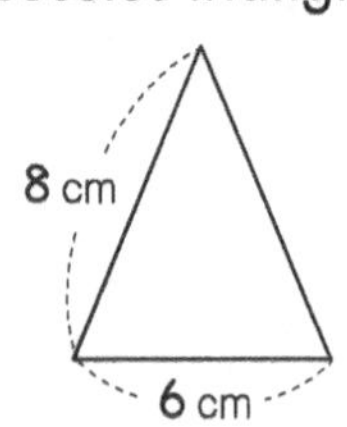

〈Ans.〉

(2) [Square]

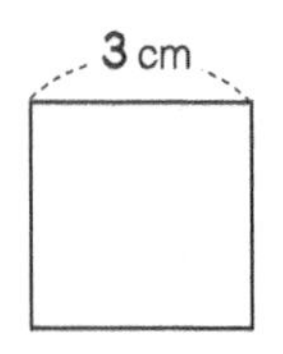

〈Ans.〉

(3) [Rectangle]

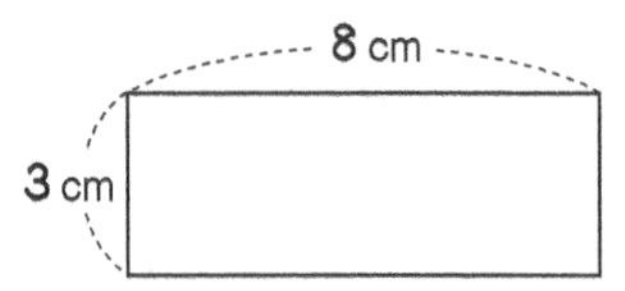

〈Ans.〉

(4) [Regular pentagon]

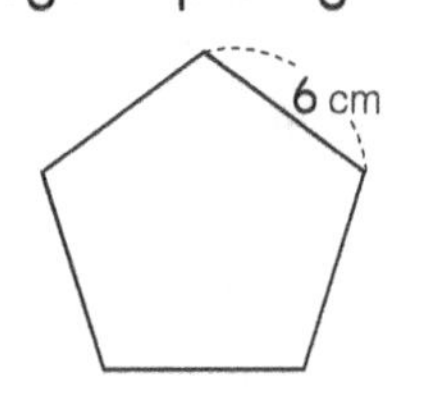

〈Ans.〉

(5) [Regular octagon]

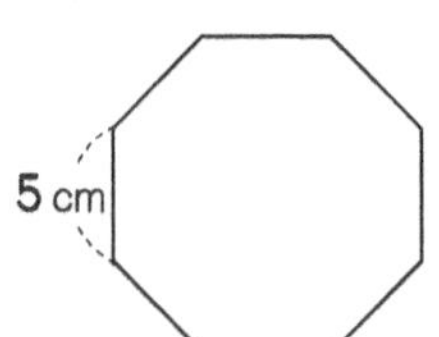

〈Ans.〉

(6) [Parallelogram]

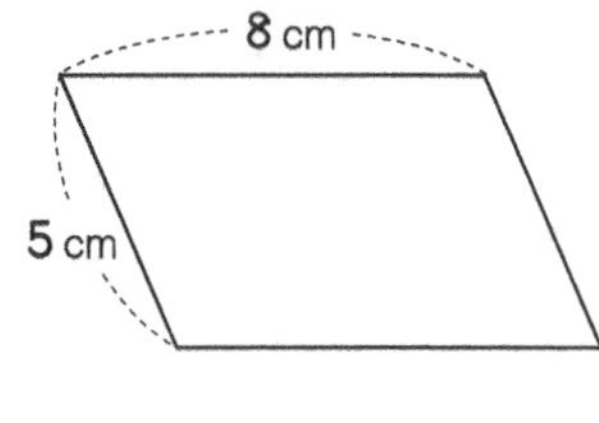

〈Ans.〉

2 Find the perimeter of the shaded region in each shape below in terms of π.

(1)(2) 7 points per question (3) 8 points

Example

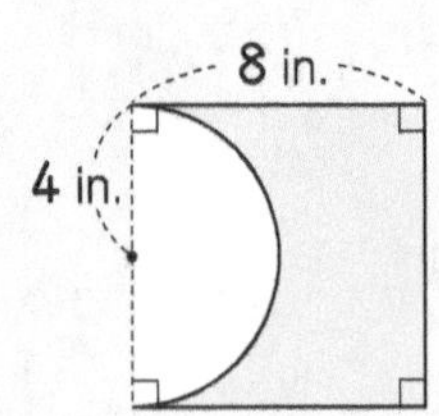

$8 \times 3 + 8\pi \times \frac{1}{2}$

$= (24 + 4\pi)$ in.

(1)

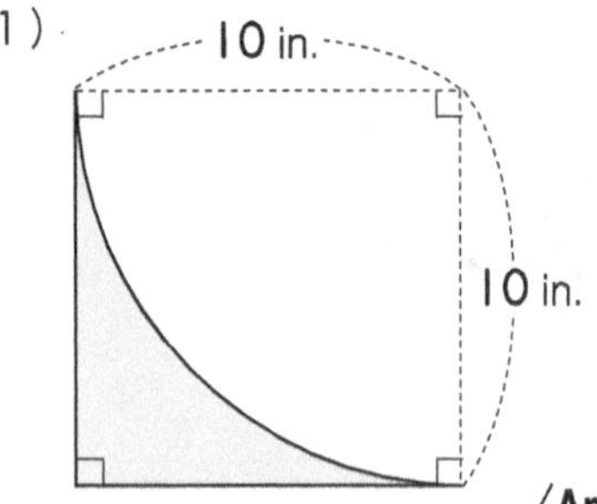

〈Ans.〉

(2)

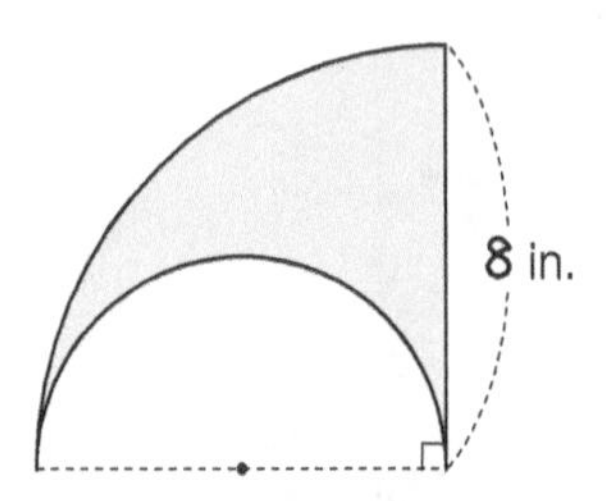

〈Ans.〉

(3)

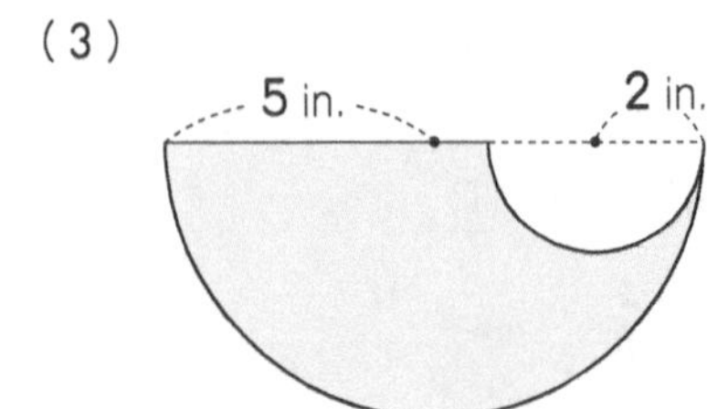

〈Ans.〉

3 Find the area and the perimeter of the shaded region in each shape below in terms of π.

10 points per question

(1)

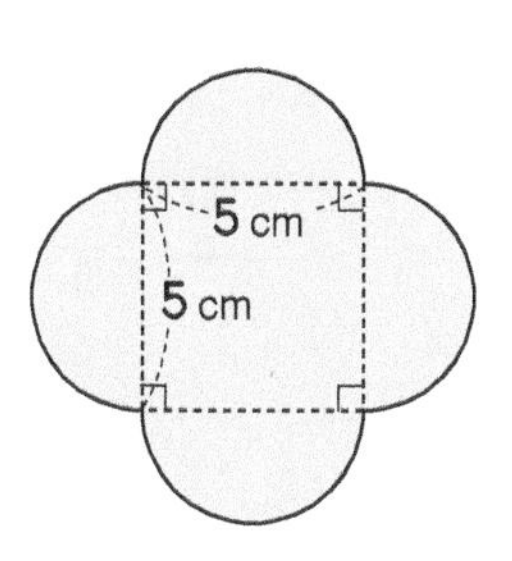

〈Ans.〉 Area : ____________________

Perimeter : ____________________

(2)

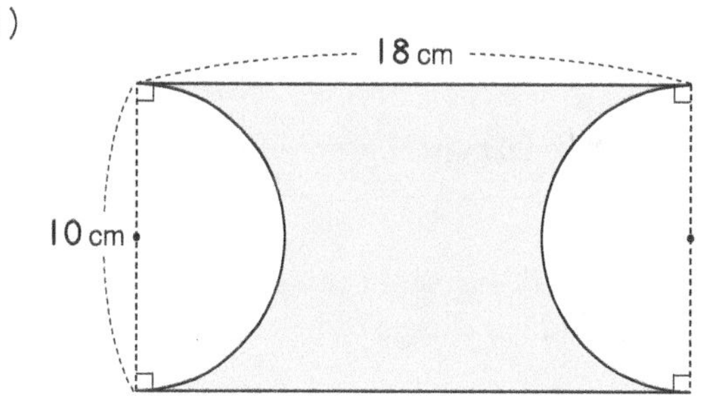

〈Ans.〉 Area : ____________________

Perimeter : ____________________

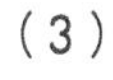

(3)

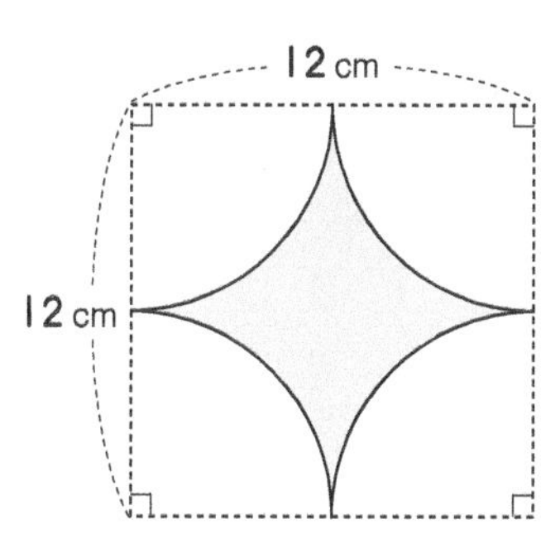

〈Ans.〉 Area : ____________________

Perimeter : ____________________

(4)

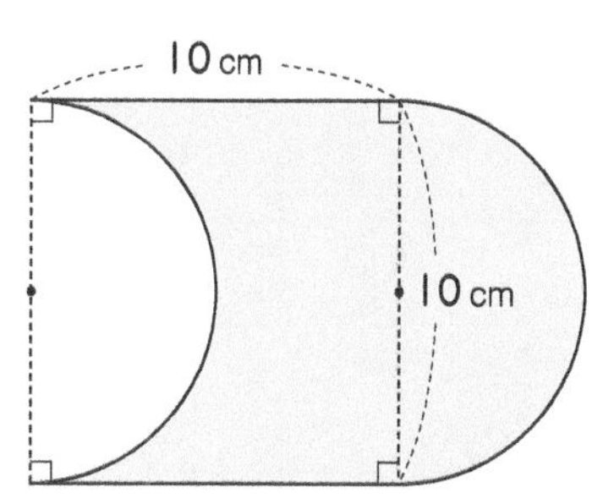

〈Ans.〉 Area : ____________________

Perimeter : ____________________

(5)

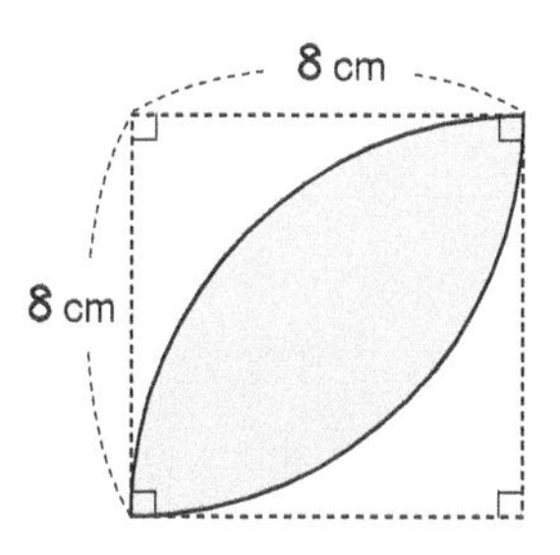

p.96

〈Ans.〉 Area : ____________________

Perimeter : ____________________

(6)

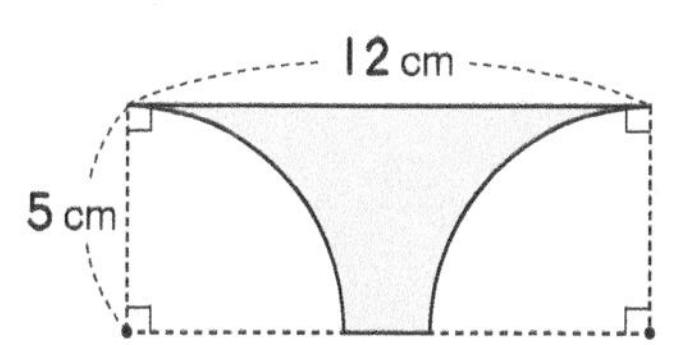

p.96

〈Ans.〉 Area : ____________________

Perimeter : ____________________

23 Area of Polygons

Level ★★

Date / / Name

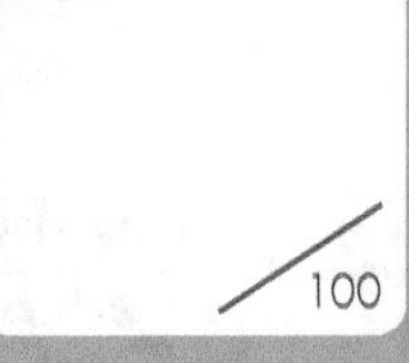

■ The Answer Key is on page 91.

Don't forget!

- The line segment connecting opposite vertices of a quadrilateral is called the **diagonal**.
- In a pentagon, you can draw two diagonals from one vertex.

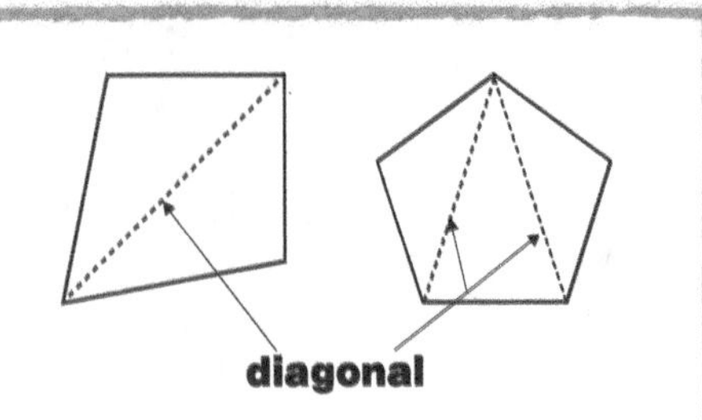

1 **Find the area of each shape below.** 9 points per question

(1)

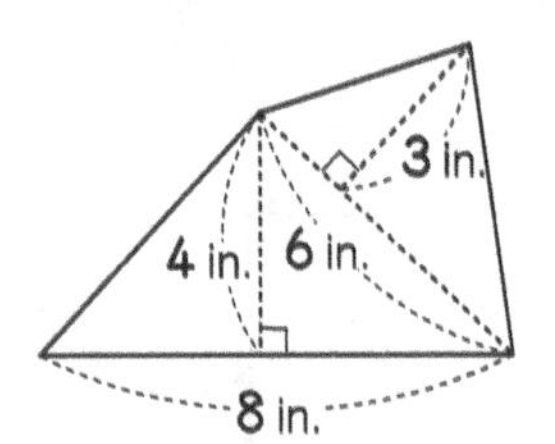

⟨Ans.⟩

(2)

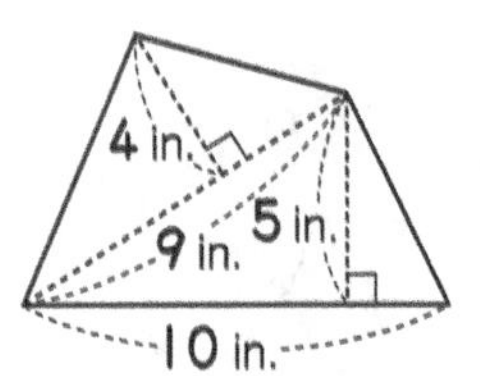

⟨Ans.⟩

(3)

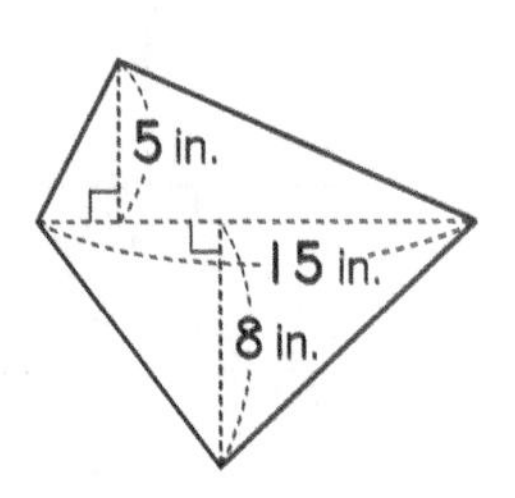

⟨Ans.⟩

(4)

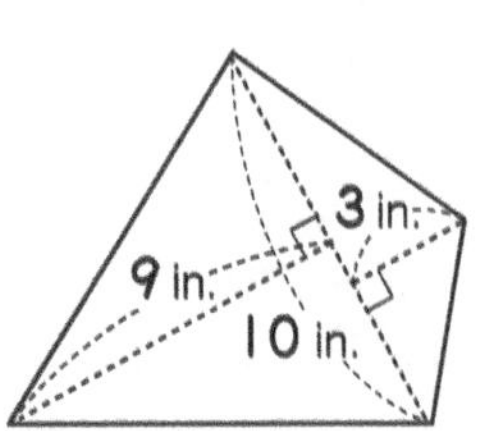

⟨Ans.⟩

(5)

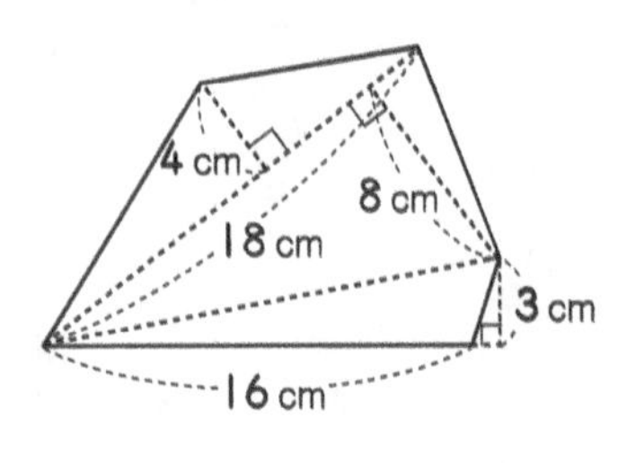

⟨Ans.⟩

(6)

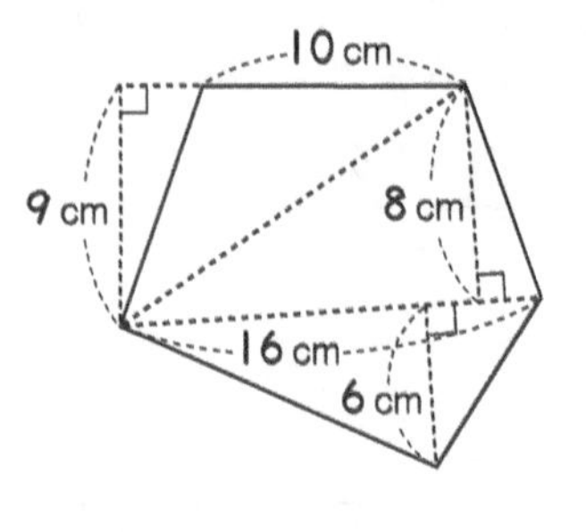

⟨Ans.⟩

2 **Answer the questions below about the polygons pictured on the left.** 9 points per question

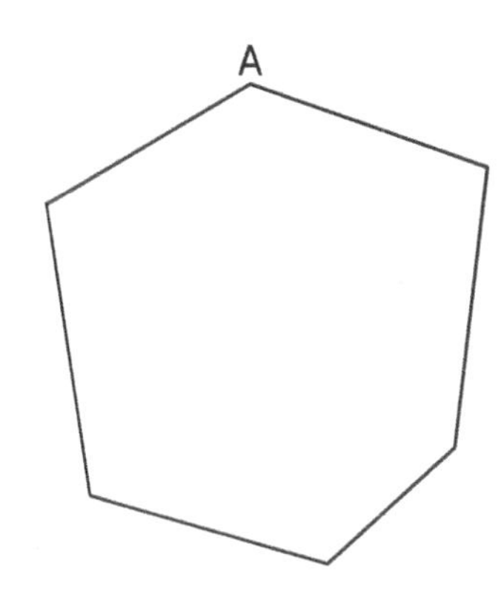

(1) How many diagonals can be drawn from one vertex A of the hexagon?

〈Ans.〉

(2) From the result of (1), how many triangles are there?

〈Ans.〉

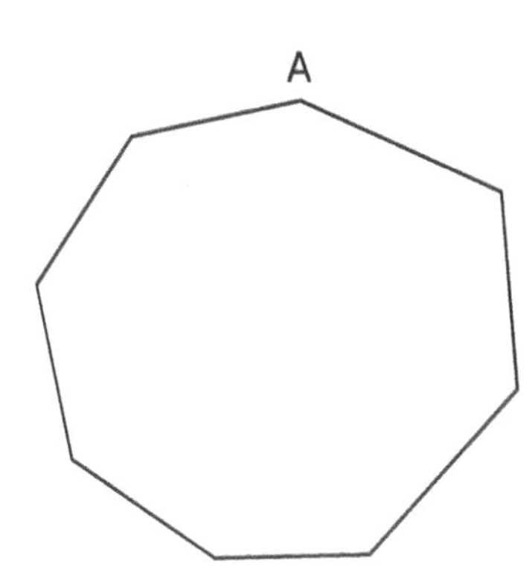

(3) How many diagonals can be drawn from one vertex A of the octagon?

〈Ans.〉

(4) From the result of (3), how many triangles are there?

〈Ans.〉

Don't forget!

- A polygon can be divided into several triangles by drawing a diagonal from one vertex.
- The area can be obtained by dividing a polygon into triangles and measuring the length of the sides and heights of the triangle.

Find the area of hexagon below. 10 points

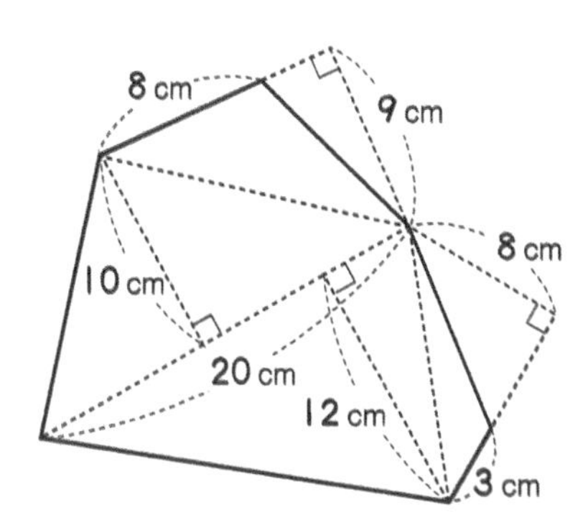

〈Ans.〉

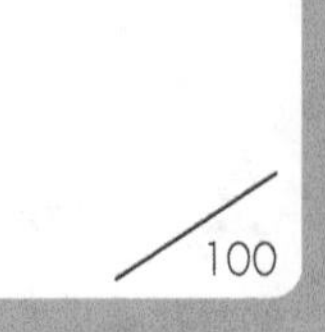

24 Rectangular Prisms 1 (Figure 1)

Level ★★

Date / / Name

Score /100

■ The Answer Key is on page 91.

Don't forget!

A solid made up of only rectangles or squares is called a **rectangular prism**.

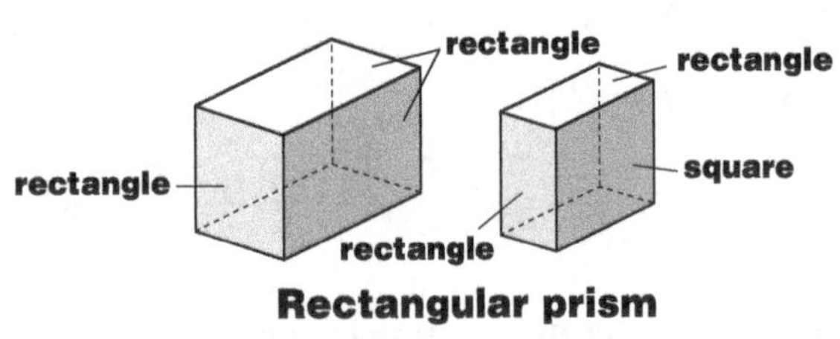

Rectangular prism

A solid made up of only six congruent squares is called a **cube**.

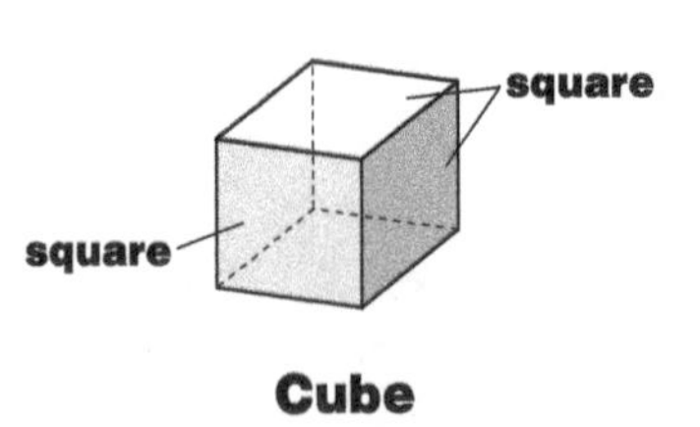

Cube

1 **The figures below are bound by different faces. Judging by their faces, which of the shapes are rectangular prisms? Which ones are cubes?**

20 points for completion

A
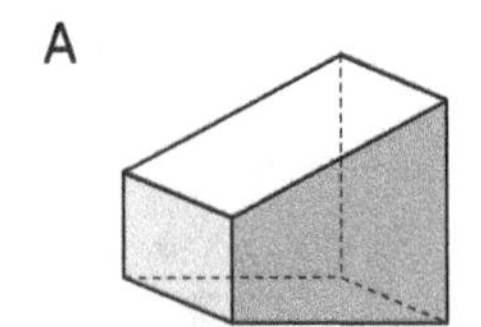

B
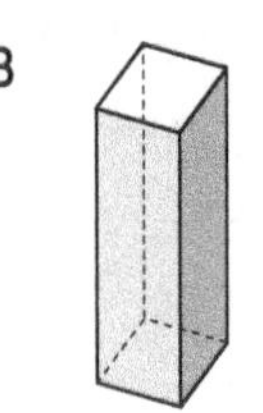

C
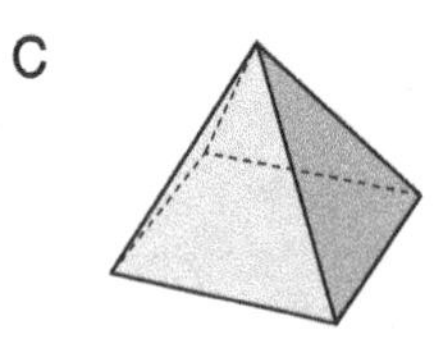

D
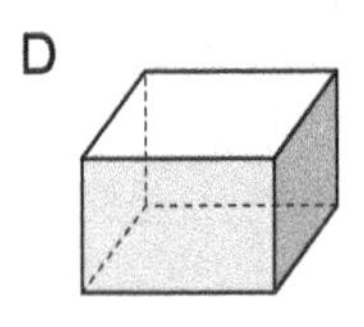

E
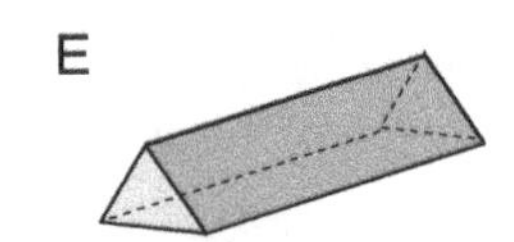

F
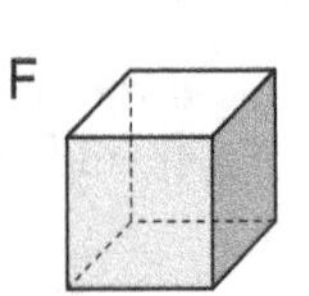

G
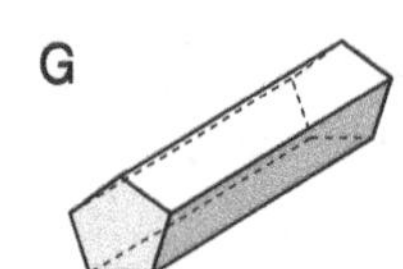

H
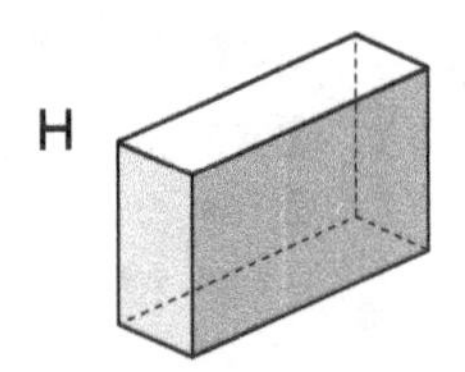

〈Ans.〉 Rectangular prism : ______ , ______ , ______

〈Ans.〉 Cube : ______

2 **Fill the table on the right with the number of faces, edges, and vertices of the rectangular prism and cube pictured below.**

20 points for completion

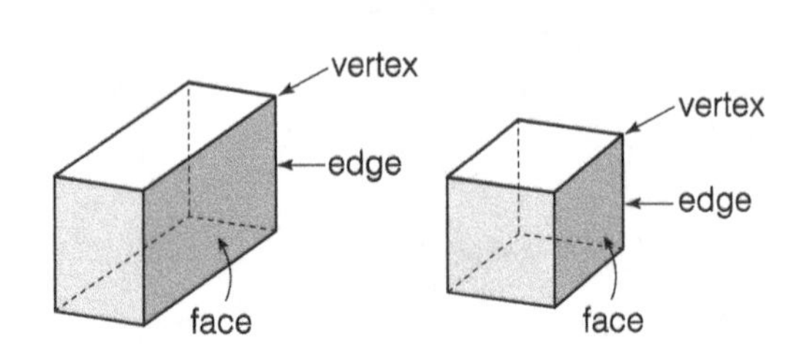

	Rectangular prism	Cube
Number of faces		
Number of edges		
Number of vertices		

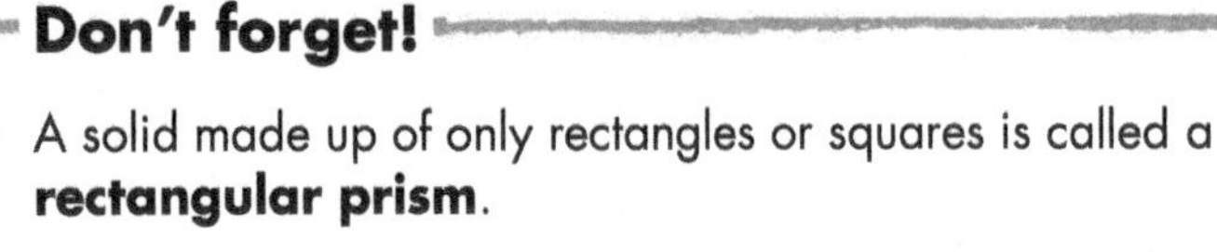

3 The rectangular prism on the right has edges of 5 cm, 7 cm, and 2 cm.

5 points per question

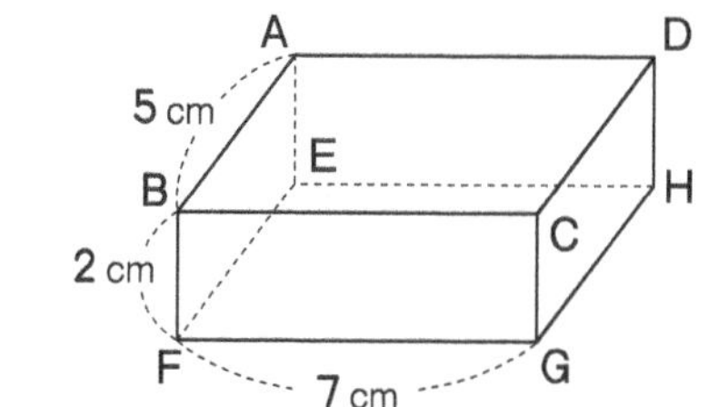

(1) Which edges are equal to edge AB?

〈Ans.〉 edge ______, edge ______, edge ______

(2) Which edges are equal to edge AD?

〈Ans.〉 edge ______, edge ______, edge ______

(3) Which edges are equal to edge AE?

〈Ans.〉 edge ______, edge ______, edge ______

(4) How many faces have a length of 2 cm and a width of 5 cm?

〈Ans.〉 ______

(5) How many faces have a length of 2 cm and a width of 7 cm?

〈Ans.〉 ______

(6) How many faces have a length of 5 cm and a width of 7 cm?

〈Ans.〉 ______

(7) How many edges have a length of 5 cm?

〈Ans.〉 ______

(8) Which edges have a length of 2 cm?

〈Ans.〉 edge ______, edge ______, edge ______, edge ______

4 A cube whose edges are all the same length is a special type of rectangular prism. The cube on the right has edges that are 6 cm long. Answer the questions below.

10 points per question

(1) Which edges are equal to edge AB?

〈Ans.〉

______, ______, ______

______, ______, ______, ______

______, ______, ______, ______

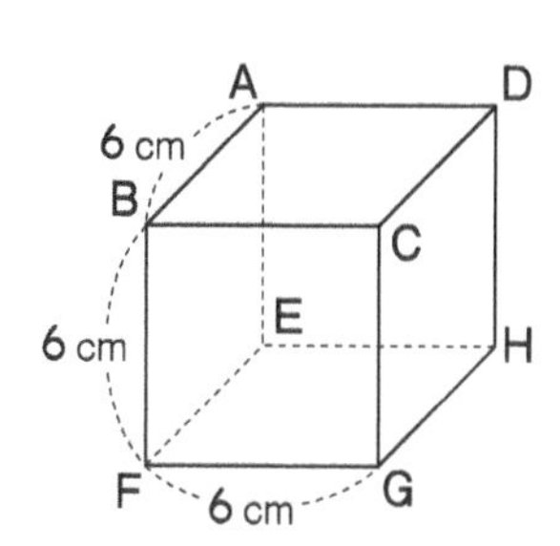

(2) How many faces have a length and a width of 6 cm?

〈Ans.〉 ______

25 Rectangular Prisms 2 (Figure 2)

Level ★★

Date / /

Name

Score /100

■ The Answer Key is on page 91.

1 **Answer the questions below about the rectangular prism on the right.** 7 points per question

(1) Edge AB and edge AD are perpendicular lines.
Which edges are perpendicular to edge AB through vertex A?

〈**Ans.**〉 edge AD , edge

(2) Which edges are perpendicular to edge BF?

〈**Ans.**〉 edge , edge , edge , edge

(3) Edges AB and DC are parallel. Which edges are parallel to edge AB?

〈**Ans.**〉 edge DC , edge , edge

(4) Which edges are parallel to edge BC?

〈**Ans.**〉 edge , edge , edge

2 **Answer the questions below about the rectangular prism on the right.** 8 points per question

(1) Edge AE intersects face *a* perpendicularly.
Which edges intersect face *a* perpendicularly?

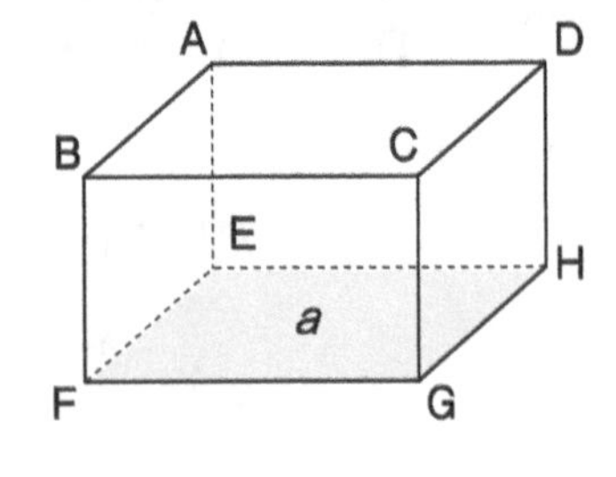

〈**Ans.**〉 edge AE , edge , edge , edge

(2) Which edges intersect face BCGF perpendicularly?

〈**Ans.**〉 edge , edge , edge , edge

3 **Answer the questions below about the rectangular prism on the right.**

8 points per question

(1) Face BFEA intersects face *a* perpendicularly. Which other faces intersect face *a* perpendicularly?

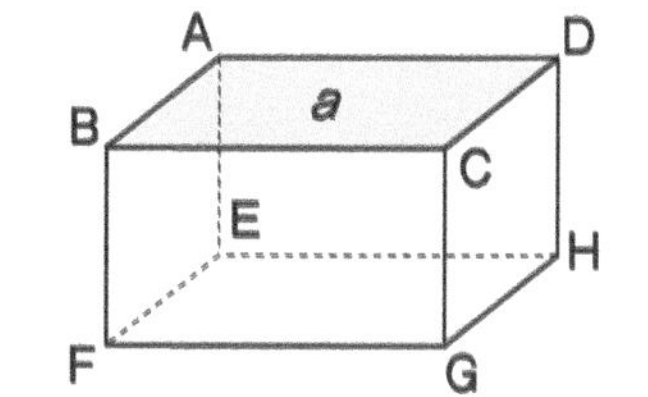

⟨**Ans.**⟩ face ________, face ________, face ________

(2) Which faces intersect face BFEA perpendicularly?

⟨**Ans.**⟩ face ________, face ________, face ________, face ________

(3) Which faces intersect face BFGC perpendicularly?

⟨**Ans.**⟩ face ________, face ________, face ________, face ________

4 **Face *a* and face EFGH are parallel in the rectangular prism on the right.**

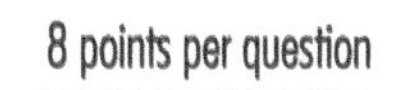

(1) Which face is parallel to face BFEA?

⟨**Ans.**⟩ face ________

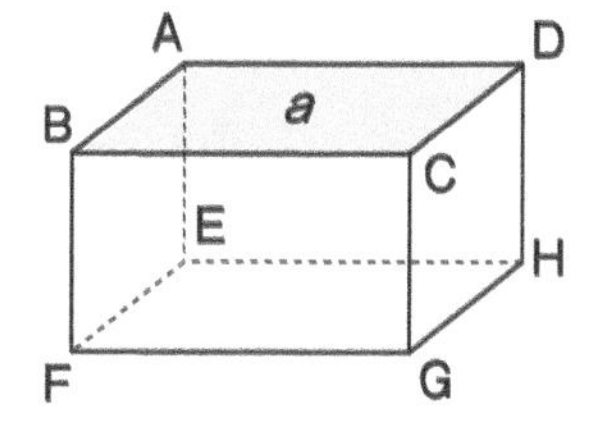

(2) Which face is parallel to face BFGC?

⟨**Ans.**⟩ face ________

5 **The edges (AB, BC, CD, and DA) of face ABCD are all parallel to face *a* in the rectangular prism on the right.**

8 points per question

(1) Which edges are parallel to face BFEA?

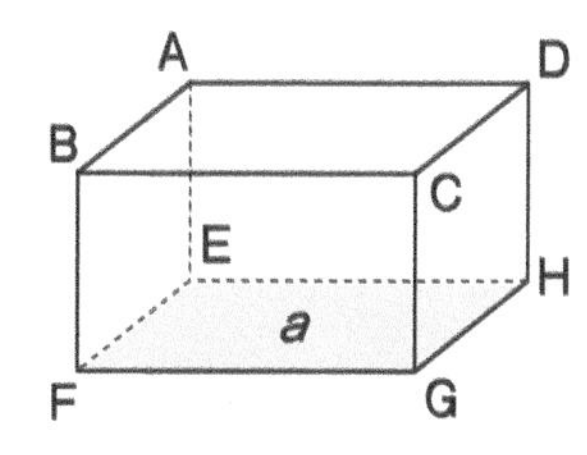

⟨**Ans.**⟩ edge ________, edge ________, edge ________, edge ________

(2) Which edges are parallel to face BFGC?

⟨**Ans.**⟩ edge ________, edge ________, edge ________, edge ________

26 Rectangular Prisms 3 (Net 1)

Level 

Date / / Name 

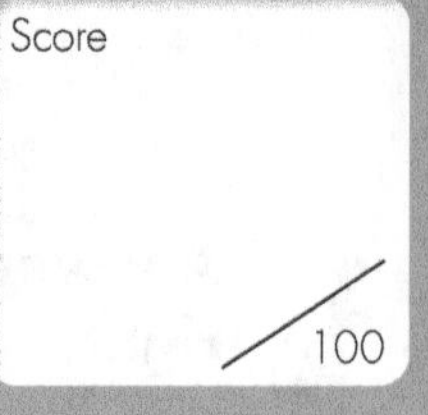

■ The Answer Key is on page 92.

1 Which nets would make the shapes on the left?

7 points per question

(1) (Built shape) (Net)

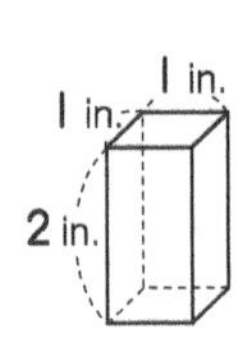

A
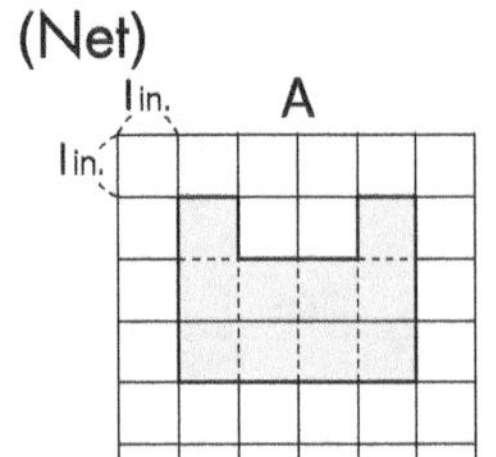

B
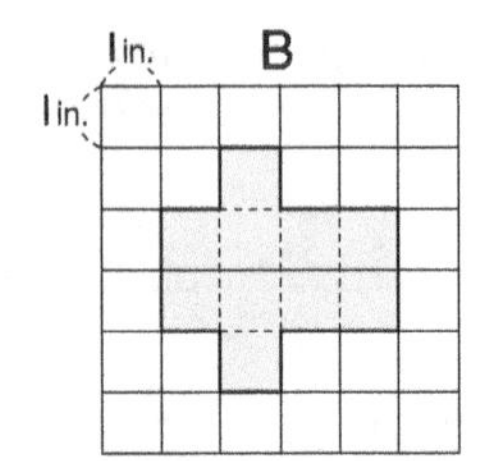

C
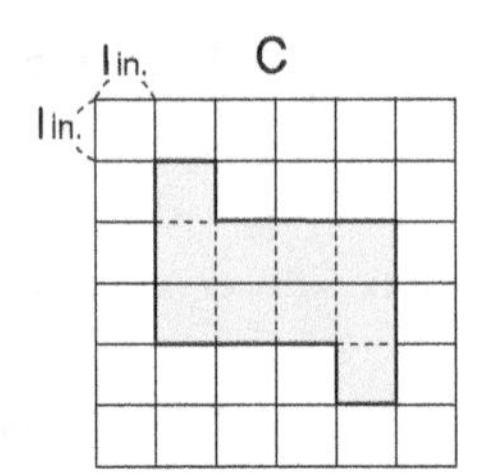

⟨Ans.⟩ ,

(2) (Built shape) (Net)

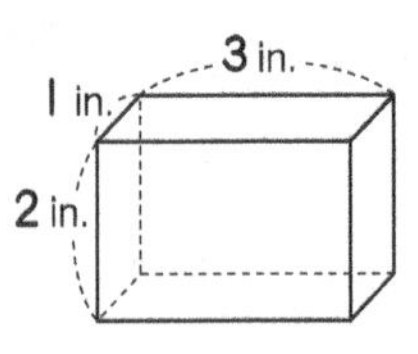

A
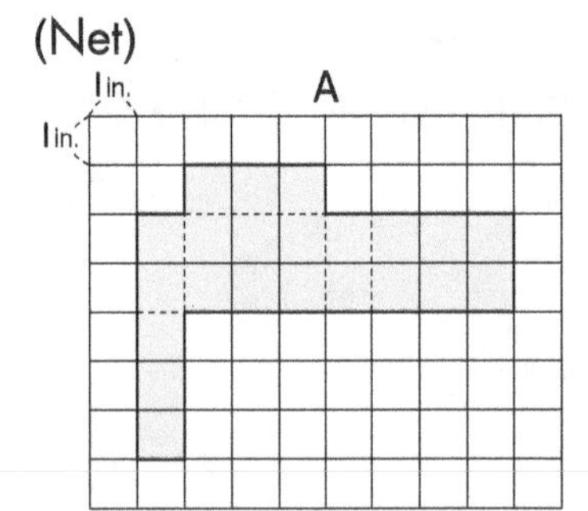

B
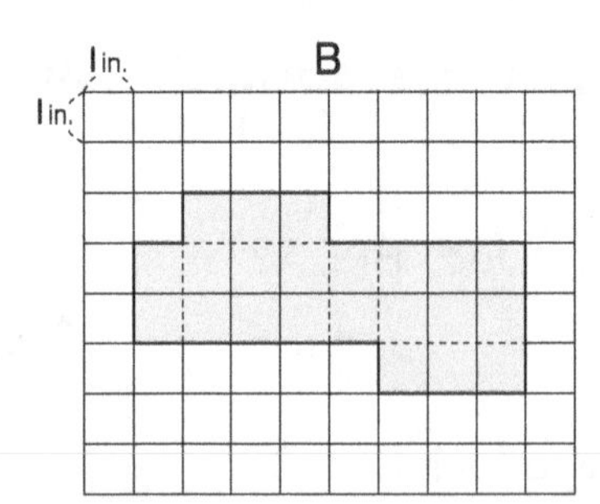

C
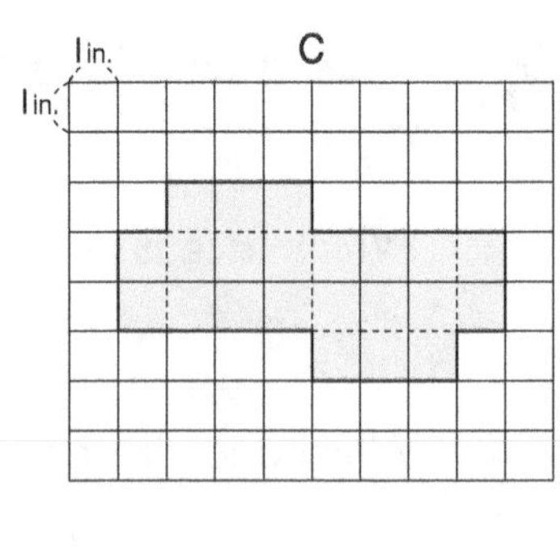

⟨Ans.⟩ ,

2 Answer the questions about the nets below.

10 points per question

A
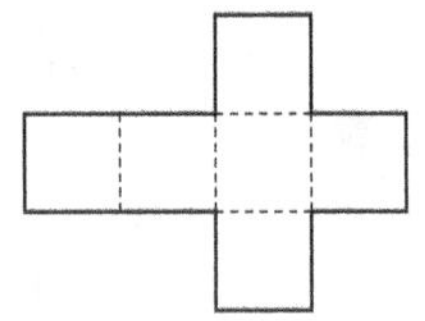

B
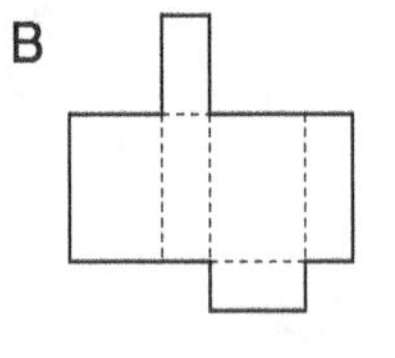

C
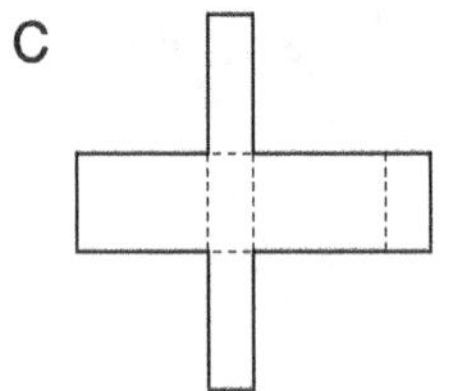

D
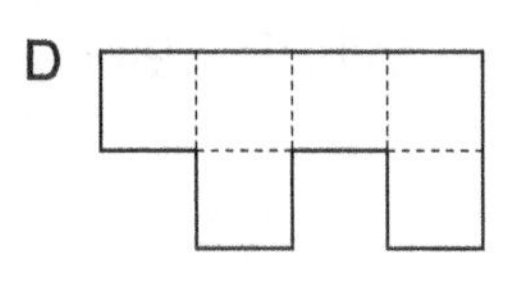

E

F
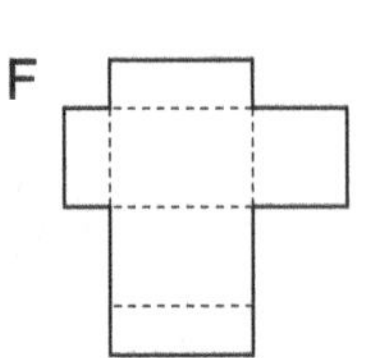

G
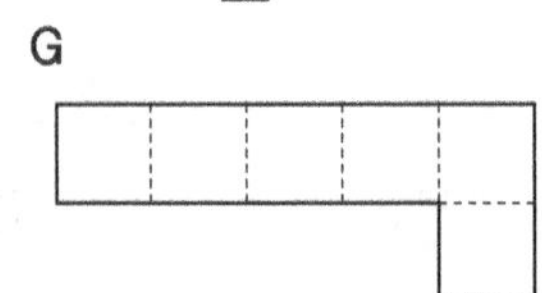

H
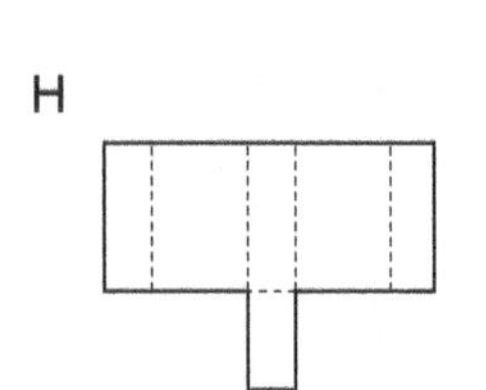

(1) Which nets would build rectangular prisms? ⟨Ans.⟩ ,

(2) Which nets would build cubes? ⟨Ans.⟩ ,

3 **The figures below show the net and the built shape for a rectangular prism with edges that are 5 cm, 8 cm, and 12 cm.**

6 points per question

(Built shape)

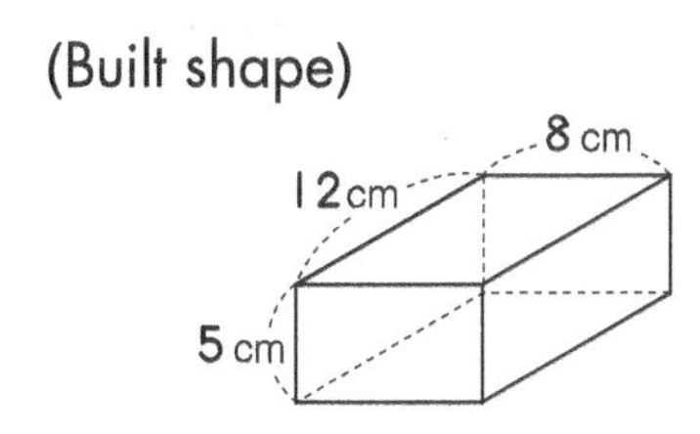

(Net)

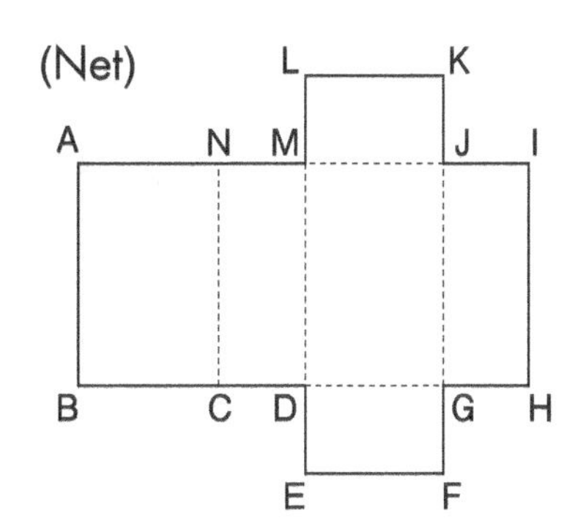

(1) How long is edge AB?

〈Ans.〉

(2) How long is edge FG?

〈Ans.〉

(3) How long is edge DG?

〈Ans.〉

(4) When you fold up the net, which edge lines up with edge AB?

〈Ans.〉 edge

(5) Which edge lines up with edge DE?

〈Ans.〉 edge

(6) Which edge lines up with edge GH?

〈Ans.〉 edge

(7) When you fold up the net, which vertex lines up with vertex C?

〈Ans.〉 vertex

(8) Which vertex lines up with vertex L?

〈Ans.〉 vertex

(9) Which vertex lines up with vertex A?

〈Ans.〉 vertex , vertex

(10) When you fold up the net, which face is parallel to face KJML?

〈Ans.〉 face

(11) When you fold up the net, which faces intersect face NCDM perpendicularly?

〈Ans.〉 face , face , face , face

Rectangular Prisms 4 (Net 2)

Level 

Date / /

Name

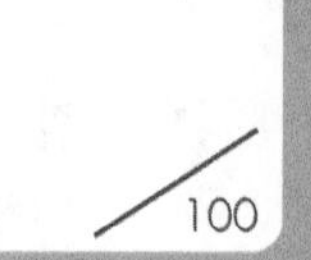

Score

/100

■ The Answer Key is on page 92.

1 **Draw a net that would make each shape on the left.**

15 points per question

(1)

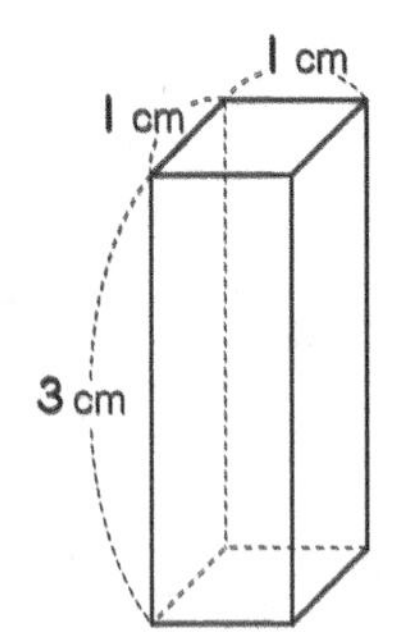

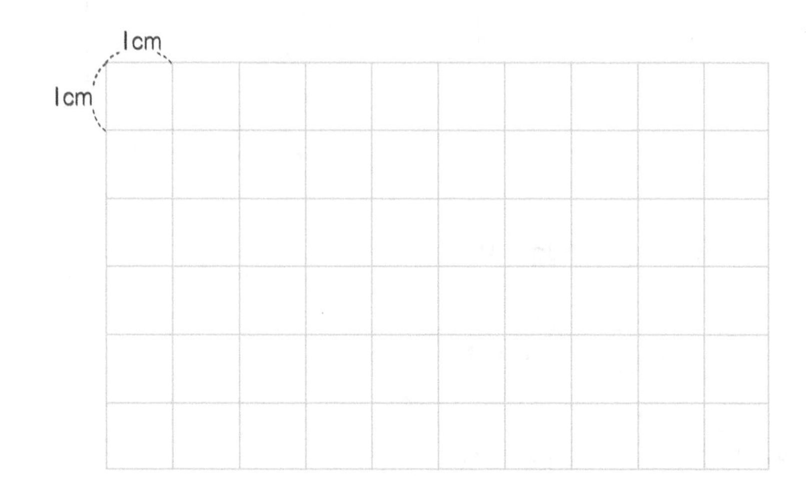

(2)

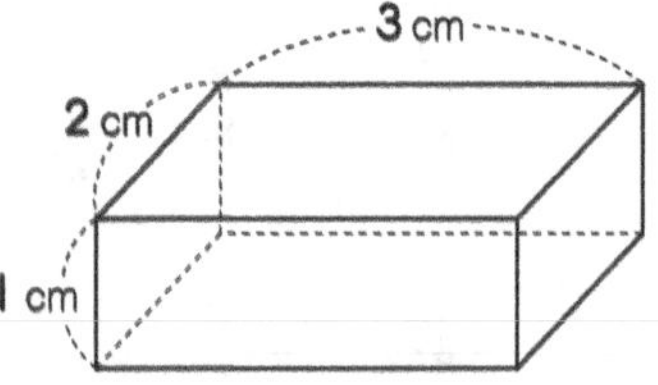

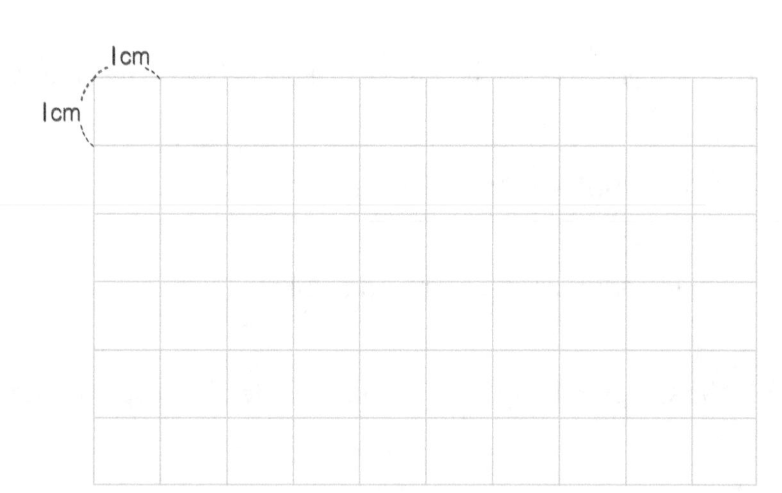

(3)

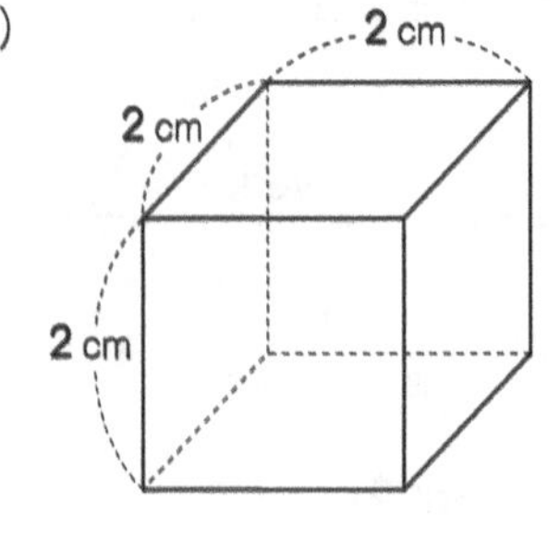

1 cm

1 cm

2 **Draw the shape that would be made by each net on the left.** 15 points per question

(1) 〔Rectangular prism〕

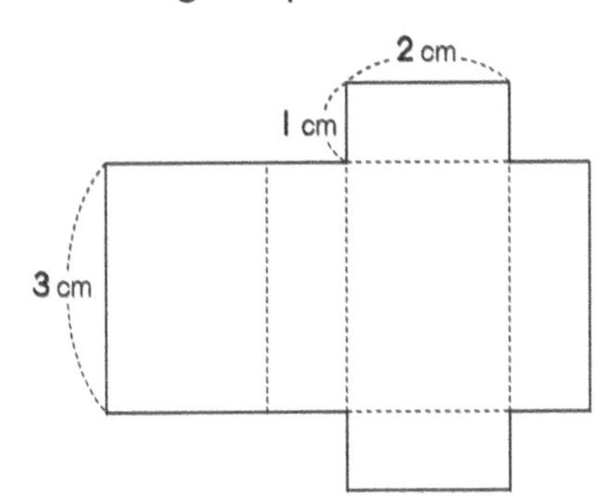

(2) 〔Cube〕

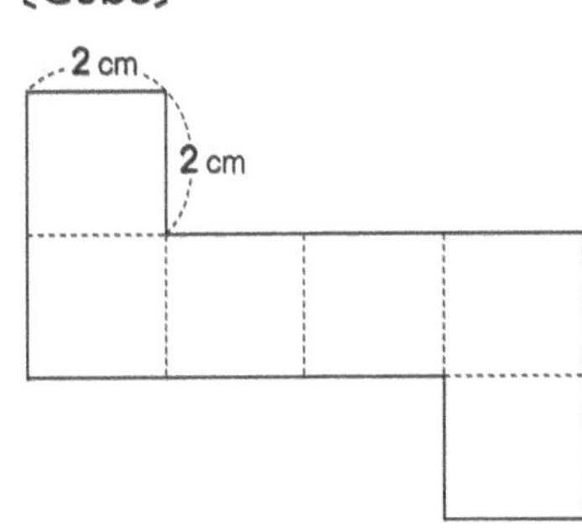

3 **Draw the shape and the net that would make a rectangular prism with edges that are 2 cm, 3 cm, and 4 cm long.** 25 points for completion

(Built shape)

(Net)

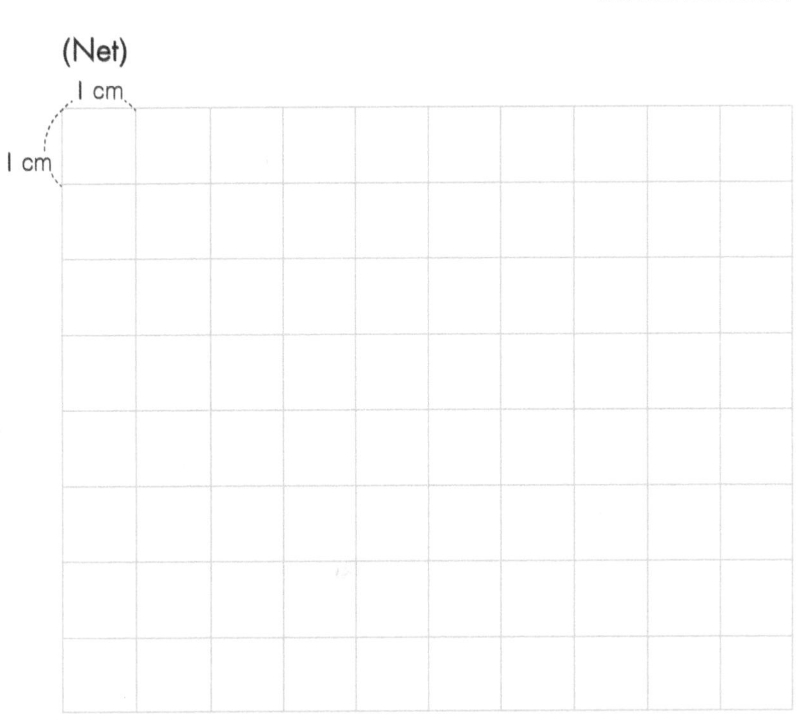

Prisms & Cylinders

Level ★★

Date / /　Name

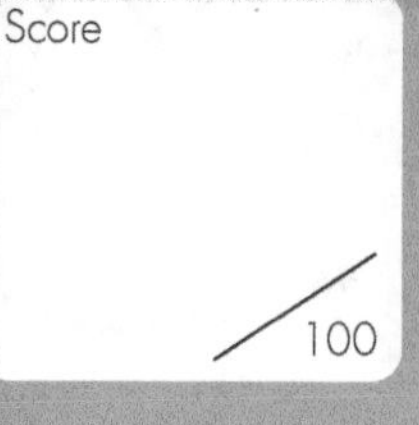

Score /100

■ The Answer Key is on page 92.

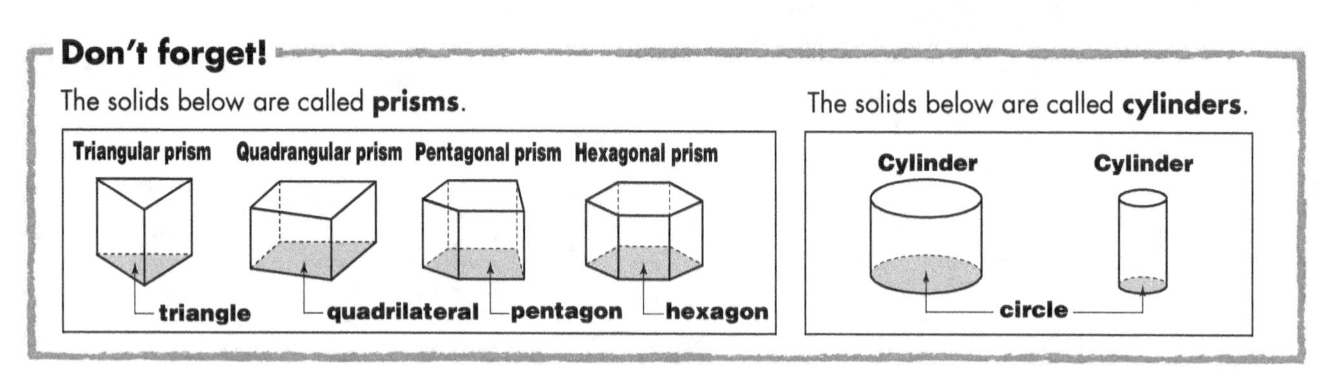

Don't forget!

The solids below are called **prisms**.

The solids below are called **cylinders**.

1 What is the name of each solid below?

4 points per question

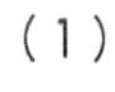

(1)

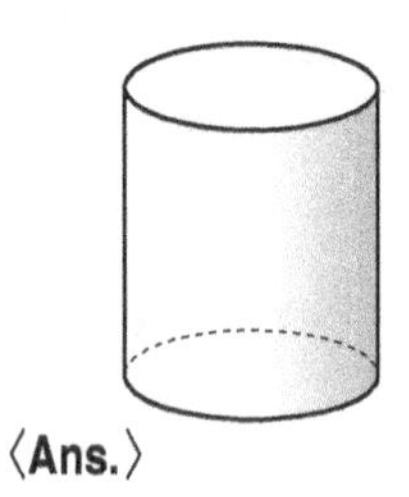

⟨Ans.⟩

(2)

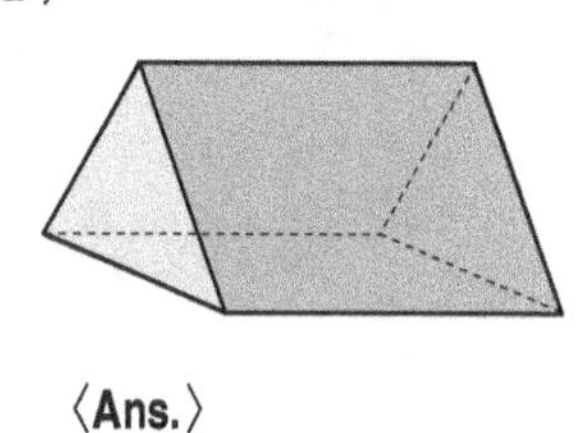

⟨Ans.⟩

(3)

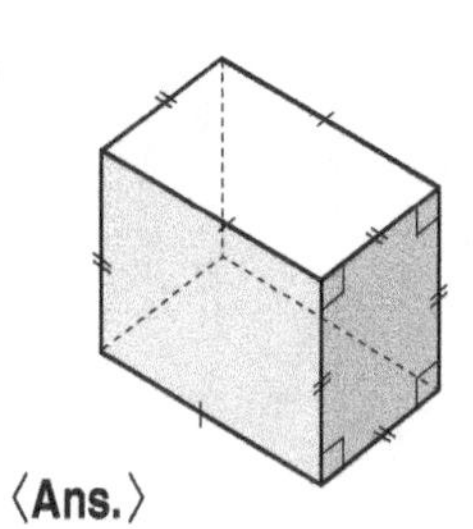

⟨Ans.⟩

(4)

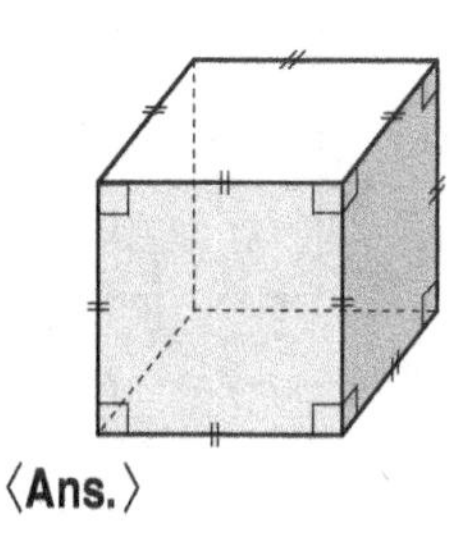

⟨Ans.⟩

(5)

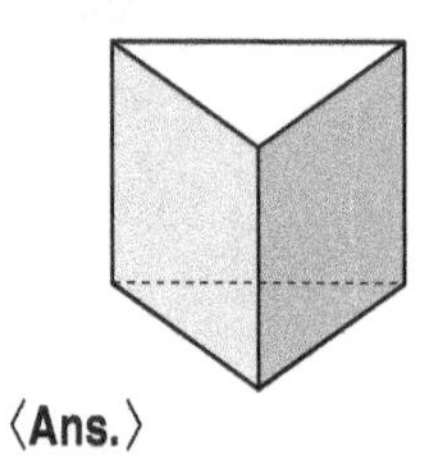

⟨Ans.⟩

(6)

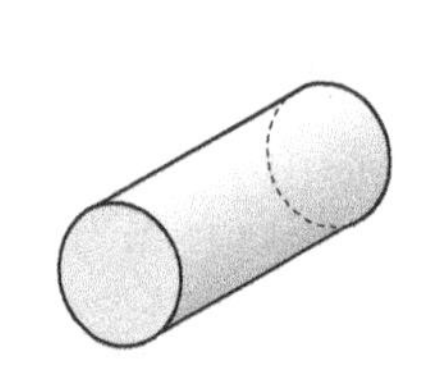

⟨Ans.⟩

(7)

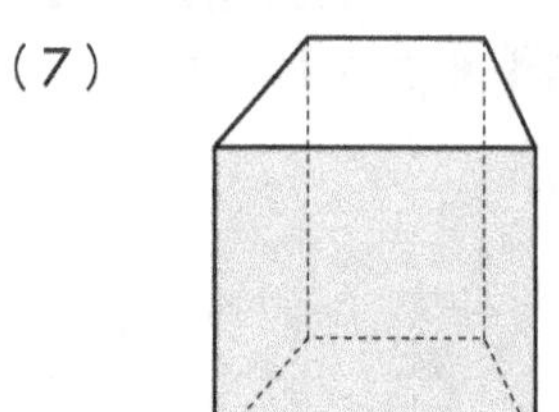

⟨Ans.⟩

(8)

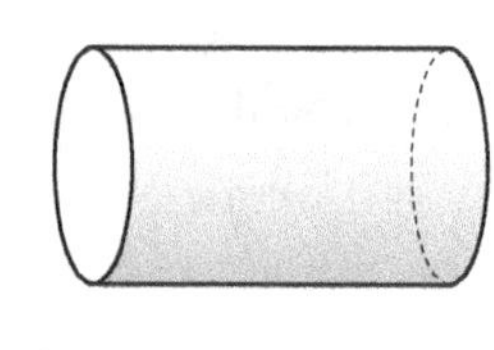

⟨Ans.⟩

(9)

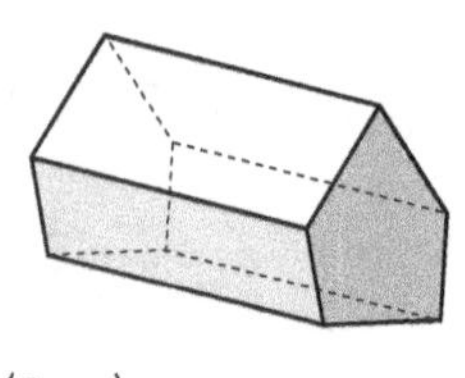

⟨Ans.⟩

(10)

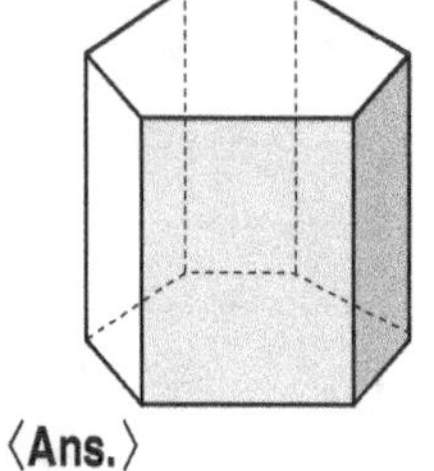

⟨Ans.⟩

(11)

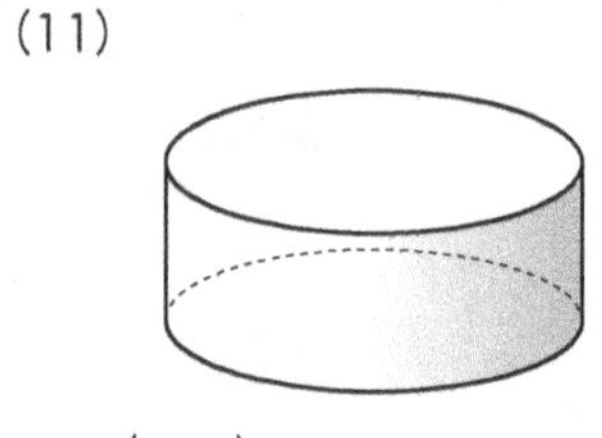

⟨Ans.⟩

(12)

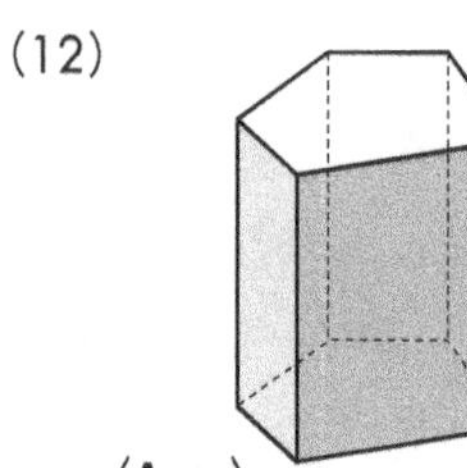

⟨Ans.⟩

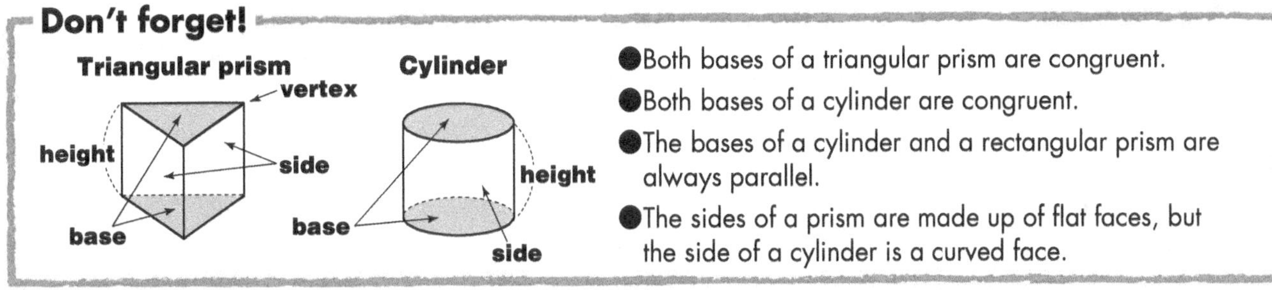

2 **Answer the questions about the triangular prism below.**

4 points per question

(1) What shape is each base?

〈Ans.〉

(2) How many bases are there?

〈Ans.〉

(3) What shape is each side?

〈Ans.〉

(4) How many sides are there?

〈Ans.〉

(5) How many vertices are there?

〈Ans.〉

3 **Answer the questions about the hexagonal prism below.**

4 points per question

(1) What shape is each base?

〈Ans.〉

(2) How many bases are there?

〈Ans.〉

(3) What shape is each side?

〈Ans.〉

(4) How many sides are there?

〈Ans.〉

(5) How many vertices are there?

〈Ans.〉

4 **Answer the questions about the cylinder below.**

6 points per question

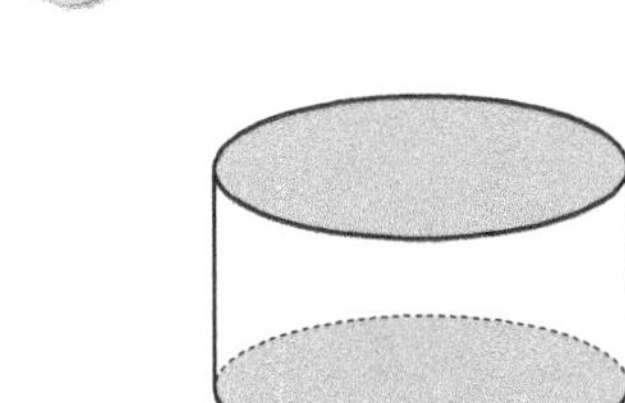

(1) What shape is each base, and how many bases are there?

〈Ans.〉 Shape , Number

(2) Is the side face a flat face, or a curved face?

〈Ans.〉

Surface Area & Volume

Level ★★

Date / /

Name

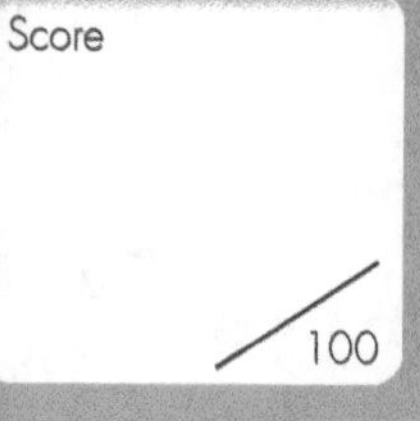

■ The Answer Key is on page 92.

Don't forget!

- The area of an entire side is called the **lateral area**.
- The area of one base is called the **area of the base**.
- The sum of the areas of all faces (= surfaces) of a solid is called the **surface area**.
  (The surface area of a prism = area of the base × 2 + lateral area.)

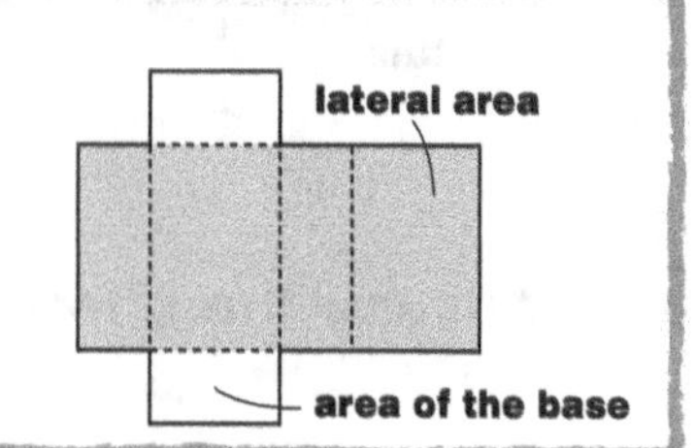

1 Find the surface area of the cubes and the rectangular prisms below.

8 points per question

(1)

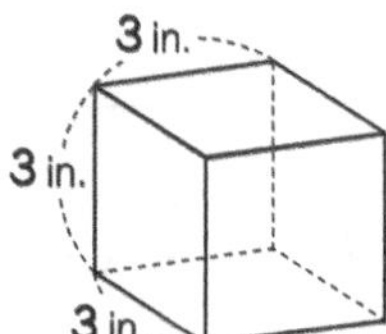

〈Ans.〉

(2)

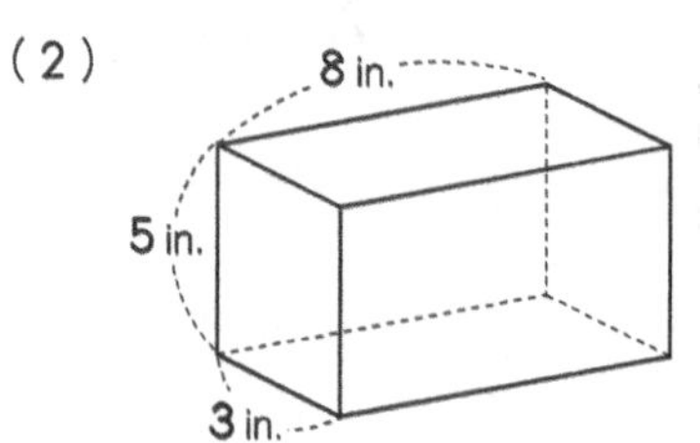

〈Ans.〉

(3)

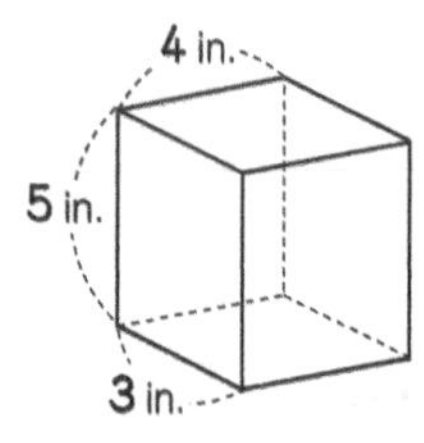

〈Ans.〉

(4)

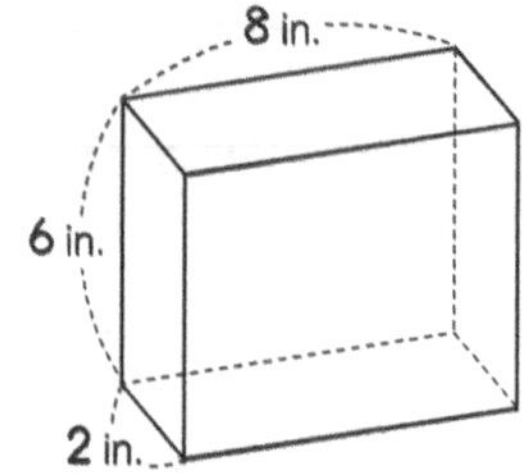

〈Ans.〉

2 Find the volume of the solids below.

8 points per question

(1)

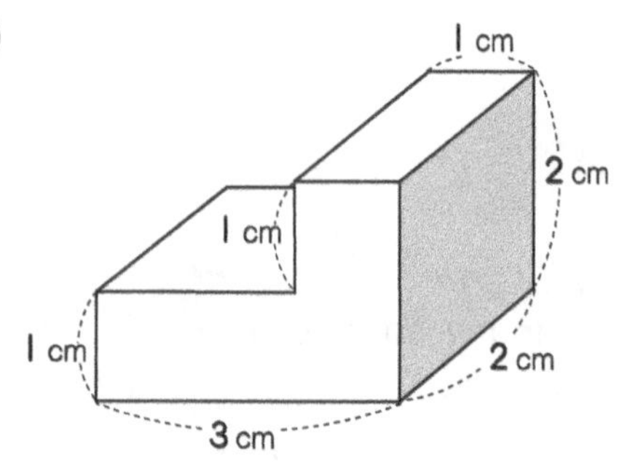

〈Ans.〉

(2)

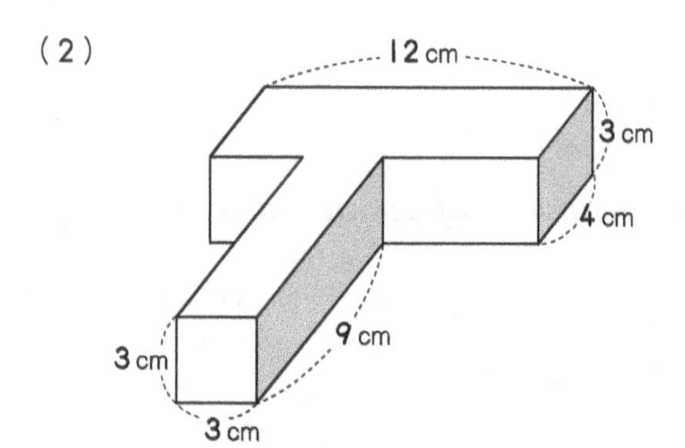

〈Ans.〉

3 Find the volume of the solids below.

9 points per question

(1)

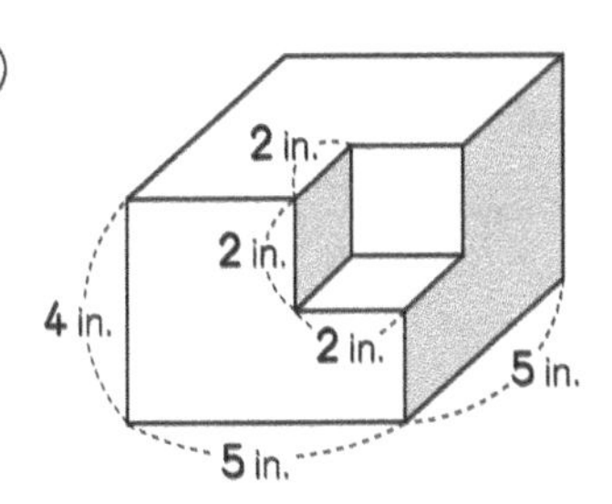

〈Ans.〉

(2)

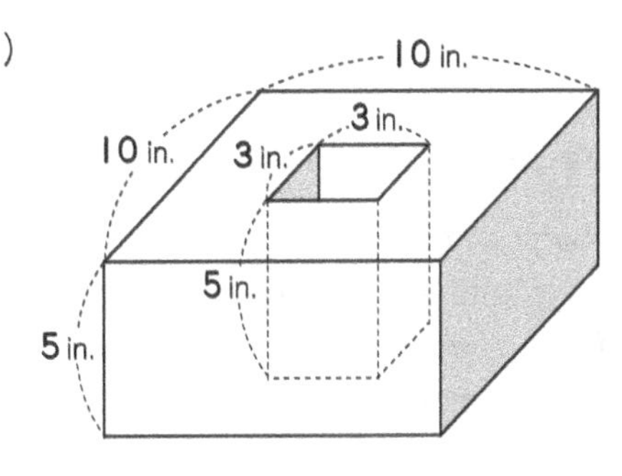

〈Ans.〉

(3)

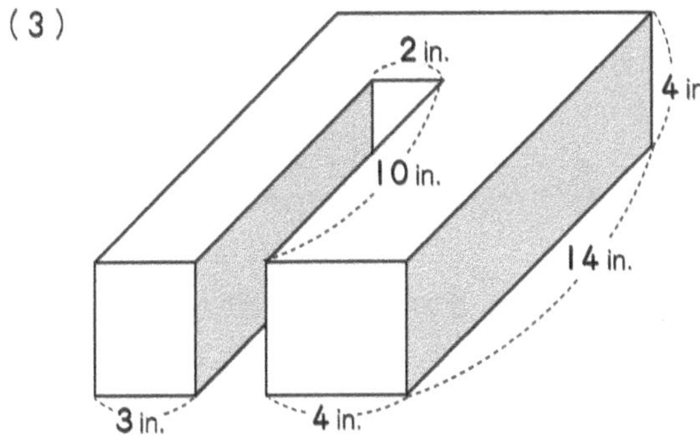

p.96

〈Ans.〉

(4)

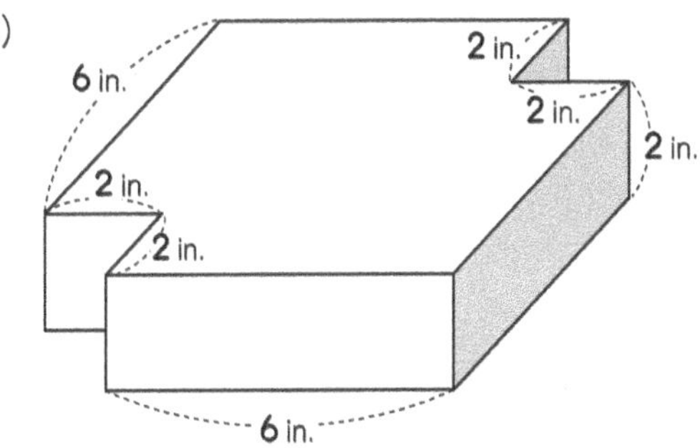

〈Ans.〉

4 Find the surface area of the solids below.

8 points per question

(1)

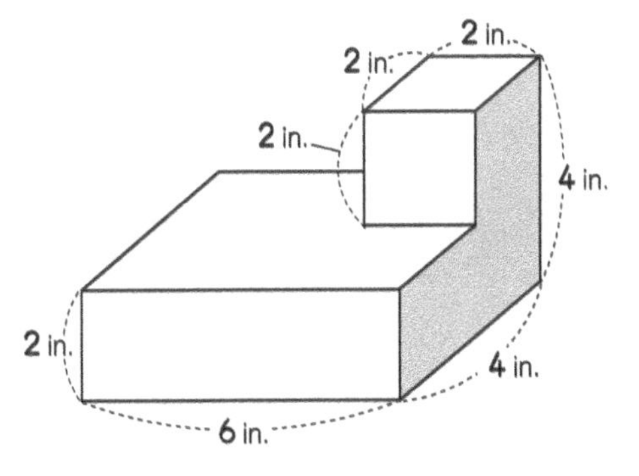

p.96

〈Ans.〉

(2)

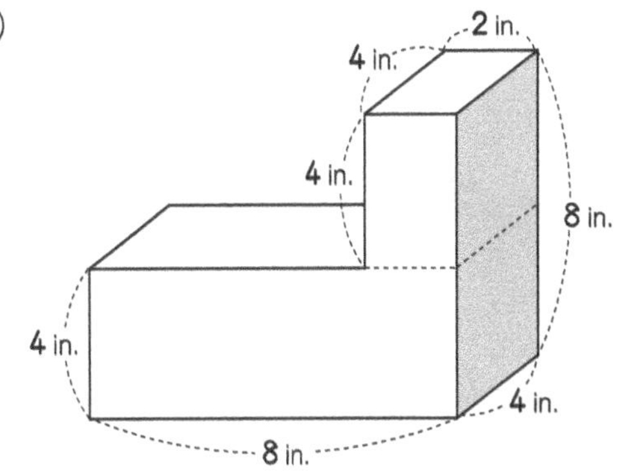

〈Ans.〉

30 Geometric Notation 1 (Line, Ray, & Line Segment)

Level ★★

Date / / Name

■The Answer Key is on page 92.

Don't forget!

- Although numerous lines can pass through one point, there is only one line that can pass through two points.
- A **line** has no endpoints and extends infinitely in both directions.
- A **ray** is a portion of a line that has one endpoint, but extends infinitely in a single direction.
- A **line segment** represents the bounded area of a line. So a line segment has two endpoints with a determined length.
- A line, a ray, and a line segment are expressed as follows.

Line AB ⇨ $\overleftrightarrow{AB}$ Ray AB ⇨ $\overrightarrow{AB}$ Line segment AB ⇨ $\overline{AB}$

1 **Draw the following figures. Use the example on the right as a guide.** 5 points per question

(1) Line CD

C D

(2) Line segment CD

C D

(3) Ray CD

C D

(4) Ray DC

C D

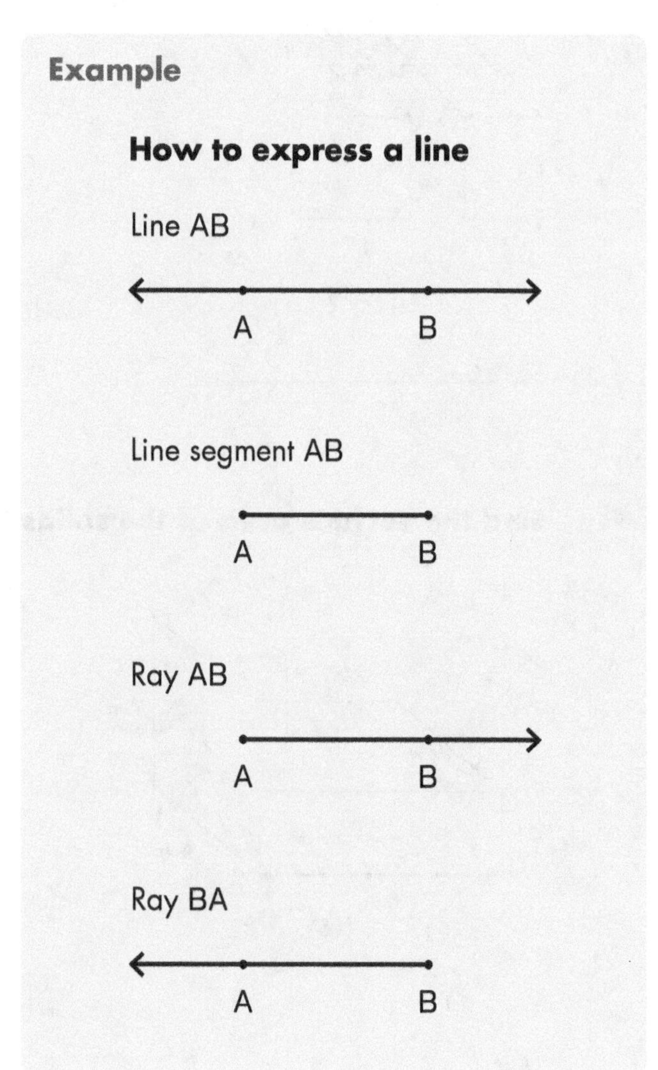

2 Use points A to J on the right to answer the questions below.

(1)-(5) 6 points per question (6)(7) 5 points per question

(1) Draw line AB.

(2) Draw line segment CD.

(3) Draw ray EF.

(4) Draw ray GH.

(5) Draw line segment IJ.

(6) Which line intersects with ray EF?

〈Ans.〉 ______________________

(7) Which ray intersects with line segment IJ?

〈Ans.〉 ______________________

A B F G C E D H I J

3 Express the following statements with the correct symbol.

5 points per question

(1) The length of line segment EF is 5 cm.

〈Ans.〉 ______ = 5 cm

(2) Line segment MN and line segment PQ have the same length.

〈Ans.〉 $\overline{MN}$= ______

(3) The length of line segment AB is three times of line segment CD.

〈Ans.〉 ______ $= 3\overline{CD}$

(4) The length of line segment GH is half of line segment CD.

〈Ans.〉 ______________________

(5) The length of line segment AB and line segment AC are equal.

〈Ans.〉 ______________________

(6) The length of line segment KL is 7 inches.

〈Ans.〉 ______________________

(7) Line segment RS is twice the length of line segment TU.

〈Ans.〉 ______________________

(8) One-half the length of line segment OP is line segment QR.

〈Ans.〉 ______________________

31 Geometric Notation 2 (Parallel & Perpendicular)

Level 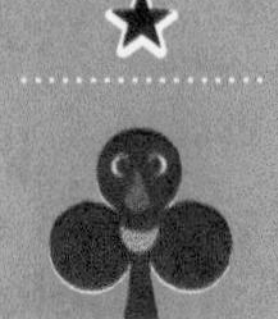

Score 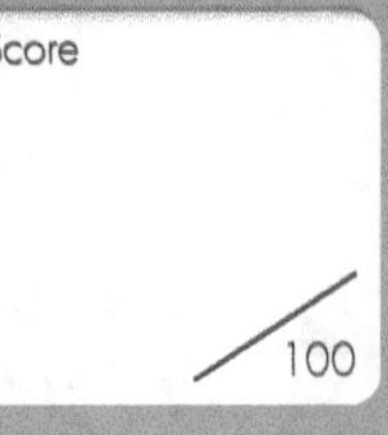

Date / / Name

■ The Answer Key is on page 93.

Don't forget!

● The positional relationship of two lines on the same plane is either ① or ②.

① **Intersect** ② **Parallel** (= Do not intersect)

● When lines intersect, and the intersection angle is a right angle, the two lines are said to be perpendicular.

● When two lines ℓ and m are parallel, it is expressed as $\ell \parallel m$, using the symbol $\parallel$.

● When two lines p and q are perpendicular, it is expressed as $p \perp q$, using the symbol $\perp$.

(Parallel) ℓ m

(Intersect) p q

Example

Line ℓ and line m are parallel. .... $\ell \parallel m$ (Do not put a symbol on the letter.)

Line AB and line CD are parallel. .... $\overleftrightarrow{AB} \parallel \overleftrightarrow{CD}$

Line segment EF and line segment GH are perpendicular. .... $\overline{EF} \perp \overline{GH}$

1 Express the following statements with the correct symbol.

5 points per question

(1) Line k and line ℓ are parallel.

⟨Ans.⟩ ____________

(2) Line MN and line PQ are parallel.

⟨Ans.⟩ ____________

(3) Line segment AB and line segment CD are parallel.

⟨Ans.⟩ ____________

2 Express the following statements with the correct symbol.

7 points per question

(1) Line g and line h are perpendicular.

⟨Ans.⟩ ____________

(2) Line AB and line CD are perpendicular.

⟨Ans.⟩ ____________

(3) Line segment EF and line segment GH are perpendicular.

⟨Ans.⟩ ____________

3 **Draw the following figures, using a protractor or a ruler.** 8 points per question

(1) $g \perp h$ (Draw line h.)

(2) $\overline{AB} \parallel \overline{CD}$ (Draw line segment CD.)

C•

A ——— B

(3) $k \parallel m$ (Draw line m.)

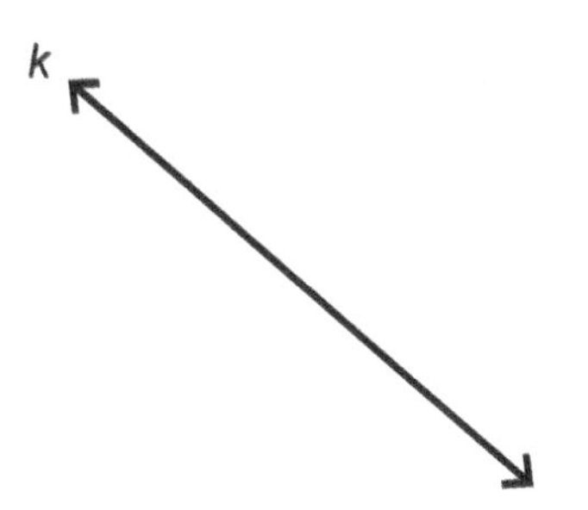

(4) $\overline{EF} \perp \overline{FG}$ (Draw line segment FG.)

(5) $\overline{AB} \perp \overline{AC}$ (Draw line segment AC.)

$\overline{AB} = \overline{AC}$

A ——— B

(6) $\overline{EF} \parallel \overline{CD}$ (Draw line segment CD.)

$\overline{CD} = 3$ cm

E ——— F

C•

(7) $\overline{PQ} \perp \overline{RQ}$ (Draw line segment RQ.)

$\overline{PQ} = 2\,\overline{RQ}$

(8) $k \parallel \ell$ (Draw line ℓ.)

$k \perp m$ (Draw line m.)

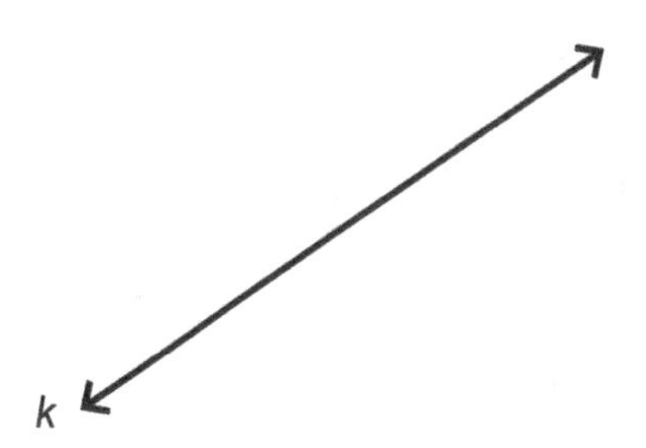

Geometric Notation 3 (Angle)

Level 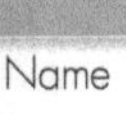

Date / /

Name

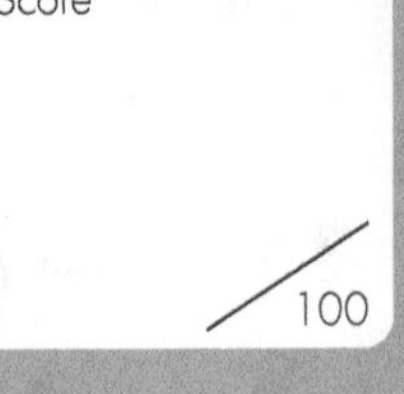

■ The Answer Key is on page 93.

Don't forget!

● As shown in the figure on the right, if the points and vertices are A, B, and C, the angle is represented by ∠ABC, or ∠B using the symbol ∠.

● ∠ABC is 30 degrees. ………… ∠ABC = 30°.

● The measure of ∠A and ∠B are equal. ………… ∠A = ∠B.

● The length of side AB and side AC are equal. ……… AB = AC.

● Triangle ABC is expressed as △ABC.

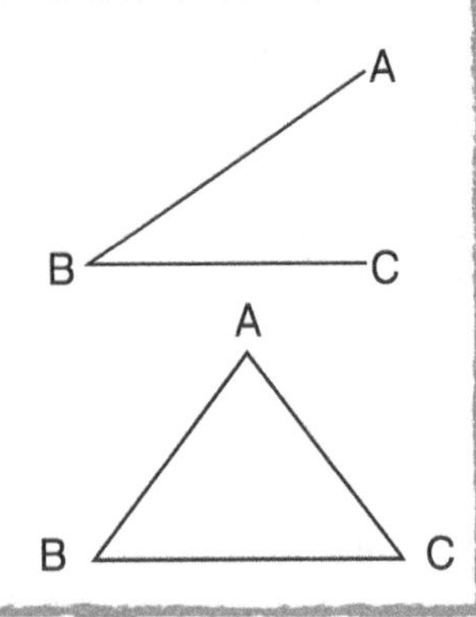

1 **Express the following statements with the correct symbol based on the isosceles triangle on the right.** 4 points per question

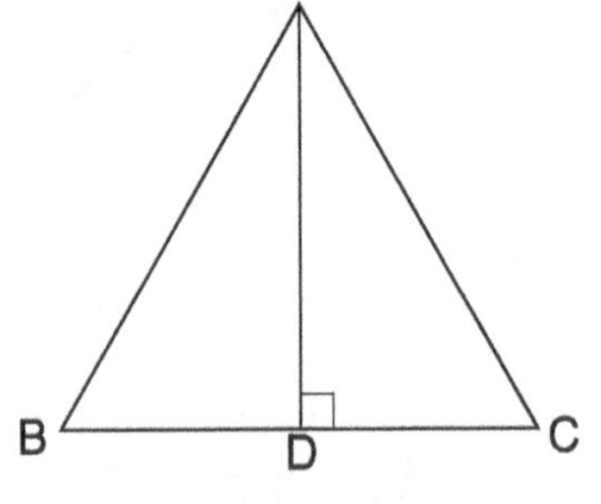

(1) The measure of ∠B and ∠C are equal. 〈Ans.〉

(2) ∠ABC is 70 degrees. 〈Ans.〉

(3) AD and BC are perpendicular. 〈Ans.〉

(4) The lengths of BD and CD are equal. 〈Ans.〉

(5) The length of BC is twice the length of BD. 〈Ans.〉

(6) The measure of ∠BAD is half the measure of ∠BAC. 〈Ans.〉

2 **Fill in the blanks below using the letters of the vertex A, B, C, and D and symbols of the figure on the right.** 5 points per question

(1) Angles①, ②, and ③ .

①  ② 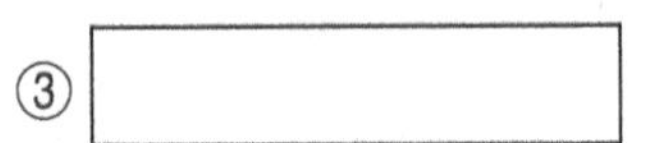 ③

(2) The equation for the sum of the interior angles of △BCD.

∠BCD + 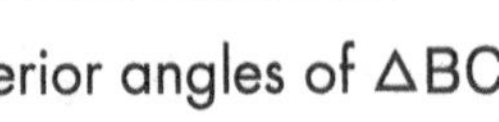 +  = 180°

(3) The equation for the sum of the interior angles of △ABD.

∠ABD + ☐ + ☐ = 180°

(4) The equation for the sum of the interior angles of quadrilateral ABCD.

∠BCD + ☐ + ☐ + ☐ = 360°

3 Express the following statements with the correct symbol based on the rectangle on the right.

4 points per question

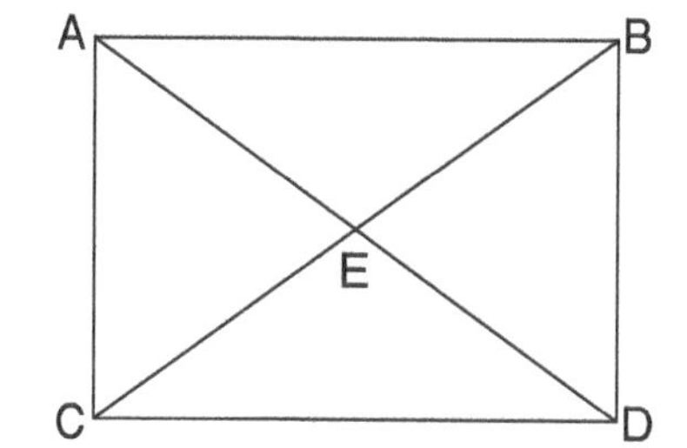

(1) AB and CD are parallel.

⟨Ans.⟩

(2) AC and BD are parallel.

⟨Ans.⟩

(3) The lengths of AB and CD are equal.

⟨Ans.⟩

(4) AC and CD are perpendicular.

⟨Ans.⟩

(5) AB and BD are perpendicular.

⟨Ans.⟩

(6) The measures of ∠BAD and ∠ADC are congruent.

⟨Ans.⟩

(7) The lengths of $\overline{AE}$ and $\overline{ED}$ are equal.

⟨Ans.⟩

(8) The lengths of $\overline{BC}$ and $\overline{AD}$ are equal.

⟨Ans.⟩

4 Express the following statements with the correct symbol based on the rhombus on the right and answer the questions below.

4 points per question

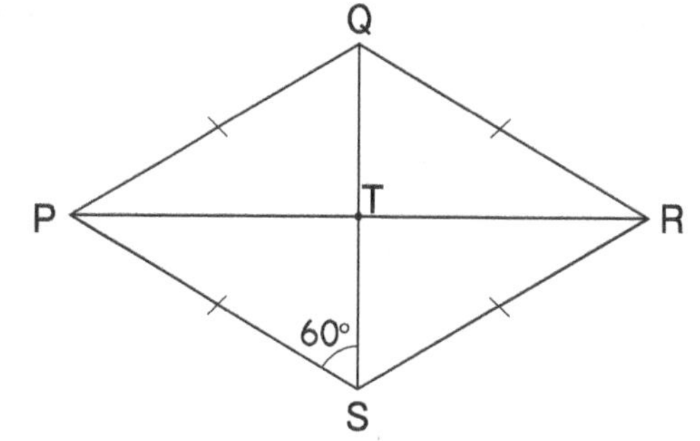

(1) The lengths of PQ and RS are equal.

⟨Ans.⟩

(2) QR and PS are parallel.

⟨Ans.⟩

(3) The lengths of PQ and PS are equal.

⟨Ans.⟩

(4) Find the measure of ∠PQS

⟨Ans.⟩

(5) $\overline{PR}$ and $\overline{QS}$ are perpendicular.

⟨Ans.⟩

(6) The lengths of $\overline{QT}$ and $\overline{TS}$ are equal.

⟨Ans.⟩

33 Geometric Notation 4 (Circle)

Date / /

Name

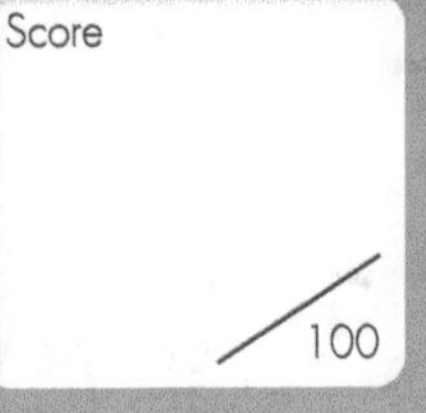

■ The Answer Key is on page 93.

Don't forget!

●The point where a line or a curve intersects is called an **intersection point**.

●The circle whose center is point A is called circle A.

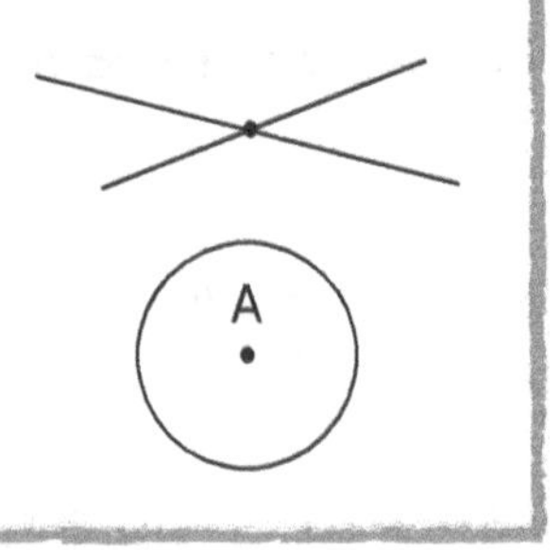

1 **Points B, C, D, and E are on the circumference of circle A. Answer the questions about the figure on the right.**

6 points per question

(1) What is the line segment of the same length as $\overline{AB}$?

⟨Ans.⟩ ______ , ______ , ______

(2) What is the intersection point of $\overline{BD}$ and $\overline{CE}$?

⟨Ans.⟩ ______

(3) Which is the diameter of circle A?

⟨Ans.⟩ ______

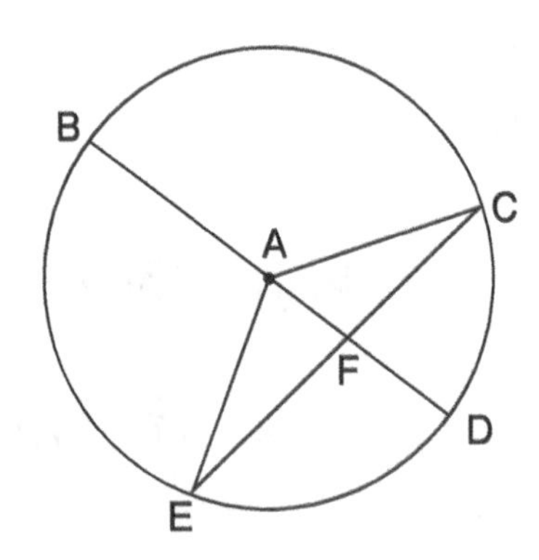

2 **△PQR is a triangle. $\overline{ST}$ is parallel to PR. Answer the questions about the figure on the right.**

6 points per question

(1) Write the intersection point X of $\overline{ST}$ and PQ.

(2) Write the intersection point Y of $\overline{ST}$ and RQ.

(3) Which angle is equal to ∠QXY?

⟨Ans.⟩ ______

(4) Which angle is equal to ∠QYX?

⟨Ans.⟩ ______

(5) Which triangle is similar to △QPR?

⟨Ans.⟩ ______

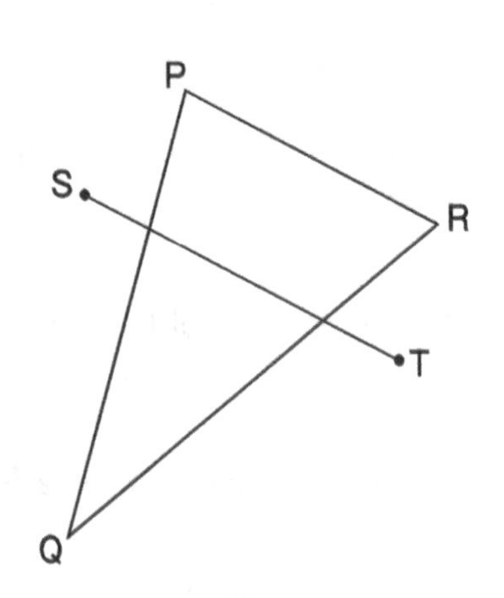

3 Circle O and circle P overlap each other. Answer the questions about the figure on the right.

6 points per question

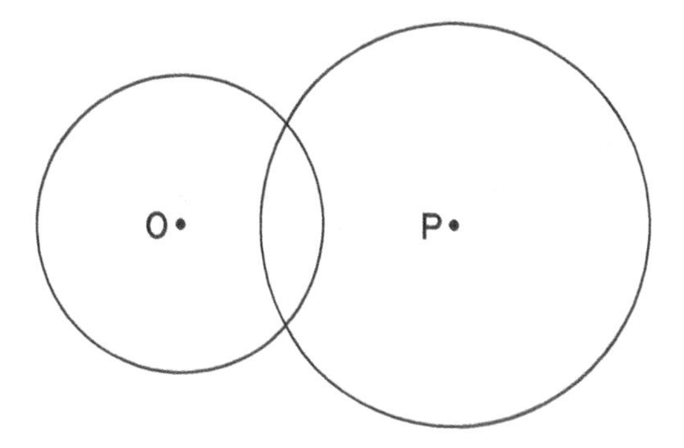

(1) Draw line g that intersects circle P without going through point P and does not intersect circle O.

(2) Label intersection points M and N, intersecting line g and circle P.

(3) Draw $\overline{PM}$ and $\overline{PN}$.

(4) What is $\overline{PM}$ and $\overline{PN}$ for circle P?

〈Ans.〉

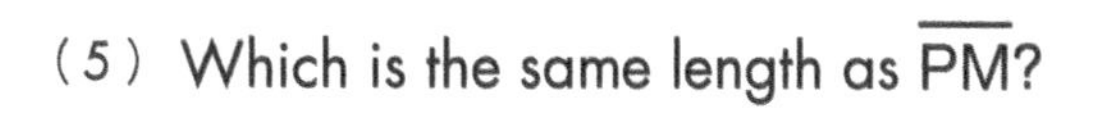

(5) Which is the same length as $\overline{PM}$?

〈Ans.〉

(6) What kind of triangle is $\triangle MPN$?

〈Ans.〉

Don't forget!

- A line segment connecting two points on the circumference is called a **chord**.
- Line segment AB connecting point A and point B on the circumference is shown as **chord AB**.
- The arc AB which is a part of the circumference is expressed as $\overset{\frown}{\mathbf{AB}}$.

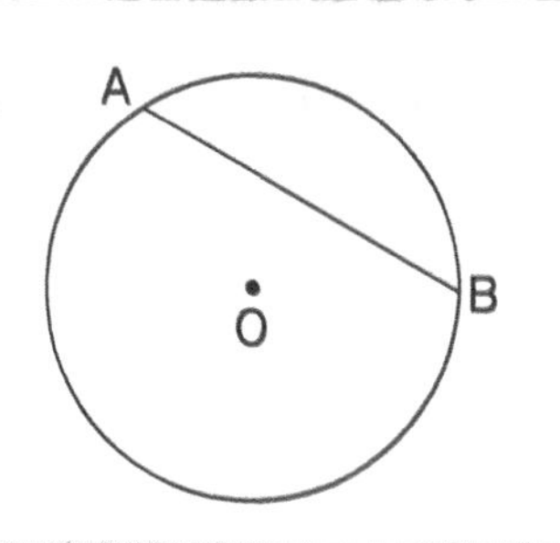

4 Answer the questions about the figure on the right.

8 points per question

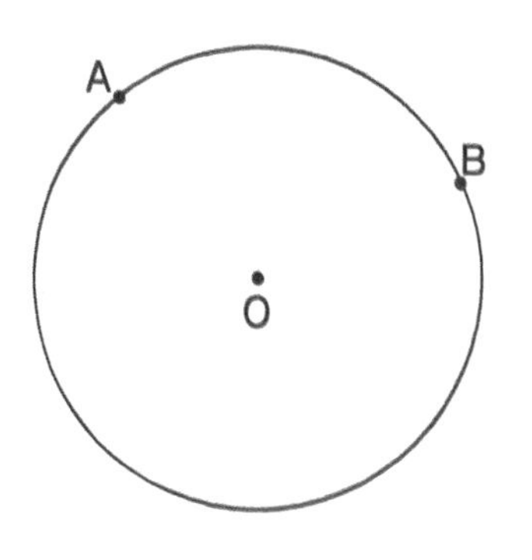

(1) Draw chord AB on circle O.

(2) Draw the longest chord through point A.

34 Construction 1 (Using Compass & Ruler)

Level 

Date / /

Name

Score 

■ The Answer Key is on page 93.

1 **There are two points C and D on the right. Using a compass and a ruler, draw and answer in the following order.** 6 points per question

(1) Draw a circle with a radius of 3 cm centered on point C and a circle with a radius of 2 cm centered on point D.

(2) Label intersection point M of $\overline{CD}$ and the line segment connecting the intersection points P and Q of circle C and circle D.

(3) Show the relationship between $\overline{CD}$ and $\overline{PQ}$ with the correct symbol.

C D

〈Ans.〉

(4) Which line segment is the same length as $\overline{PC}$?

〈Ans.〉

(5) Which line segment is the same length as $\overline{PD}$?

〈Ans.〉

(6) Which line segment is the same length as $\overline{PM}$?

〈Ans.〉

2 **There is $\overline{AB}$ with the length of 5 cm. Using a compass and a ruler, draw and answer in the following order.** 6 points per question

(1) Draw circle A with a radius of 3 cm.

(2) Draw circle B with a radius of 1 cm.

(3) Draw circle B with a radius of 2 cm next. Look at the picture you drew and check that circle A and circle B do not intersect.

A B

(4) How many centimeters should the radius of circle B be in order for circle A and circle B to intersect?

〈Ans.〉

Don't forget!

- If two circles or a circle and a straight line are in contact without intersection, it is called **tangent**.
- A straight line in contact with a curve or a circle is called a **tangent (line)**.

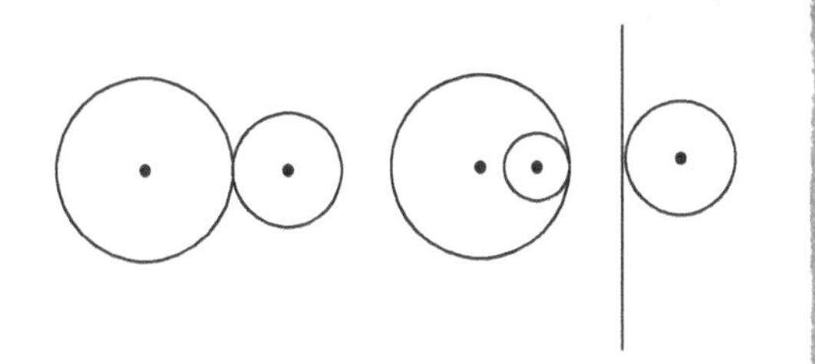

3 **There is a circle P with a radius of 2 cm on line g. Answer the questions about the figure on the right.**

5 points per question

(1) Consider point Q, 3 cm from point P on line g. When drawing circle Q tangent to circle P between PQ, how many centimeters is the radius of circle Q?

〈Ans.〉 ____________

(2) Another circle Q tangent to circle P can be drawn around point Q as the center. How many centimeters is the radius of circle Q?

〈Ans.〉 ____________

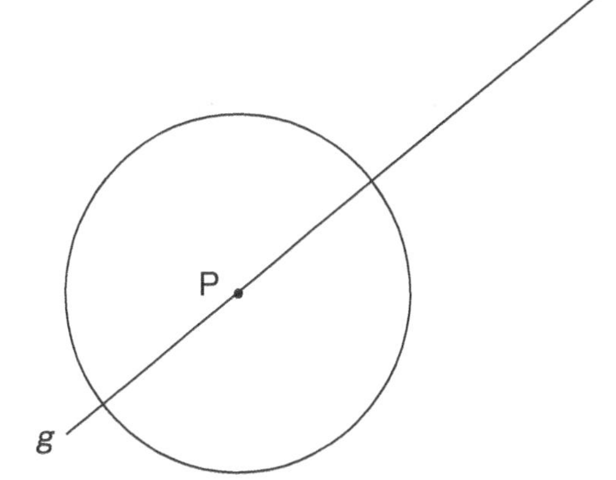

4 **There are two points, A and B with a distance of 2 cm. Answer the questions about the figure on the right.**

6 points per question

(1) Draw circle A and circle B with radius of 2 cm.

(2) Let M and N be the intersection points of circle A and circle B. Draw lines connecting A and M, B and M, and A and B.

(3) What type of triangle is △ABM?

〈Ans.〉 ____________

(4) When M and N are connected by a line, express the relationship between $\overline{MN}$ and $\overline{AB}$ with the correct symbol.

〈Ans.〉 ____________

(5) Let P be the intersection of $\overline{MN}$ and $\overline{AB}$. What is $\overline{MP}$ equal to? What is $\overline{AP}$ equal to?

〈Ans.〉 $\overline{MP}$ = ____________ , $\overline{AP}$ = ____________

A

B

35 Construction 2 (Perpendicular)

Level ★★

Date / / Name

Score /100

■ The Answer Key is on page 93.

1 **Draw and answer in the following order.** 5 points per question

(1) Draw two points A and B on line ℓ. Next, draw circle A and circle B with different radii so that two circles intersect.

(2) Draw $\overline{GH}$ by connecting the two intersection points of circle A and circle B with G and H by a line.

ℓ

(3) Draw intersection point M of $\overline{GH}$ and $\overline{AB}$.

(4) Show the relationship between $\overline{GH}$ and $\overline{AB}$ with the correct symbol. 〈**Ans.**〉

(5) Show the relationship between $\overline{GM}$ and $\overline{HM}$ with the correct symbol. 〈**Ans.**〉

2 **There is point P not on line ℓ. In the following order, draw a line perpendicular from P to ℓ.** 5 points per question

(1) Draw point A freely on line ℓ.

(2) Draw a circle with radius PA centered on point A.

(3) Draw point B freely on the line ℓ on the opposite side of the point A centered on point P.

P

(4) Draw a circle with radius PB centered on point B.

ℓ

(5) Draw point Q which is not point P among the two intersection points of circle A and circle B.

(6) When connecting point P and point Q, show the relationship between $\overline{PQ}$ and $\overline{AB}$ with the correct symbol.

〈**Ans.**〉

Don't forget!

●A perpendicular line can be drawn in the order of **2** (1) to (6).

3 **Draw a perpendicular line from point M (not on line *g*) to line *g*. Then, draw figures and lines and answer the questions in the following order.**

5 points per question

(1) Draw circle M centered on point M intersecting line *g* and find the intersection points R and S with line *g*.

(2) Draw circle R and circle S with the same radius around points R and S so that the two circles intersect.

(3) Find T which is one of the intersections of circle S and circle R on the opposite side of point M.

(4) Draw a line passing through point M and point T so that it is perpendicular to line *g*.

g

M

4 **Draw a straight line passing through point A on line *h* and perpendicular to *h* in the following order.**

5 points per question

(1) Draw a circle freely around point A. Then draw two intersection points C and D with line *h*.

(2) Draw a line crossing a circle of the same radius centered on point C and point D.

(3) Connect the intersection of circle C and circle D with a line so that it passes through point A and is a straight line perpendicular to line *h*.

h

A

5 **Draw a line passing through point P and perpendicular to line *g*.**

10 points

P

g

36 Construction 3 (Distance of Point & Line)

Level ★★

Date / / Name

Score /100

■ The Answer Key is on page 94.

1 **There is point P not on line ℓ. Find the distance between point P and line ℓ in the following order.** 6 points per question

(1) Draw line m passing point P and perpendicular to line ℓ.

(2) Mark intersection point Q on line m with line ℓ.

(3) Find the length of $\overline{PQ}$ using a ruler.

⟨Ans.⟩ ____________

ℓ

P

Don't forget!

- The length of perpendicular $\overline{PQ}$ of ① is called the distance between point P and line ℓ.
- If the two lines ℓ and m are parallel, the distance to line m is the same from any point on line ℓ, and this distance is called the distance between line ℓ and line m.

2 **Find point P, 3 cm away from line ℓ, in the following order, by answering the questions below.** 6 points per question

(1) Draw point Q freely on line ℓ and draw circle Q around point Q.

(2) Draw circle G and circle H so that the two circles intersect around the intersection points G and H of circle Q and line ℓ.

(3) Show the relationship between line ℓ and line m (connecting the two intersection points of circle G and circle H) with the correct symbol.

⟨Ans.⟩ ____________

ℓ

(4) When the intersection point of lines ℓ and m is R, find point P on line m 3 cm away from point R.

(5) How many points on line m are 3 cm away from intersection point R on line m?

⟨Ans.⟩ ____________

3 Draw parallel line *h*, 1 cm from line *g*, in the following order.

6 points per question

(1) Draw two points A and B on line *g*.

g

(2) Draw line *j* perpendicular to line *g* through point A.

(3) Draw line *k* perpendicular to line *g* through point B.

(4) Draw circle A with a radius of 1 cm around point A and find two intersection points P and Q with line *j*.

(5) Draw circle B with radius of 1 cm around point B and find two intersection points R and S with line *k*. At that time, draw point R on the same side as point P with respect to line *g* and point S on the same side as point Q.

(6) Draw a line passing points P and R and the line passing points Q and S.

(7) How many lines are 1 cm away from line *g*?

〈Ans.〉

Don't forget!

- Since non-parallel lines intersect somewhere, there is no distance between lines.

4 With reference to the answers to ③, draw a 2 cm line in parallel with line ℓ.

10 points

ℓ

37 Construction 4 (Perpendicular Bisector)

Date / / Name

Level

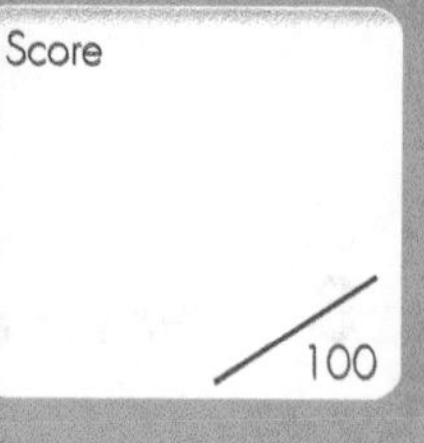
Score
100

■ The Answer Key is on page 94.

1 **There is $\overline{AB}$ on the right. Using a compass and a ruler, draw and answer in the following order.**

7 points per question

(1) Draw circle A and circle B with a radius of 2 cm centered on point A and point B. Write the relationship between point A and point B at two intersection points P and P′.

⟨**Ans.**⟩ $\overline{AP}$ = , $\overline{AP'}$ =

(2) Draw circle A and circle B with a radius of 2.5 cm centered on point A and point B. Write the relationship between point A and point B at two intersection points Q and Q′.

A B

⟨**Ans.**⟩ $\overline{AQ}$ = , $\overline{AQ'}$ =

(3) Draw circle A and circle B with a radius of 3 cm centered on point A and point B. Write the relationship between point A and point B at two intersection points R and R′.

⟨**Ans.**⟩ $\overline{AR}$ = , $\overline{AR'}$ =

(4) Draw circle A and circle B with a radius of 3.5 cm centered on point A and point B. Write the relationship between point A and point B at two intersection points S and S′.

⟨**Ans.**⟩ $\overline{AS}$ = , $\overline{AS'}$ =

(5) If S and S′ are connected by a line, points P, Q, R, P′, Q′, and R′ are aligned on $\overline{SS'}$. Are points A and B the same distance from the point on line $\overline{SS'}$, or are they different?

⟨**Ans.**⟩

(6) Show the relationship between $\overline{AB}$ and $\overline{SS'}$ with the correct symbol.

⟨**Ans.**⟩ $\overline{AB}$ $\overline{SS'}$

(7) Let M be the intersection point of $\overline{AB}$ and $\overline{SS'}$. Show the relationship between $\overline{AM}$ and $\overline{BM}$ with the correct symbol.

⟨**Ans.**⟩ $\overline{AM}$ $\overline{BM}$

Don't forget!

- If point M on $\overline{AB}$ and $\overline{AM} = \overline{BM}$, then point M is called the **midpoint**.
- Midpoint M of $\overline{AB}$ bisects $\overline{AB}$.

A M B

Don't forget!

- A straight line passing through the midpoint of a line segment and perpendicular to the line segment is called a **perpendicular bisector**.
- All points on the perpendicular bisector of $\overline{AB}$ have the same distance from point A and point B.

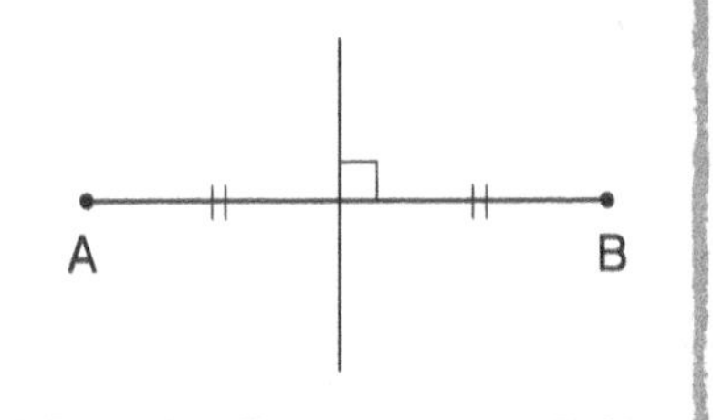

2 **With reference to the problems of ① on Pg.74, draw a perpendicular bisector of $\overline{PQ}$.** 9 points

3 **There is chord AB and chord CD on the circumference of circle O on the right. Answer the following questions in order.** 7 points per question

(1) Draw a perpendicular bisector of chord AB.

(2) Draw a perpendicular bisector of chord CD.

(3) At which point of circle O do the two perpendicular bisectors intersect?

〈Ans.〉 ____________

B
C
O
A
D

4 **There are three points A, B, and C on the right. Answer in the following order.** 7 points per question

(1) Draw a perpendicular bisector of $\overline{AB}$ connecting two points A and B.

(2) Draw a perpendicular bisector of $\overline{BC}$ connecting two points B and C.

(3) Label intersection point P of the two perpendicular bisectors, and then draw a circle P of radius $\overline{PA}$.

B
A
C

※Confirm point B and point C are on circle P.

38 Construction 5 (Angle Bisector)

Level ★★

Date / / Name

Score /100

■ The Answer Key is on page 94.

1 The figure on the right is ∠AOB. Draw ray OE using the following steps. 6 points per question

(1) Draw a circle around vertex O. Under the condition that the radius is shorter than $\overline{OA}$, and in the area that intersects $\overline{OA}$ and $\overline{OB}$.

(2) Find intersection point C of circle O, side $\overline{OA}$ and intersection point D of side $\overline{OB}$.

(3) Draw two circles of the same radius centered on C and D, and let E be one of two intersection points.

(4) Draw $\overrightarrow{OE}$.

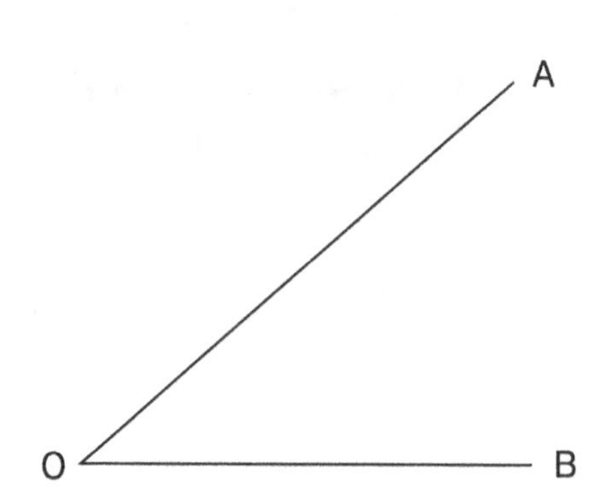

Don't forget!

● Ray OE written in 1 bisects ∠AOB.

● A ray which bisects an angle like this is called an **angle bisector**.

$\angle AOE = \angle BOE = \frac{1}{2} \angle AOB$

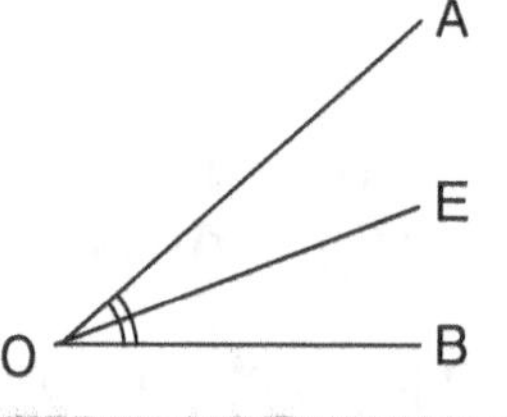

2 Using the method in 1, draw the bisector of each angle below. 8 points per question

(1)

X
O Y

(2)

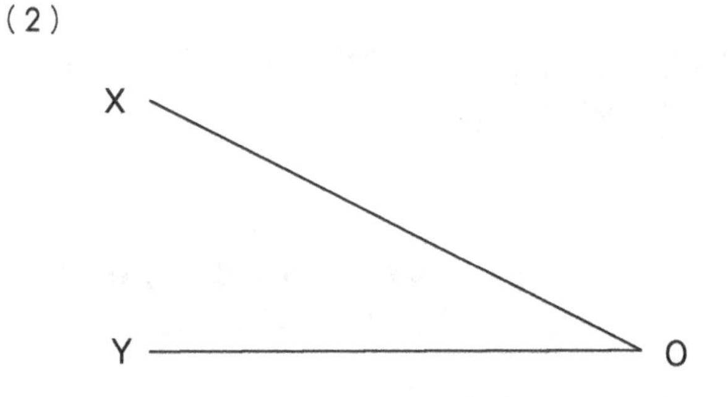

(3)

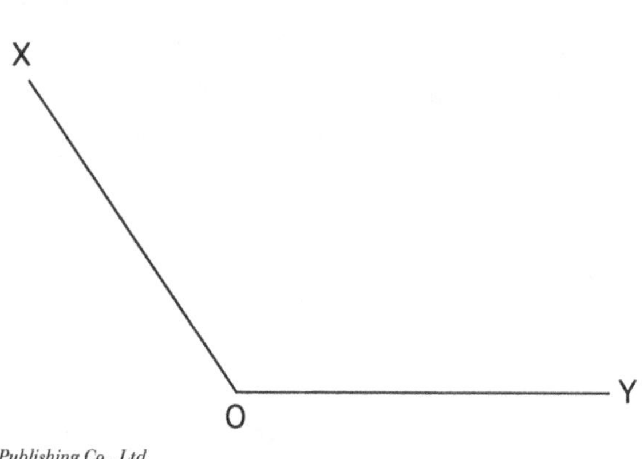

(4)

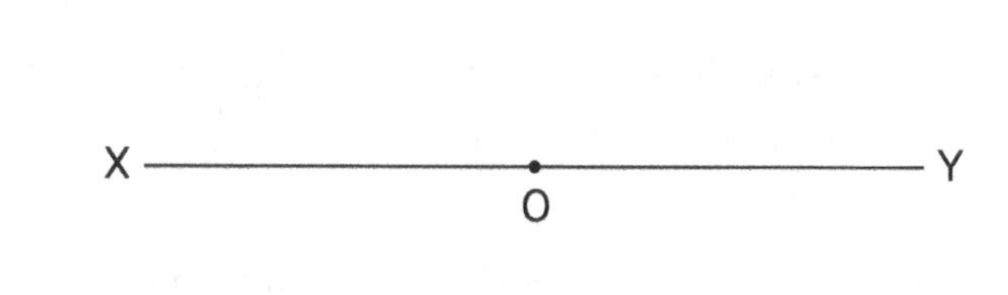

3 Draw the rays dividing ∠XOY into four angles below.

10 points per question

(1)

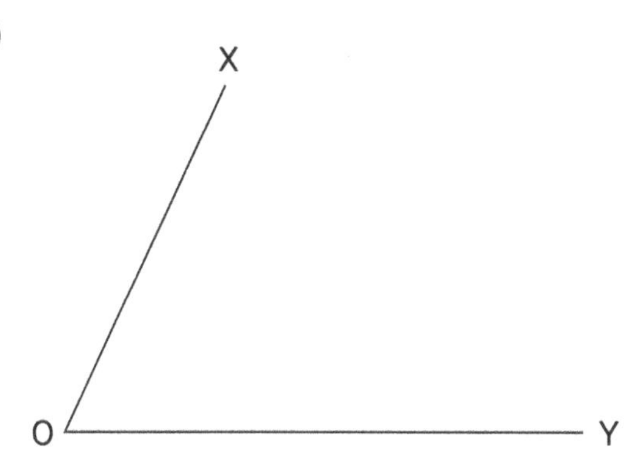

(2)

4 There is ray CD with ∠BCD = 40° drawn from point C on line AB. Answer the questions below.

4 points per question

(1) Draw bisecters CE of ∠ACD and CF of ∠BCD.

(2) Find the measure of ∠ECF.

⟨**Ans.**⟩ ________________

(3) Given ∠BCD = 60°, find the measure of ∠ECF.

⟨**Ans.**⟩ ________________

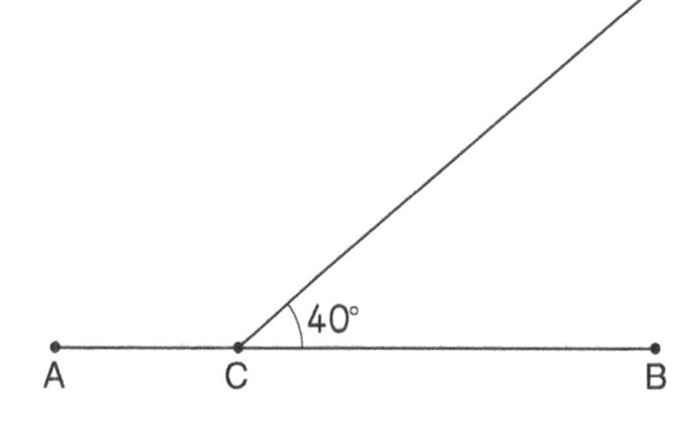

5 Draw a 45° ray on $\overline{AB}$ below using the following steps.

6 points per question

(1) Draw $\overline{AC}$ perpendicular to $\overline{AB}$.

(2) Draw the bisector of ∠CAB.

39 Construction 6 (Application 1)

Level ★★★

Date / /

Name

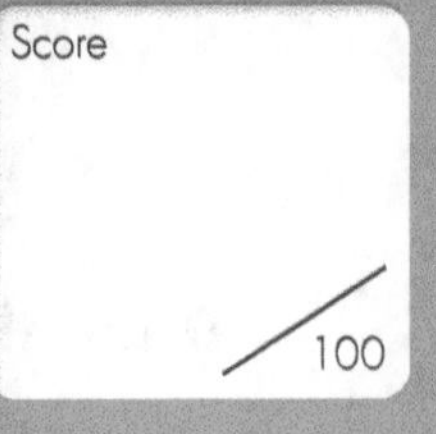

■ The Answer Key is on page 95.

1 **Draw equilateral triangle ABC with $\overline{AB}$ as one side in the following order.**

6 points per question

(1) Draw circle A and circle B with $\overline{AB}$ as its radius.

(2) Find intersection point C of circle A and circle B.
One of two intersection points can be set as C.

(3) Since △ABC is an equilateral triangle, what is side AB equal to?

A ———— B

〈**Ans.**〉 AB = ______ = ______

(4) What is ∠ABC equal to? , and find the measure.

〈**Ans.**〉 ∠ABC = ______ = ______ = ______ °

2 **Draw square PQRS with $\overline{PQ}$ as one side in the following order.**

6 points per question

(1) Draw a perpendicular line passing through point Q on $\overline{PQ}$.

(2) Draw circle Q with a radius of $\overline{PQ}$ and find intersection point R with perpendicular line of $\overline{PQ}$.

(3) Draw circle P with radius PQ around point P and circle R with radius RQ around point R.

(4) Find intersection point S of circle P and circle R.

P
|
Q

※Confirm rectangle PQRS is a square.

3 **The figure on the right is isosceles triangle ABC. Draw in the following order.**

7 points per question

(1) Draw a perpendicular line from vertex A to side BC in △ABC with side AB equal to side AC.

(2) Find intersection point H of the perpendicular line and side BC.

(3) Draw circle H with radius BH around point H and confirm point C is on the circumference of circle H.

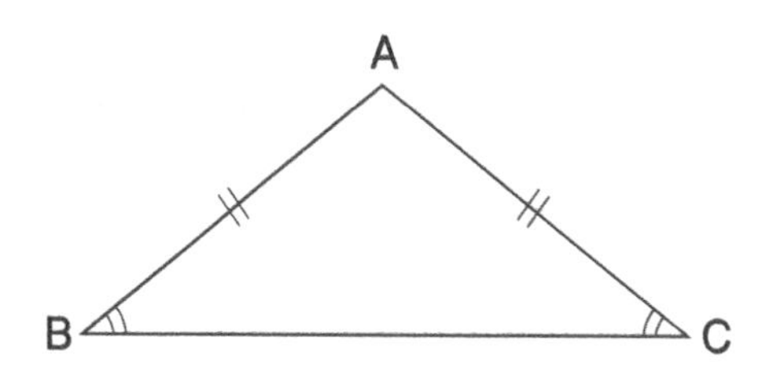

※Confirm $\overline{BH}$ and $\overline{CH}$ are equal.

4 **The figure on the right is isosceles triangle ABC. Draw in the following order.**

7 points per question

(1) Draw a bisector of angle A of the isosceles triangle.

(2) Find intersection point H of bisector A and side BC.

(3) Draw circle H with radius BH around point H and confirm point C is on the circumference of circle H.

※Confirm $\overline{BH}$ and $\overline{CH}$ are equal.

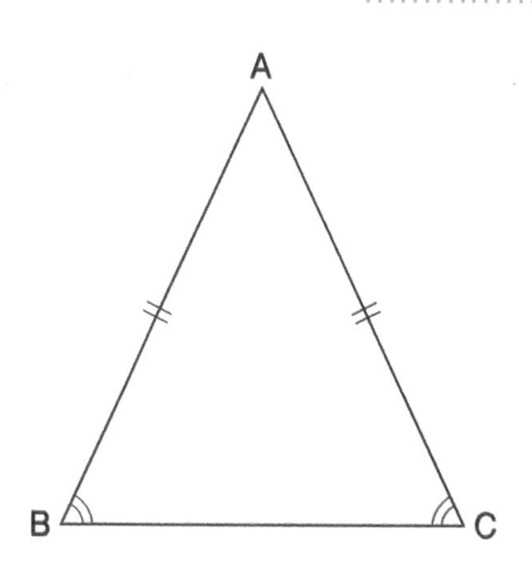

Don't forget!

●From **3** and **4**, it can be seen that the angle bisector of vertex A sandwiched between equal sides of the isosceles triangle coincides with the perpendicular from vertex A to side BC.

5 **Point A is outside circle O below. Draw point P and point P' which are on the circumference of circle O and are at the same distance from point A and point O.**

10 points for completion

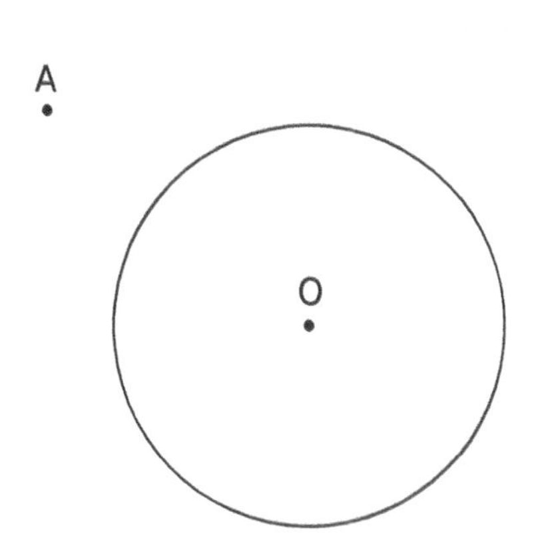

40 Construction 7 (Application 2)

Level ★★★

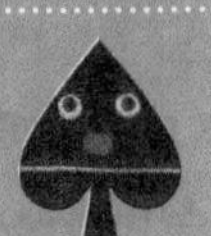

Date / /

Name

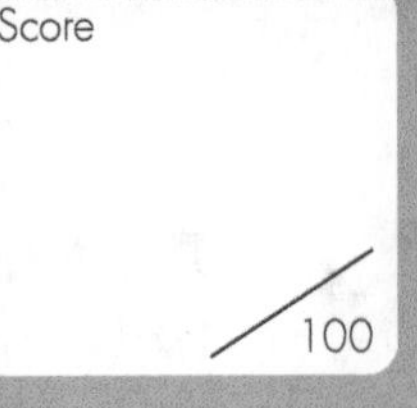

■ The Answer Key is on page 95.

1 **Draw and answer in the following order. Using a compass and a ruler.** 5 points per question

(1) To the right are point O and line g. Draw circle O with a radius of 2 cm.

(2) Label intersection points A and B of circle O and line g.

(3) What is the triangle that connects $\overline{OA}$, $\overline{OB}$ and chord AB?

g

O

⟨Ans.⟩ ______

(4) Find another intersection point P of circle O with the straight line passing through point A and point O.

(5) What is $\overline{AP}$ on circle O?

⟨Ans.⟩ ______

(6) How many centimeters is $\overline{AP}$?

⟨Ans.⟩ ______

(7) What is the longest chord of circle O?

⟨Ans.⟩ ______

(8) Draw $\overline{PB}$ connecting point P and point B.

(9) Show the relationship between $\overline{PB}$ and $\overline{AB}$ with the correct symbol.

⟨Ans.⟩ ______

(10) What type of triangle is $\triangle PBA$?

⟨Ans.⟩ ______

2 **Draw $\overrightarrow{AC}$ with a 30° angle to $\overline{AB}$ using the equilateral triangle drawing method.** 7 points

A ———— B

3 **There are three points A, B, and C on the right. Draw and answer in the following order.**

5 points per question

(1) Draw a perpendicular bisector of $\overline{AB}$.

(2) Draw a straight line passing at the same distance from point B and point C.

(3) Draw intersection point P of the straight lines of (1) and (2).

(4) Draw a circle with radius PA centered at point P.

B•

•C

A•

※Confirm circle P passes through point B and point C.

4 **Using $\triangle$ABC on the right, draw and answer in the following order.**

5 points per question

(1) Draw a perpendicular bisector on each side of $\triangle$ABC.

(2) Confirm the three straight lines intersect at one point P.

(3) Draw circle P with the radius of $\overline{PA}$.

※Confirm there are points B and C on the circumference of circle P.

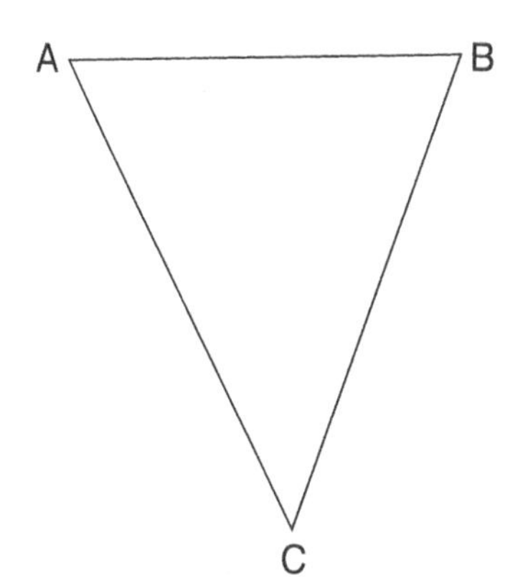

5 **Using $\triangle$DEF below, draw a bisector of the three angles and confirm that the three straight lines intersect at one point.**

8 points for completion

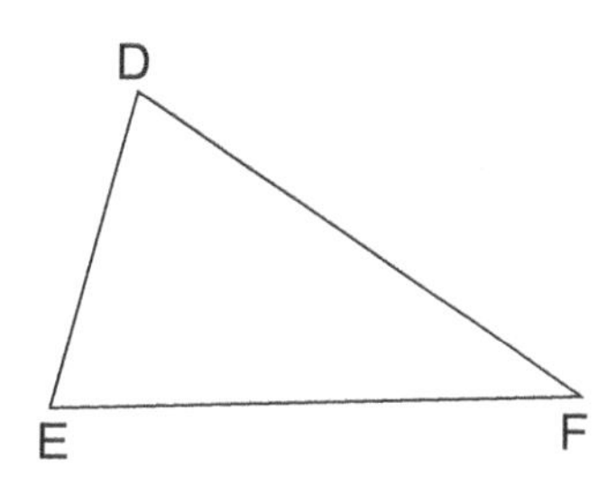

Review 1

Date / /

Name

Level ★★★

Score /100

■ The Answer Key is on page 95.

1 **There are five points A, B, C, D, and E on the circumference of circle O on the right. Answer the questions below.** 7 points per question

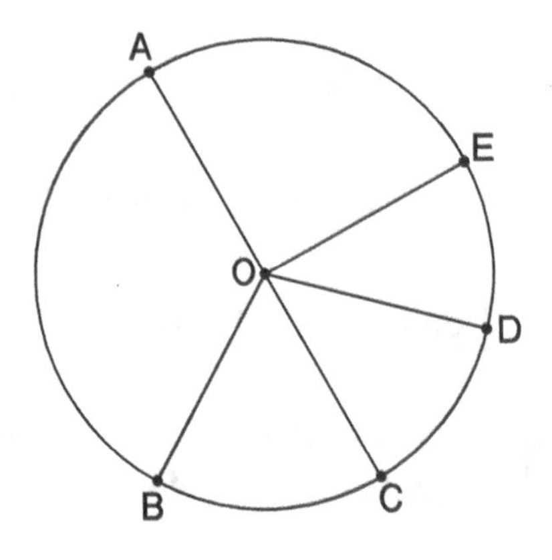

(1) What is the center angle of $\overset{\frown}{AB}$?

⟨Ans.⟩ ______

(2) What is the common arc of $\overset{\frown}{BCD}$ and the $\overset{\frown}{CDE}$?

⟨Ans.⟩ ______

(3) Given $\angle COD = \angle DOE$, what is the relationship between $\overset{\frown}{CD}$ and $\overset{\frown}{DE}$?

⟨Ans.⟩ ______

2 **Using the semicircle O with a radius of 10 cm on the right, answer the questions below in terms of π.** 7 points per question

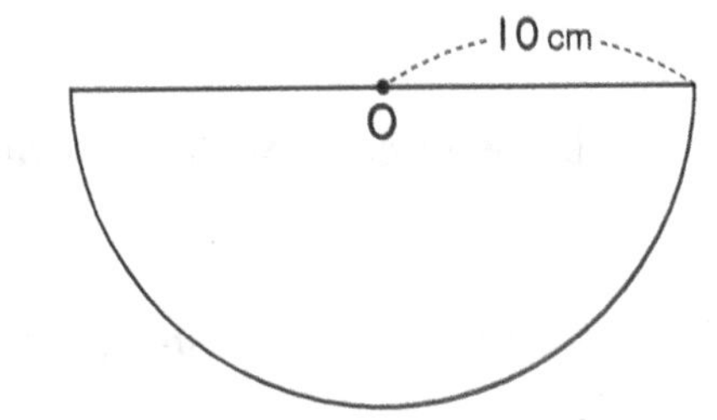

(1) Find the perimeter of semicircle O.

⟨Ans.⟩ ______

(2) Find the area of semicircle O.

⟨Ans.⟩ ______

3 **Find the length of the perimeter and the area of the sectors below.** 10 points per question

(1)

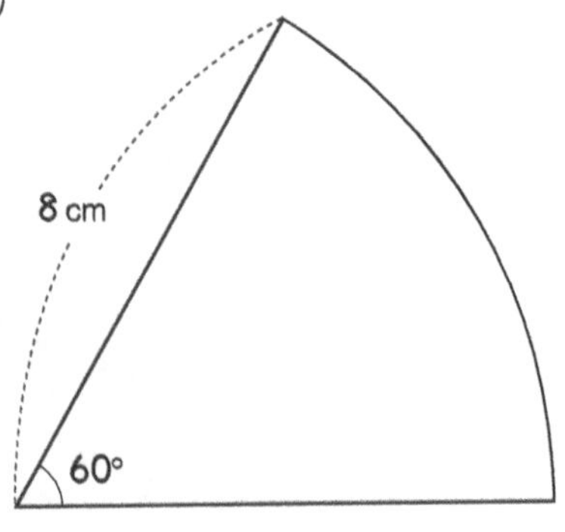

⟨Ans.⟩ Perimeter : ______

Area : ______

(2)

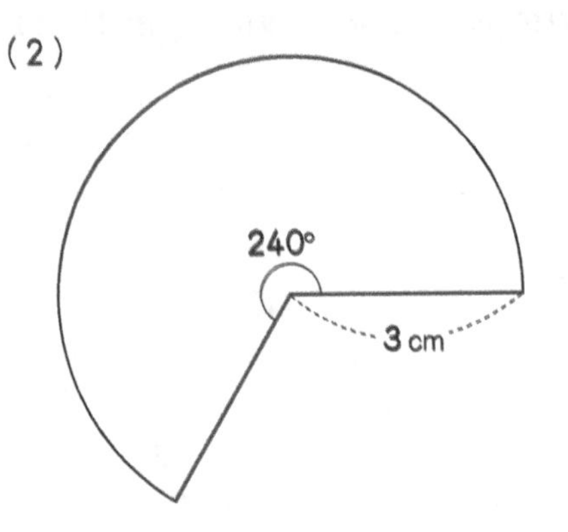

⟨Ans.⟩ Perimeter : ______

Area : ______

4 **Answer the following questions about circle O on the right.** 5 points per question

(1) Find the length of $\overset{\frown}{BC}$.

⟨Ans.⟩

(2) Find the length of $\overset{\frown}{CD}$.

⟨Ans.⟩

(3) Find the measure of ∠DOE.

⟨Ans.⟩

(4) Find the circumference of circle O.

⟨Ans.⟩

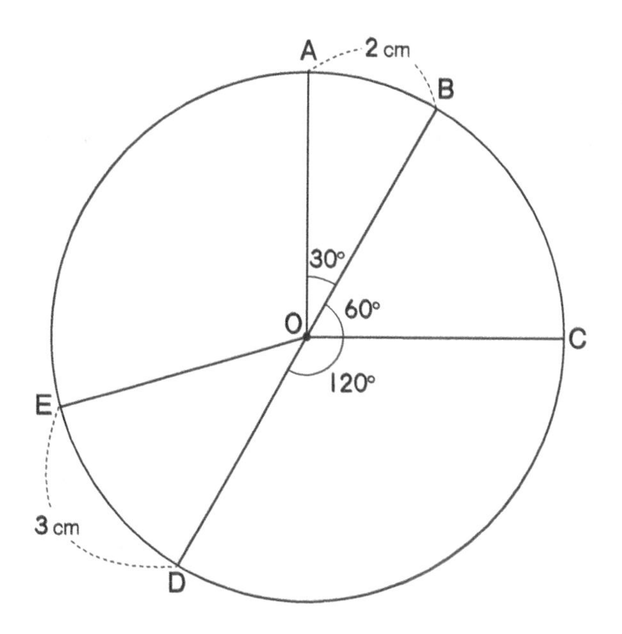

5 **Circle O, on the right, has a radius of 8 cm. Answer the questions below.** 5 points per question

(1) Find the area of circle O in terms of π.

⟨Ans.⟩

(2) Find the ratio of the area of sector AOB to the area of circle O in fractions.

⟨Ans.⟩

(3) Find the area of sector AOB in terms of π.

⟨Ans.⟩

(4) Find the ratio of the area of sector BOC to the area of sector AOC in fractions.

⟨Ans.⟩

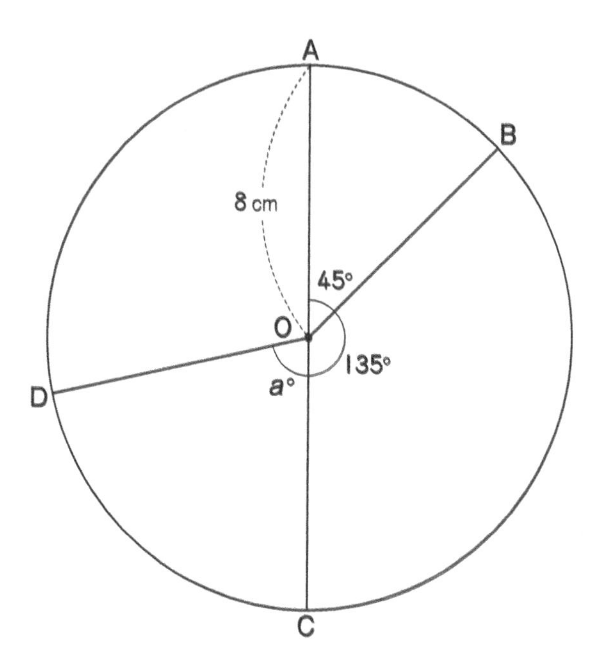

(5) Find the ratio of the area of sector DOC to the area of circle O in terms of a in fractions.

⟨Ans.⟩

Review 2

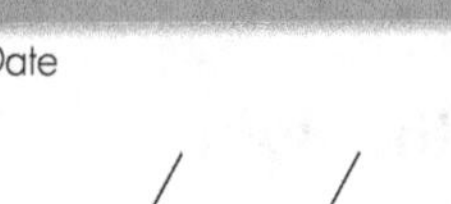

Name

■ The Answer Key is on page 95.

1 Find the area of the figures below.

5 points per question

(1)

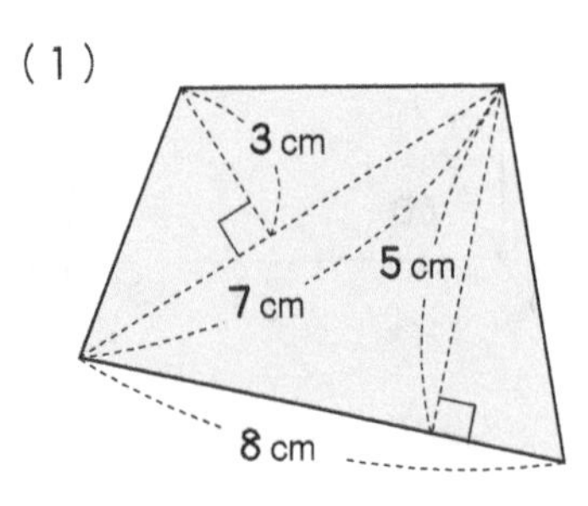

⟨Ans.⟩

(2)

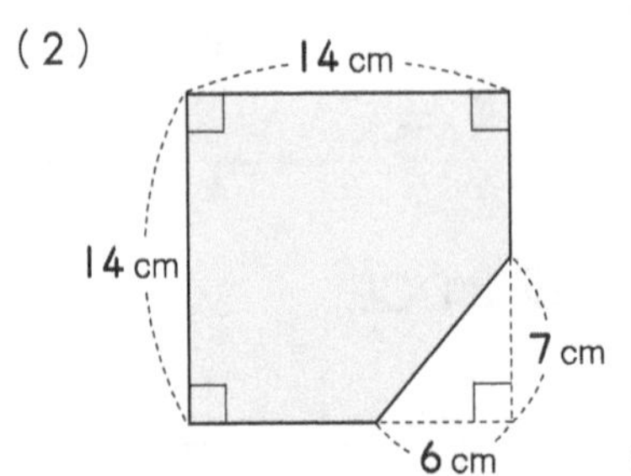

⟨Ans.⟩

2 Find the area of the trapezoids below.

5 points per question

(1)

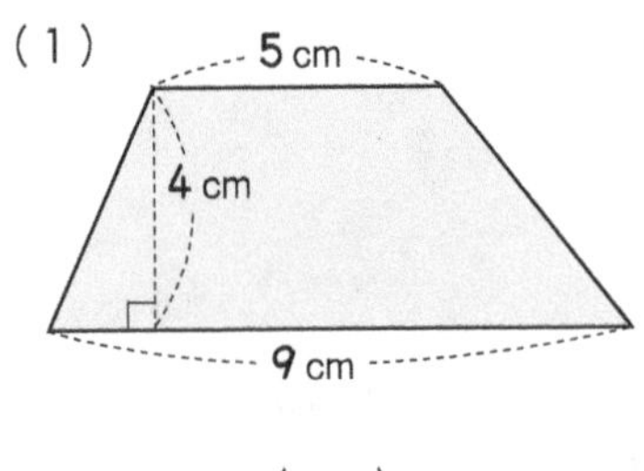

⟨Ans.⟩

(2)

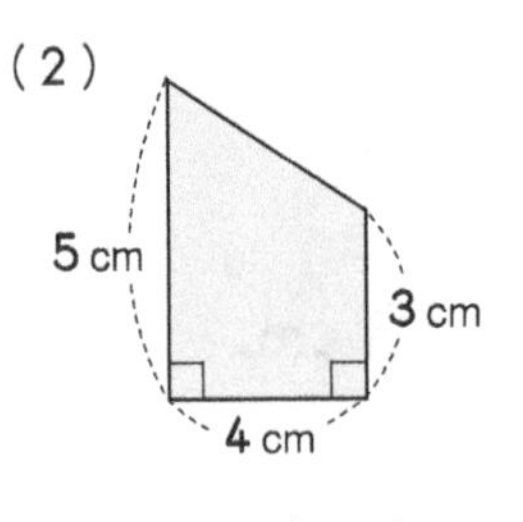

⟨Ans.⟩

(3)

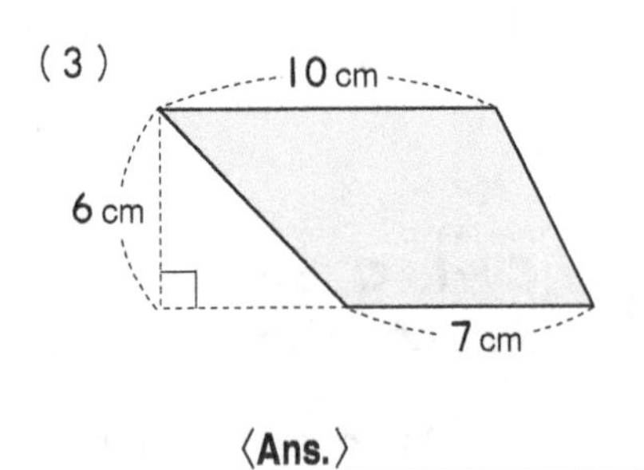

⟨Ans.⟩

3 Find the area of the figures below.

5 points per question

(1)

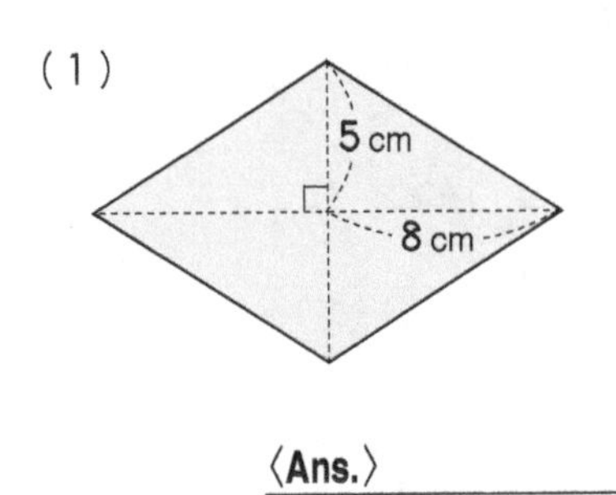

⟨Ans.⟩

(2)

15 cm

20 cm

⟨Ans.⟩

(3)

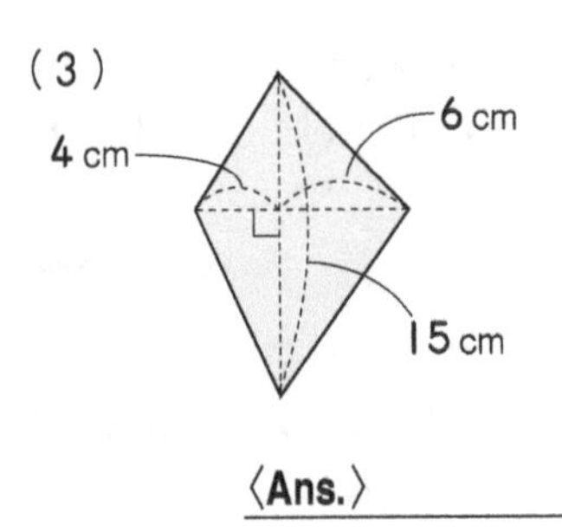

⟨Ans.⟩

4 Find the area of the figures below using x.

5 points per question

(1)

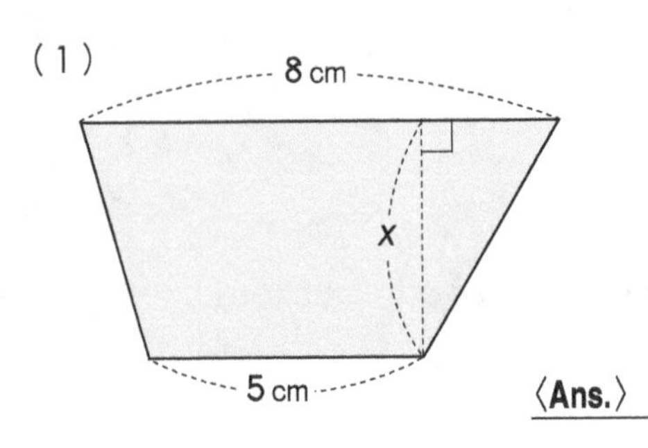

⟨Ans.⟩

(2)

6 cm

7 cm

x

x

x

8 cm

⟨Ans.⟩

5 Find the length of the perimeter and the area of the figures below.

10 points per question

(1)

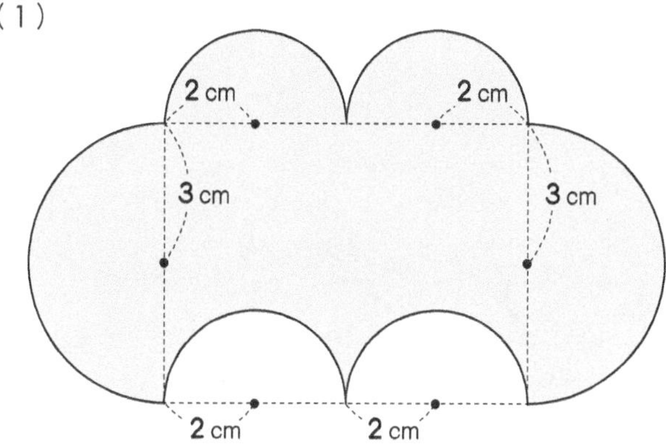

〈Ans.〉 Perimeter :

Area :

(2)

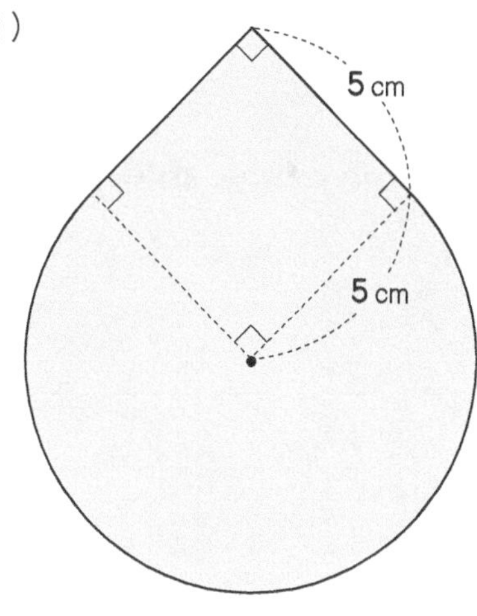

p.96

〈Ans.〉 Perimeter :

Area :

6 Answer the following questions about the built shape and the net of the rectangular prism on the right.

5 points per question

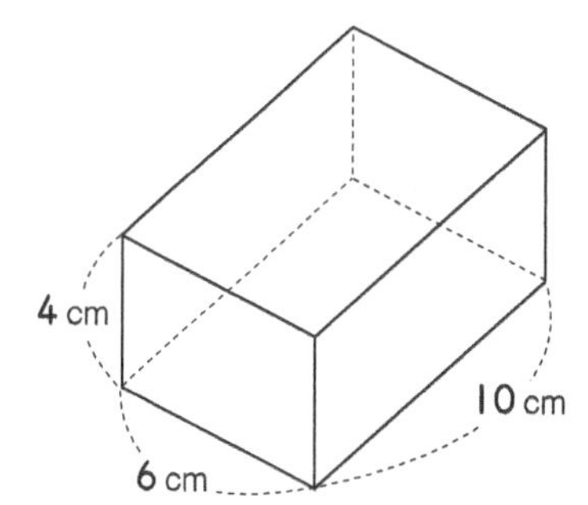

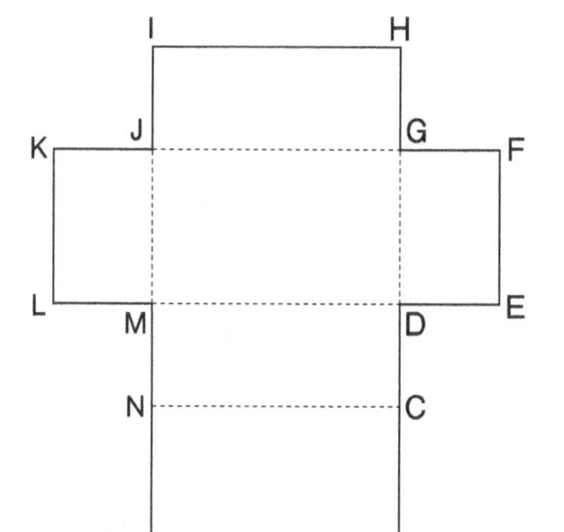

(1) Find the length of edge AB.

〈Ans.〉

(2) Find the length of edge BC.

〈Ans.〉

(3) Find the length of edge IJ.

〈Ans.〉

(4) Which edge lines up with edge FG?

〈Ans.〉

(5) Which edge lines up with edge KL?

〈Ans.〉

(6) Which surface (= face) is parallel to surface (= face) ABCN?

〈Ans.〉

43 Review 3

Date / /

Name

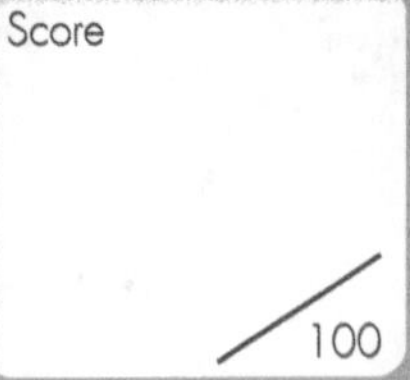

■ The Answer Key is on page 96.

1 Find the volume of the solids below.

7 points per question

(1)

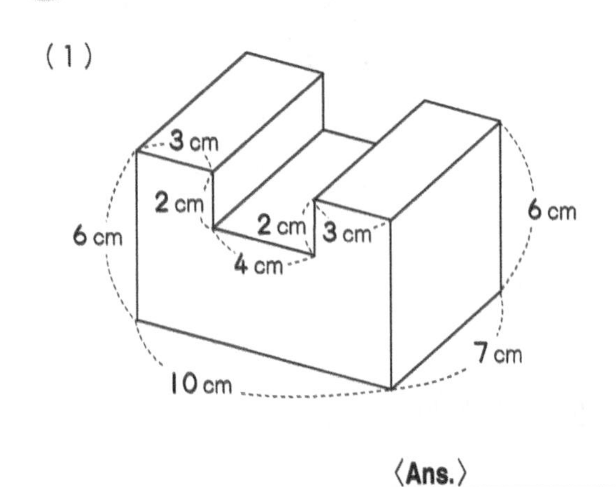

〈Ans.〉

(2)

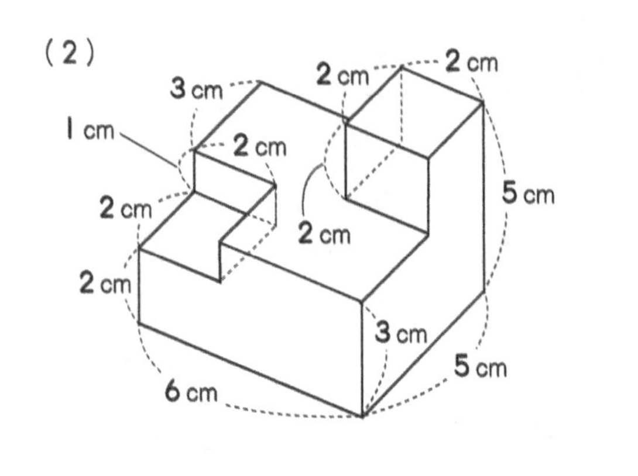

〈Ans.〉

2 Find the surface area of the solids below.

7 points per question

(1)

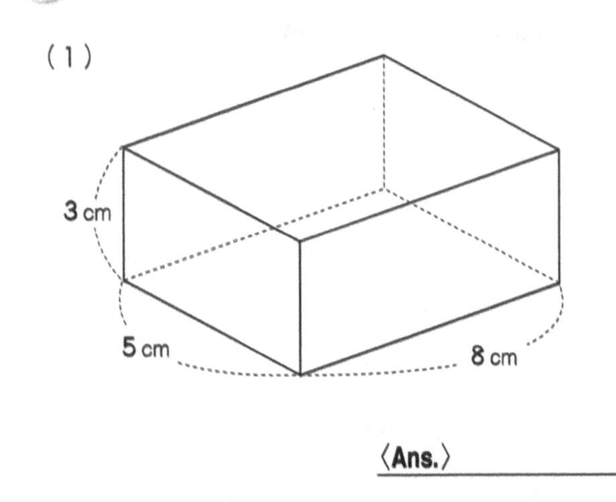

〈Ans.〉

(2)

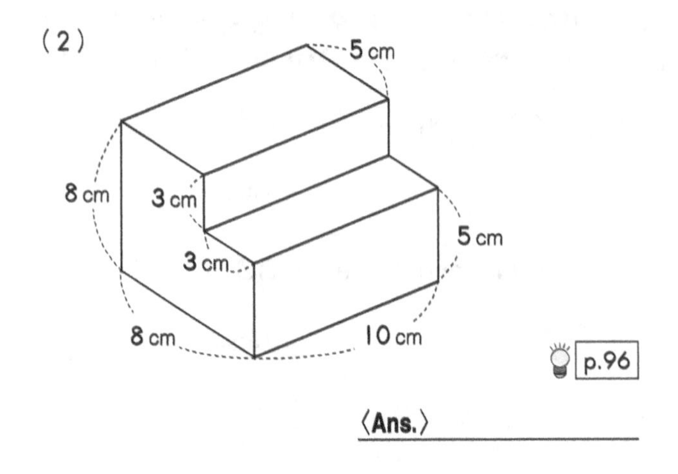

p.96

〈Ans.〉

3 Answer the appropriate words in the blanks.

8 points per question

(1) Given $\overline{AM} = \overline{MB}$ in $\overline{AB}$, point M is the ☐ of $\overline{AB}$.

〈Ans.〉

(2) Given $\overline{AB}$ is the perpendicular bisector of $\overline{CD}$ and the intersection point of $\overline{AB}$ and $\overline{CD}$ is point P, $\overline{CP} \perp$ ☐ and $\overline{CP} =$ ☐ .

〈Ans.〉 ,

(3) Given $\overline{PQ}$ and $\overline{RS}$ bisect each other and intersect at point O, $\overline{PO} =$ ☐ , $\overline{RO} =$ ☐ .

〈Ans.〉 ,

4 Lines ℓ and m are parallel to each other. Answer the questions below.

8 points per question

(1) Draw a perpendicular line from point P and point Q on line ℓ, respectively, to line m.

(2) Fill the appropriate words and numbers in the blanks.

When the intersection point with line m is point M and point N respectively, the perpendicular $\overline{PM}$ is the distance between point ______ and line m.

Since line ℓ and line m are parallel, $\overline{PM}$ = ______

As described above, when the distances from all the points on line ℓ to line m are the same, this length is called the ______ between line ℓ and line m.

5 Using the isosceles triangle with $\overline{CA} = \overline{CB}$ on the right, answer the questions below.

8 points per question

(1) Draw the perpendicular bisector of base AB.

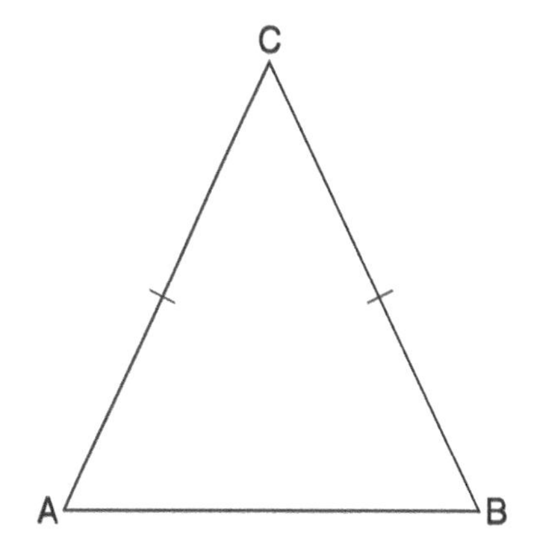

(2) Which angle is bisected by the perpendicular bisector of base AB?

⟨Ans.⟩ ______

6 Using the isosceles triangle with $\overline{AB} = \overline{AC}$ on the right, answer the questions below.

8 points per question

(1) Draw the bisector of ∠BAC.

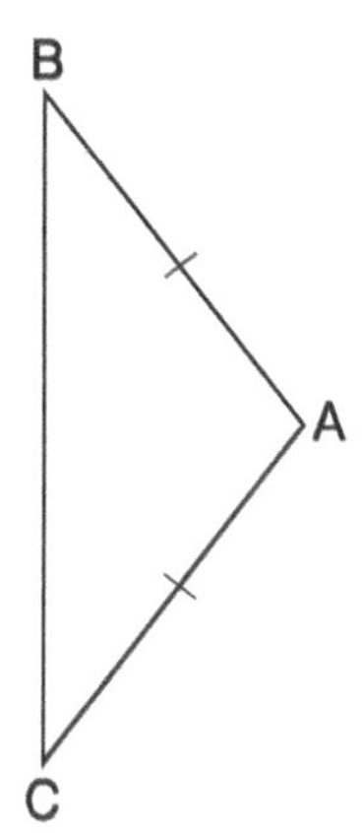

(2) Where does the bisector of ∠BAC pass through base BC?

⟨Ans.⟩ ______

Answer Key Grades 6-8 Intro to Geometry Workbook I

1 Triangles & Quadrilaterals Review pp 2, 3

1 (1) A, H, K (2) D, F, M (3) C, L, O (4) B, N

2 (1) C, G, I (2) A, D, J (3) B, E, F, H

3 (1) A, C, E (2) B, G, H

2 Quadrilaterals Review pp 4, 5

1 (1) A, H (2) C, I

2 D, F

3 (1) A, B, D, E
(2) A, B
(3) A, B, D, E
(4) A, E

4 (1) Trapezoid
(2) Parallelogram
(3) Rectangle
(4) Rhombus
(5) Square

3 Polygons Review pp 6, 7

1 (1) (Regular) octagon (2) (Regular) decagon
(3) (Regular) hexagon (4) (Regular) pentagon
(5) (Equilateral) triangle

2 (1) × (2) ○ (3) ○ (4) × (5) ×
(6) ○ (7) × (8) ○ (9) × (10) ×

3 (1) 36° (2) 72°

4 (1)

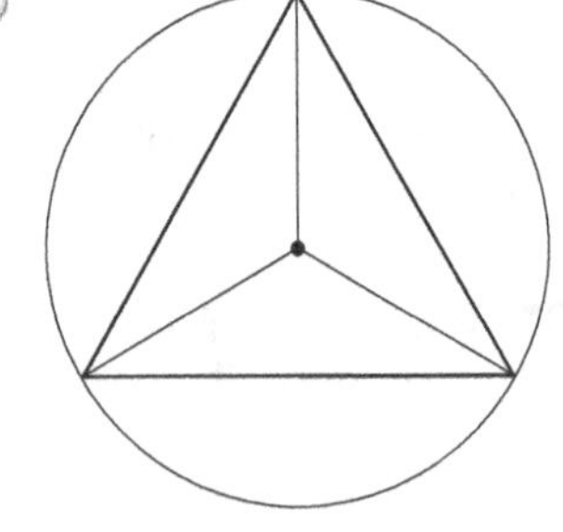

(2)

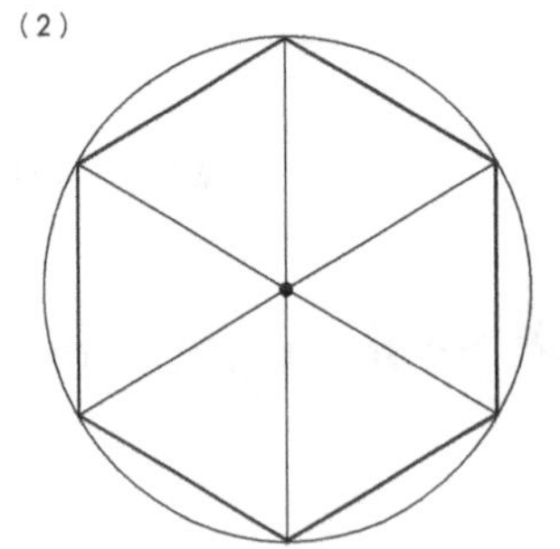

(3)

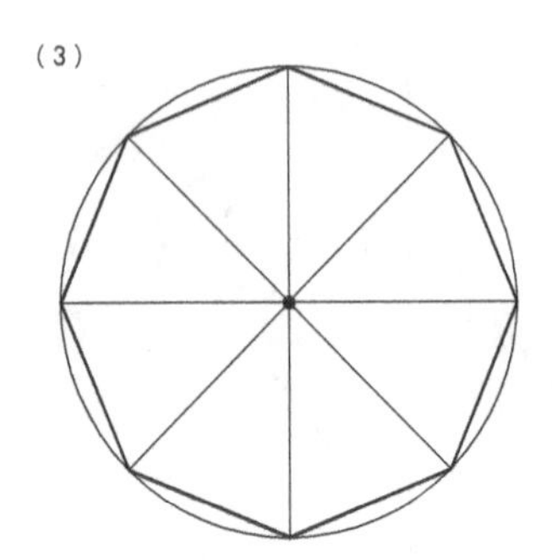

4 Angles Review 1 pp 8, 9

1 A = 30° B = 60° C = 90° D = 45°
E = 45° F = 90°

2 (1) 180° (2) 360° (3) 60° (4) 140°
(5) 270° (6) 120°

3 (1) C (2) D (3) 180°

4 (1) A = 110° B = 70° (2) A = 50° B = 130°
(3) A = 160° B = 20°

5 (1) A = 45° B = 30° (2) A = 75° B = 35°
(3) A = 50° B = 40° C = 40°

5 Angles Review 2 pp 10, 11

1 (1) 35° (2) 32° (3) 130° (4) 25°
(5) 65° (6) 60° (7) 20° (8) 60°

2 (1) 115° (2) 55° (3) 105° (4) 125°

3 (1) A = 45° B = 90° (2) A = 30° B = 120°
(3) A = 54° B = 72° (4) A = 60° B = 60°

6 Angles Review 3 pp 12, 13

1 (1) *e* (2) *f* (3) *g* (4) *h*

2 (1) *h*, *ℓ* (2) *e*, *i* (3) *ℓ* (4) *i*

3 (1) *c*, *e*, *g* (2) *d*, *f*, *h*
(3) *a*, *e*, *g* (4) *b*, *f*, *h*

4 (1) 60° (2) 120° (3) 120° (4) 120°

5 (1) 80° (2) 100° (3) 100°

7 Drawing Lines Review pp 14, 15

1 (1)

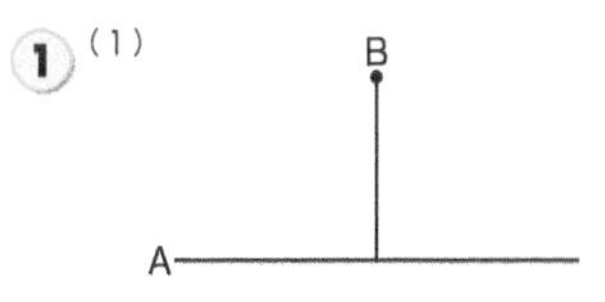

(2)

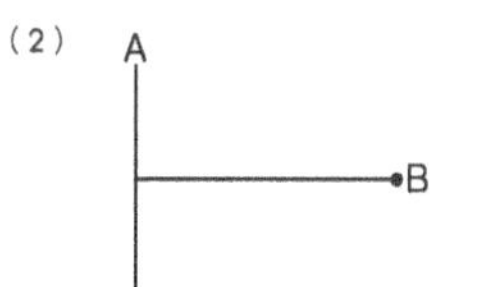

(3)

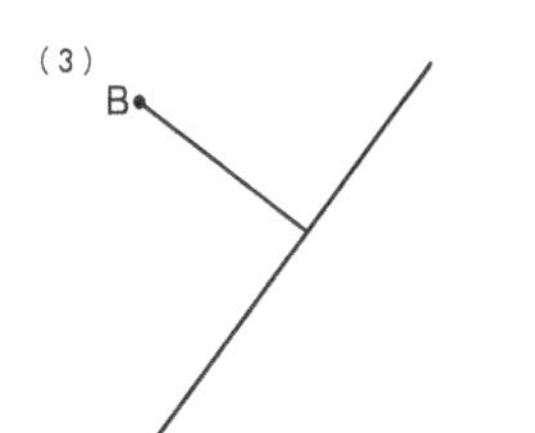

(4)

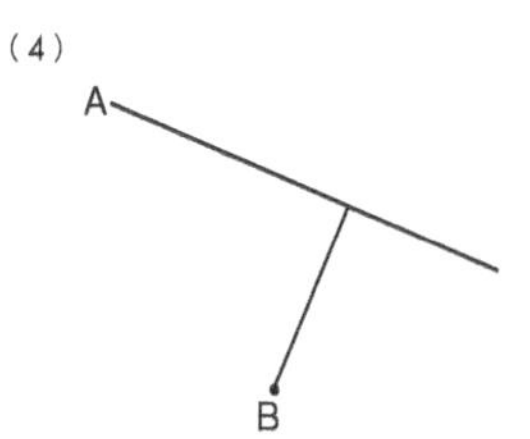

2 (1)

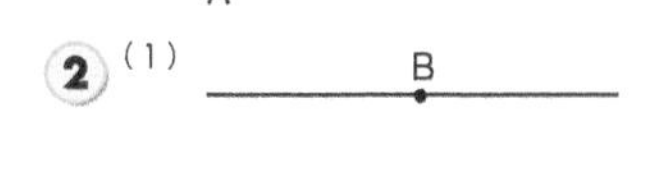

(2)

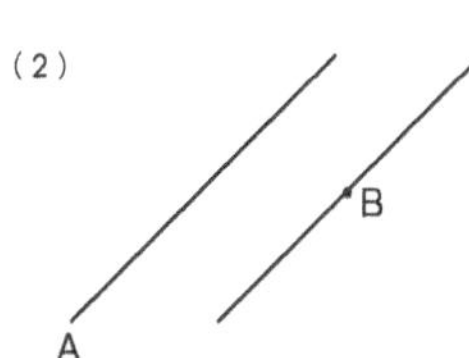

(3)

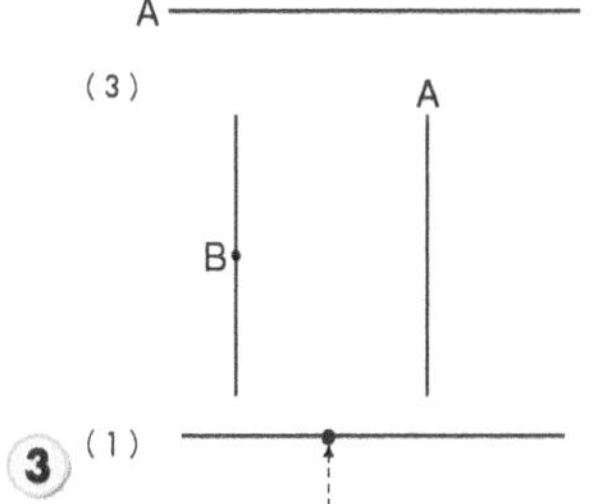

3 (1)

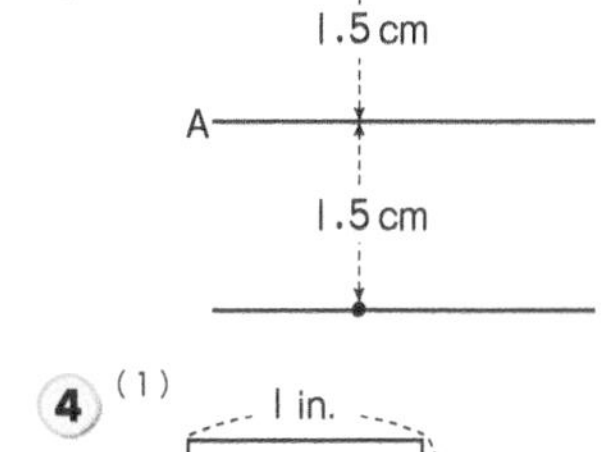

(2)

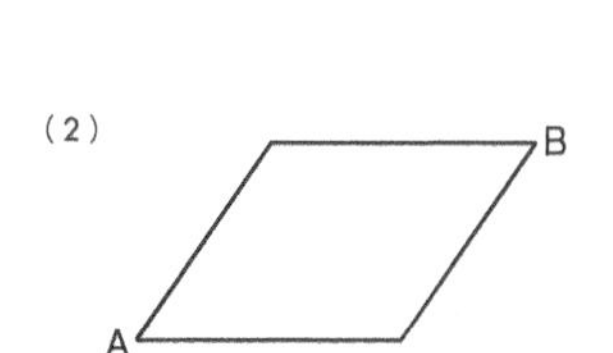

4 (1) 1 in. 1 in.

(2)

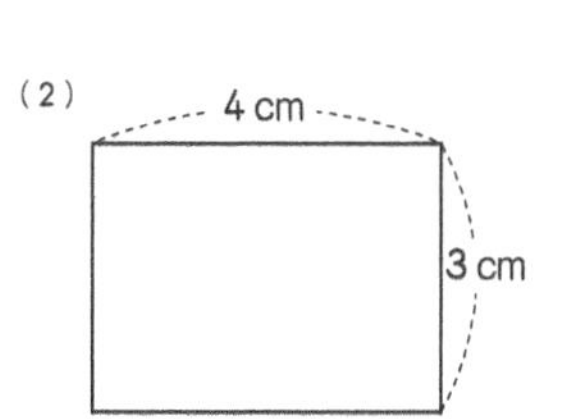

5 (1)

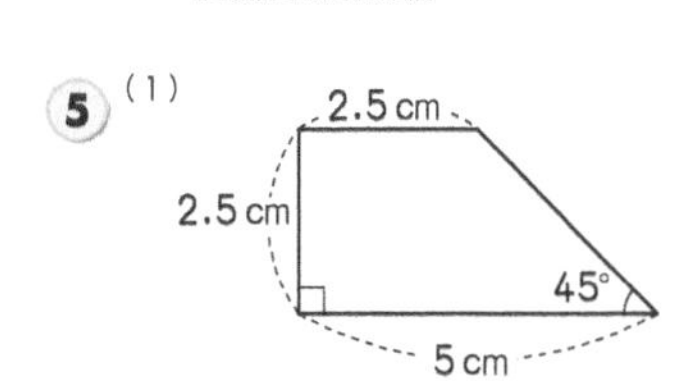

(2)

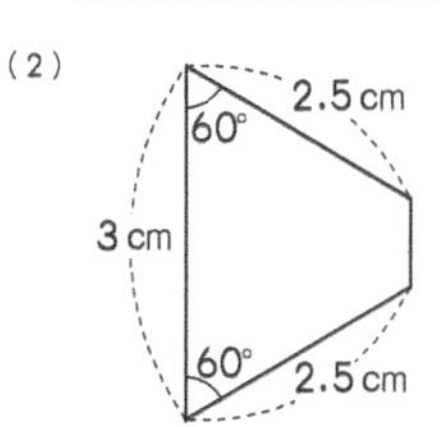

8 Drawing Quadrilaterals Review pp 16, 17

1 (1) 3 cm 45° 4 cm

(2)

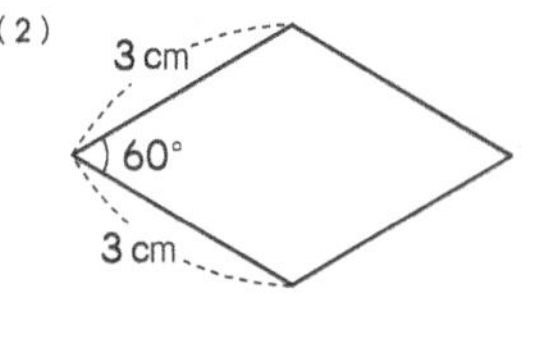

(3)

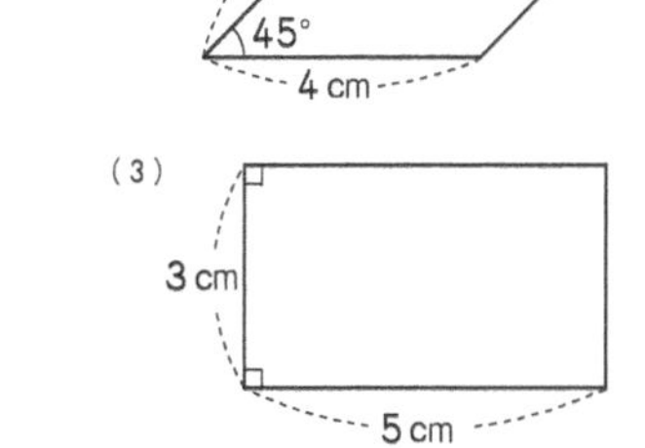

2 (1)

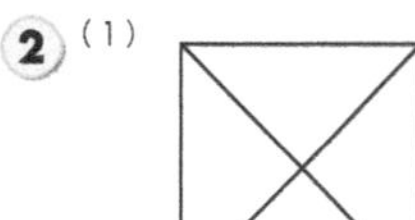

(2)

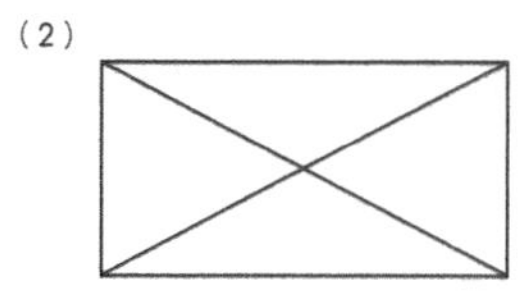

(3)

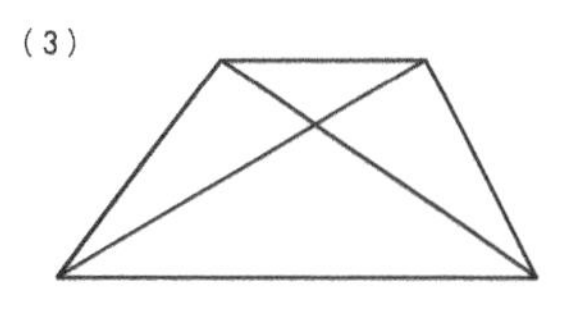

(4)

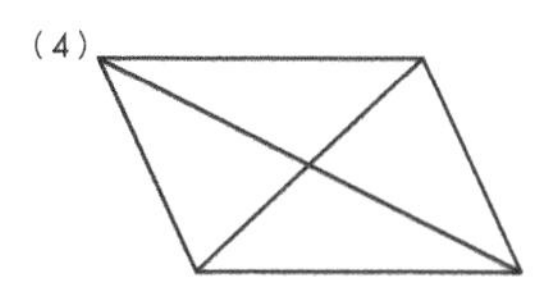

(5)

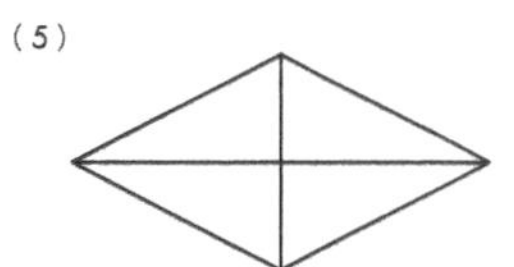

3 (1) A, B (2) A, E (3) A, B, D, E

4 (1) Parallelogram (2) Square (3) Rectangle

9 Area Review pp 18, 19

1 (1) 16 in.^{2} (2) 15 in.^{2} (3) 50 in.^{2}
(4) 70 in.^{2} (5) 16 in.^{2} (6) 75 in.^{2}

2 (1) 15 in.^{2} (2) 20 in.^{2} (3) 30 in.^{2} (4) 24 in.^{2}

3 (1) 43 in.^{2} (2) 37 in.^{2} (3) 36 in.^{2} (4) 112 in.^{2}

4 (1) 56 in.^{2} (2) 52 in.^{2} (3) 48 cm^2 (4) 225 cm^2

10 Circles & Perimeter Review pp 20, 21

1 (1) 6 in. (2) 6 in. (3) 3 in.

2 (1) 16 in. (2) 8 in. (3) 4 in.

3 (1) 10 cm (2) 6 cm (3) 8 cm

4 (1) 22 cm (2) 20 cm (3) 36 cm (4) 28 cm
(5) 32 cm

5 (1) 32 cm (2) 20 cm (3) 80 cm (4) 32 cm

11 Volume & Net Review pp 22, 23

1 (1) 27 in.^{3} (2) 495 cm^3 (3) 84 ft.^{3} (4) 24 m^3
(5) 576 in.^{3} (6) 54 cm^3 (7) 2.16 ft.^{3} (8) 8 cm^3

2 (1) *c* (2) *a* (3) *b*

3 (1) F (2) E

4 *b*, *d*

12 Congruent Figures Review
pp 24, 25

1 A and I, C and H, F and J

2 C, J, K

3 (1) D (2) E (3) F (4) DE (5) EF
(6) FD (7) D (8) E (9) F

4 (1) B (2) GF (3) E
(4) 2 cm (5) 3 cm (6) 60°

13 Scale Drawing Review
pp 26, 27

1 C, G, H

2 C, F, G

3 (1) Vertex A: D
Vertex B: E
Vertex C: F
(2) Side AB: DE
Side BC: EF
Side CA: FD

4 (1) Vertex A: E
Vertex B: F
Vertex C: G
(2) Side AB: EF
Side BC: GH
Side CA: HE

14 Line Symmetry Review
pp 28, 29

1 A, B, C, E, G, H, J

2 A, C, D, E, F, H, I, J

3 (1) 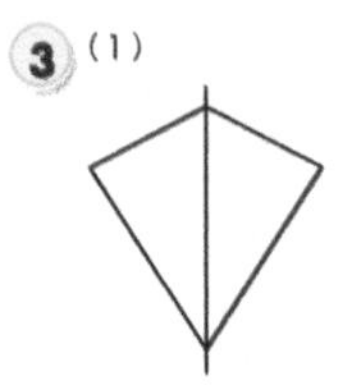 (2) 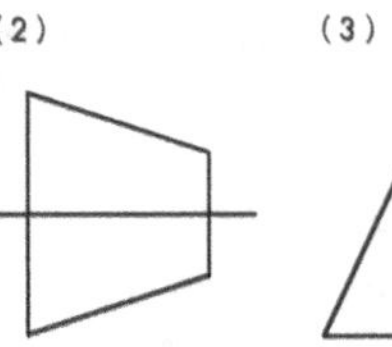 (3) 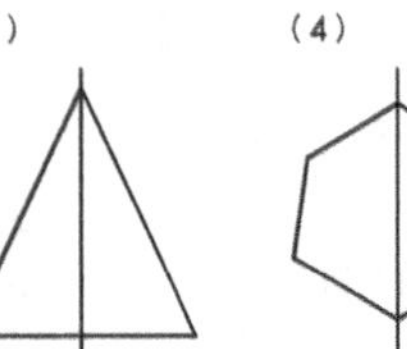 (4)

4 (1) 2 (2) 2 (3) 3 (4) 4 (5) 5
(6) 2 (7) 2 (8) 1

15 Rotational Symmetry Review
pp 30, 31

1 B, C

2 B, C, F, H

3 (1) E (2) F (3) C (4) ED (5) GH

4 (1)

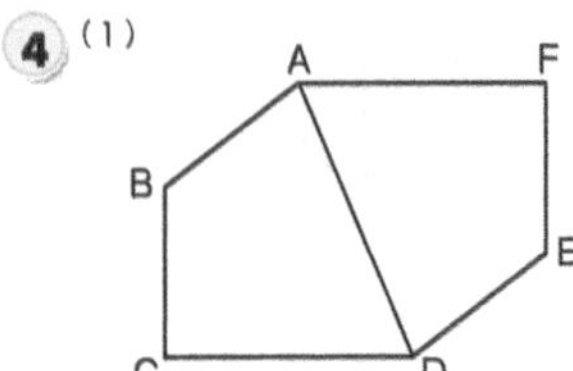

(2)–(4)

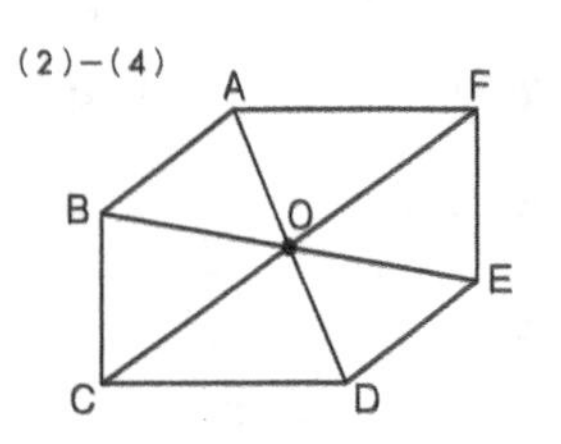

5

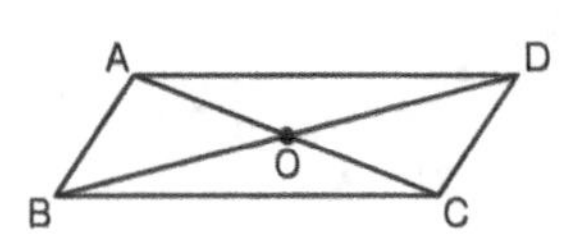

16 Line Symmetry & Rotational Symmetry Review
pp 32, 33

1 (1)

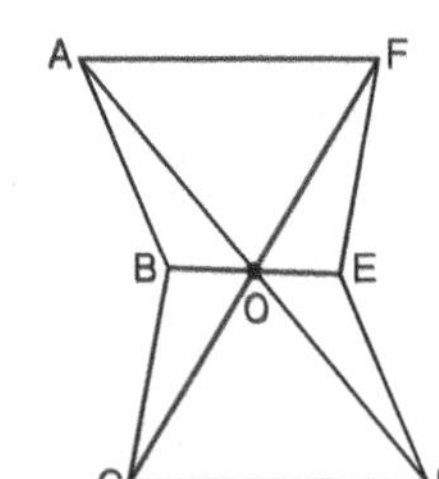

(2) D (3) E
(4) BC (5) FA
(6) F (7) OC
(8) OE

2 (1)

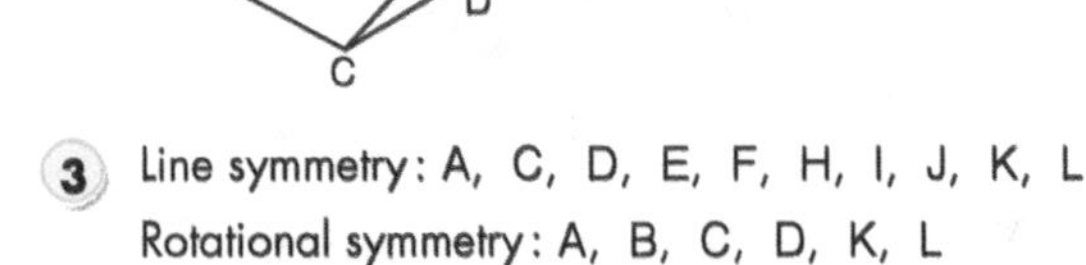

(2) ED
(3) EDO

3 Line symmetry: A, C, D, E, F, H, I, J, K, L
Rotational symmetry: A, B, C, D, K, L

4 Line symmetry: A, B, H, I, M, T, U, X
Rotational symmetry: H, I, N, S, X, Z

17 Circle 1 (Circumference)
pp 34, 35

1 (1) 12.56 cm (2) 18.84 cm (3) 25.12 cm
(4) 31.4 cm (5) 43.96 in. (6) 37.68 in.

2 (1) 56.52 in. (2) 25.7 in. (3) 41.12 in.
(4) 28.56 in. (5) 41.4 cm (6) 33.12 cm
(7) 55.12 cm (8) 55.12 cm (9) 22.85 cm
(10) 31.4 cm

18 Circle 2 (Area)
pp 36, 37

1 (1) 28.26 in.^{2} (2) 78.5 in.^{2}
(3) 113.04 in.^{2} (4) 200.96 in.^{2}

2 (1) 25.12 cm^2 (2) 6.28 cm^2
(3) 19.625 cm^2 (4) 28.26 cm^2

3 (1) 37.68 cm^2 (2) 58.875 cm^2
(3) 21.5 cm^2 (4) 13.76 cm^2
(5) 7.065 in.^{2} (6) 17.27 in.^{2}
(7) 14.13 in.^{2} (8) 37.68 in.^{2}

19 Circle 3 (Arc & Sector)
pp 38, 39

1 (1) $\frac{1}{4}$ Ans. 9.42 cm (2) $\frac{1}{2}$ Ans. 15.7 cm

(3) $\frac{3}{4}$ Ans. 28.26 cm (4) $\frac{1}{3}$ Ans. 6.28 cm

(5) $\frac{1}{6}$ Ans. 6.28 cm (6) $\frac{1}{8}$ Ans. 4.71 cm

2 A, D, F, G, I, K

3 (1) 21.42 cm (2) 20.56 cm (3) 24.56 cm

20 Circle 4 (Area of Sector)
pp 40, 41

1 (1) $\frac{1}{2}$ Ans. 39.25 cm² (2) $\frac{1}{4}$ Ans. 12.56 cm²

(3) $\frac{3}{4}$ Ans. 150.72 cm²

(4) $\frac{120}{360}=\frac{1}{3}$ Ans. 84.78 cm²

(5) $\frac{60}{360}=\frac{1}{6}$ Ans. 18.84 cm²

(6) $\frac{45}{360}=\frac{1}{8}$ Ans. 6.28 cm²

2 (1) $\frac{1}{4}$ (2) $\frac{2}{3}$ (3) $\frac{3}{4}$ (4) $\frac{3}{8}$

(5) $\frac{1}{12}$ (6) $\frac{7}{12}$ (7) $\frac{1}{24}$ (8) $\frac{5}{24}$

3 (1) 23.55 cm² (2) 188.4 cm² (3) 3.14 cm²
(4) 50.24 cm² (5) 37.68 cm²

4 (1) 6.28 cm (2) 25.12 cm (3) 9.42 cm
(4) 6.28 cm

21 Circle 5 (Use π for pi)
pp 42, 43

1 (1) 8π in. (2) 12π in. (3) 10π in.

2 (1) 9π cm² (2) 25π cm² (3) 36π cm²

3 (1) 6π in. (2) 5π in. (3) 6π in.

4 (1) 5π cm² (2) 75π cm² (3) 8π cm²

22 Perimeter & Area of Shapes
pp 44, 45

1 (1) 22 cm (2) 12 cm (3) 22 cm
(4) 30 cm (5) 40 cm (6) 26 cm

2 (1) $(20+5\pi)$ in. (2) $(8+8\pi)$ in. (3) $(6+7\pi)$ in.

3 (1) Area: $(25+12.5\pi)$ cm² Perimeter: 10π cm
(2) Area: $(180-25\pi)$ cm² Perimeter: $(36+10\pi)$ cm
(3) Area: $(144-36\pi)$ cm² Perimeter: 12π cm
(4) Area: 100 cm² Perimeter: $(20+10\pi)$ cm
(5) Area: $(32\pi-64)$ cm² Perimeter: 8π cm
(6) Area: $\left(60-\frac{25}{2}\pi\right)$ cm² Perimeter: $(14+5\pi)$ cm

23 Area of Polygons
pp 46, 47

1 (1) 25 in.² (2) 43 in.² (3) 97.5 in.²
(4) 60 in.² (5) 132 cm² (6) 157 cm²

2 (1) 3 (2) 4 (3) 5 (4) 6

3 268 cm²

24 Rectangular Prisms 1 (Figure 1)
pp 48, 49

1 Rectangular prism: B, D, H
Cube: F

2

	Rectangular prism	Cube
Number of faces	6	6
Number of edges	12	12
Number of vertices	8	8

3 (1) edge DC, edge EF, edge HG
(2) edge BC, edge EH, edge FG
(3) edge BF, edge CG, edge DH
(4) 2 (5) 2 (6) 2 (7) 4
(8) edge AE, edge BF, edge CG, edge DH

4 (1) BC, CD, DA
AE, BF, CG, DH
EF, FG, GH, HE
(2) 6

25 Rectangular Prisms 2 (Figure 2)
pp 50, 51

1 (1) edge AD, edge AE
(2) edge BA, edge BC, edge FE, edge FG
(3) edge DC, edge EF, edge HG
(4) edge AD, edge EH, edge FG

2 (1) edge AE, edge BF, edge CG, edge DH
(2) edge AB, edge DC, edge EF, edge HG

3 (1) face CGHD, face AEHD, face BFGC
(2) face ABCD, face EFGH, face AEHD, face BFGC
(3) face ABCD, face EFGH, face ABFE, face CGHD

4 (1) face CGHD (2) face AEHD

5 (1) edge CD, edge GH, edge CG, edge DH
(2) edge AE, edge DH, edge AD, edge EH

26 Rectangular Prisms 3 (Net 1) pp52,53

1 (1) B, C (2) A, B

2 (1) B, C (2) A, E

3 (1) 12 cm (2) 5 cm
(3) 8 cm (4) edge IH
(5) edge DC (6) edge GF
(7) vertex E (8) vertex N
(9) vertex I, vertex K (10) face DEFG
(11) face ABCN, face MDGJ, face LMJK, face DEFG

27 Rectangular Prisms 4 (Net 2) pp54,55

1 (1)

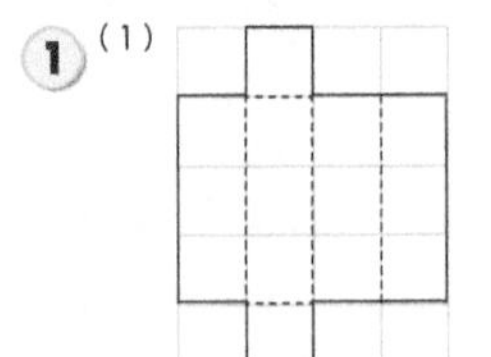

(2)

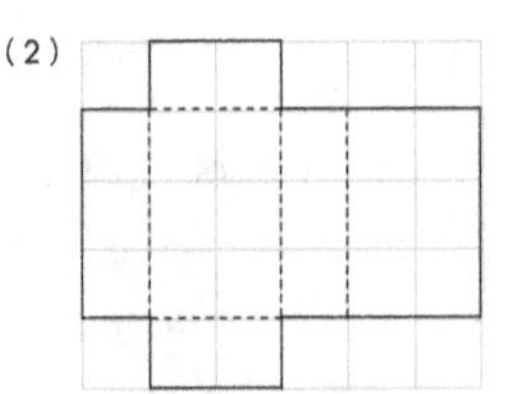

(3)

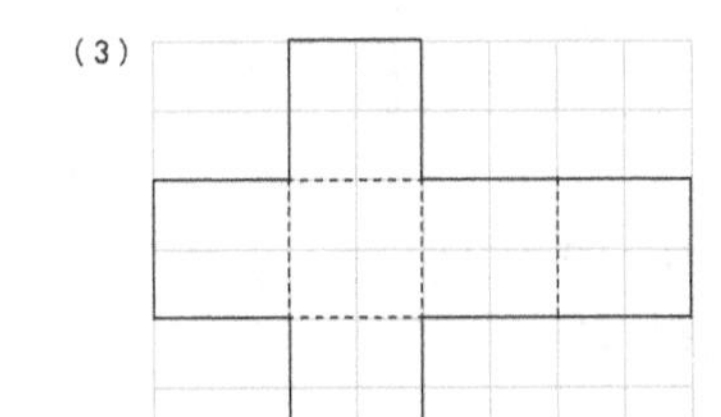

※There are several other answers.

2 (1)

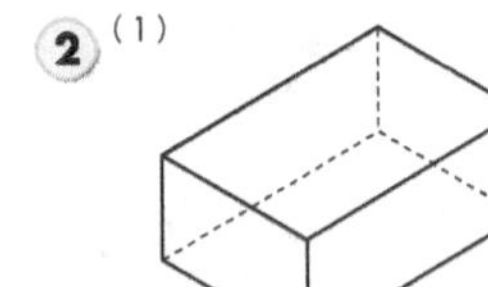

(2)

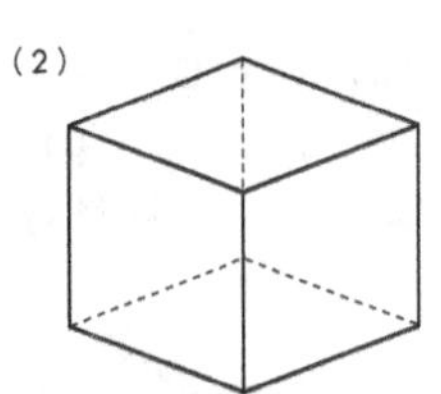

3 (Built shape)

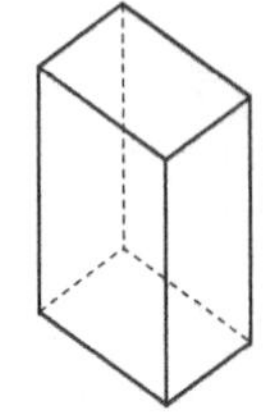

(net)

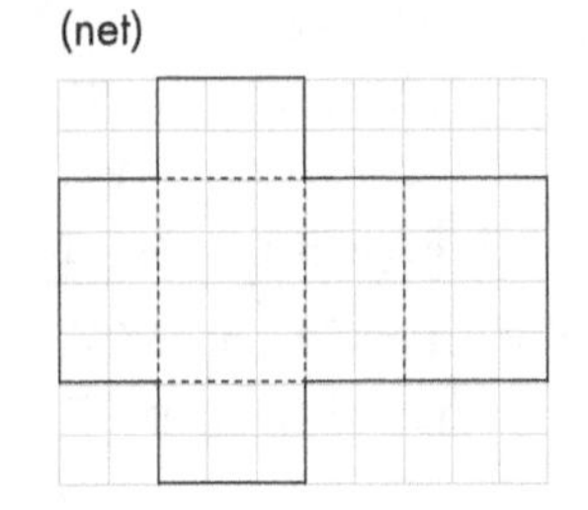

28 Prisms & Cylinders pp56,57

1 (1) Cylinder (2) Triangular prism
(3) Rectangular prism (4) Cube
(5) Triangular prism (6) Cylinder
(7) Quadrangular prism (8) Cylinder
(9) Pentagonal prism (10) Hexagonal prism
(11) Cylinder (12) Pentagonal prism

2 (1) triangle (2) 2 (3) rectangle
(4) 3 (5) 6

3 (1) hexagon (2) 2 (3) rectangle
(4) 6 (5) 12

4 (1) Shape: circle, Number: 2
(2) a curved face

29 Surface Area & Volume pp58,59

1 (1) 54 in.^{2} (2) 158 in.^{2} (3) 94 in.^{2} (4) 152 in.^{2}

2 (1) 8 cm^3 (2) 225 cm^3

3 (1) 92 in.^{3} (2) 455 in.^{3} (3) 424 in.^{3} (4) 112 in.^{3}

4 (1) 104 in.^{2} (2) 208 in.^{2}

30 Geometric Notation 1 (Line, Ray, & Line Segment) pp60,61

1 (1) (2) (3) (4)

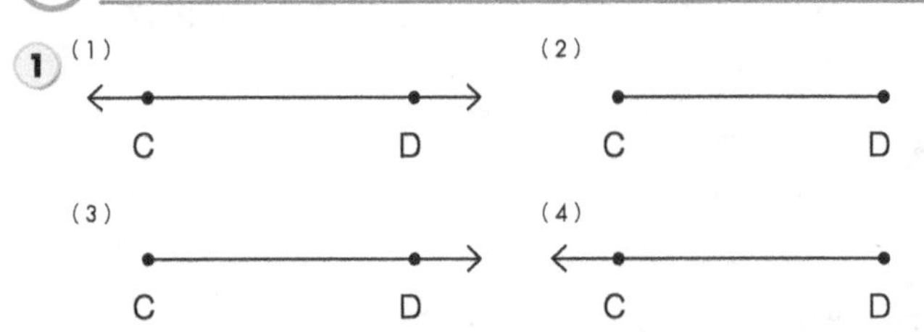

2 (1) – (5)

(6) line AB (7) ray GH

3 (1) $\overline{EF}=5$ cm　(2) $\overline{MN}=\overline{PQ}$　(3) $\overline{AB}=3\overline{CD}$

(4) $\overline{GH}=\frac{1}{2}\overline{CD}$　(5) $\overline{AB}=\overline{AC}$　(6) $\overline{KL}=7$ in.

(7) $\overline{RS}=2\overline{TU}$　(8) $\overline{QR}=\frac{1}{2}\overline{OP}$

31 Geometric Notation 2 (Parallel & Perpendicular) pp62,63

1 (1) $k \parallel \ell$　(2) $\overleftrightarrow{MN} \parallel \overleftrightarrow{PQ}$　(3) $\overline{AB} \parallel \overline{CD}$

2 (1) $g \perp h$　(2) $\overleftrightarrow{AB} \perp \overleftrightarrow{CD}$　(3) $\overline{EF} \perp \overline{GH}$

3 (1)

(2)

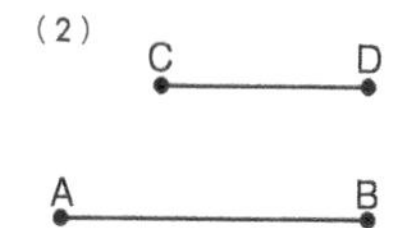

(3)

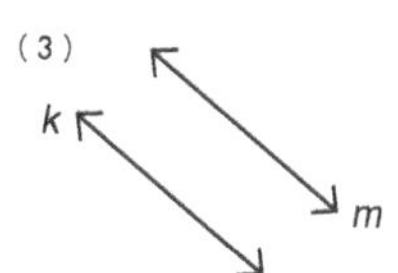

(4)

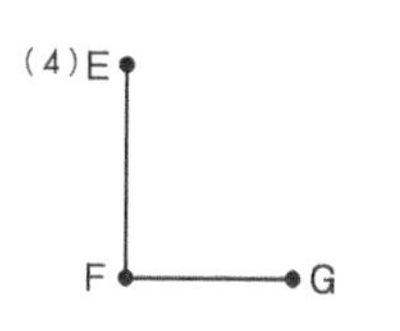

(5)

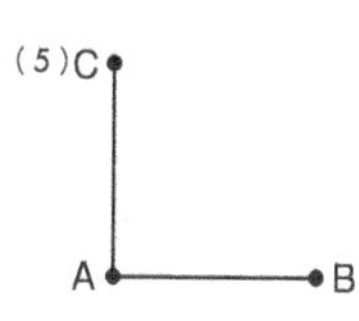

(6)

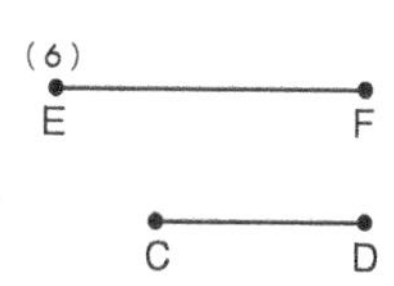

(7)

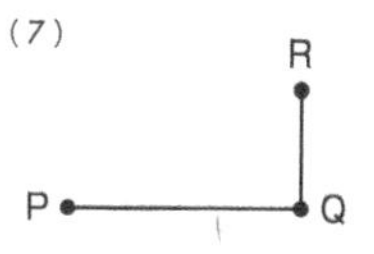

(8)

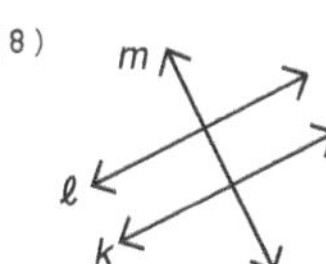

32 Geometric Notation 3 (Angle) pp64,65

1 (1) ∠B = ∠C　(2) ∠ABC = 70°　(3) AD ⊥ BC

(4) BD = CD　(5) BC = 2BD

(6) $\angle BAD = \frac{1}{2}\angle BAC$

2 (1) ①∠BAD　②∠BDC　③∠DBC

(2) ∠CDB, ∠DBC　(3) ∠BDA, ∠DAB

(4) ∠CDA, ∠DAB, ∠ABC

3 (1) AB ∥ CD　(2) AC ∥ BD　(3) AB = CD

(4) AC ⊥ CD　(5) AB ⊥ BD　(6) ∠BAD = ∠ADC

(7) $\overline{AE}=\overline{ED}$　(8) $\overline{BC}=\overline{AD}$

4 (1) PQ = RS　(2) QR ∥ PS　(3) PQ = PS

(4) 60°　(5) $\overline{PR} \perp \overline{QS}$　(6) $\overline{QT}=\overline{TS}$

33 Geometric Notation 4 (Circle) pp66,67

1 (1) $\overline{AC}$, $\overline{AD}$, $\overline{AE}$　(2) F　(3) $\overline{BD}$

2 (1) (2)

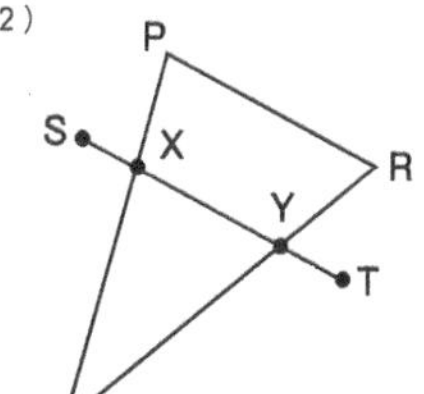

(3) ∠QPR (∠XPR)

(4) ∠QRP (∠YRP)

(5) △QXY

3 (1)–(3)

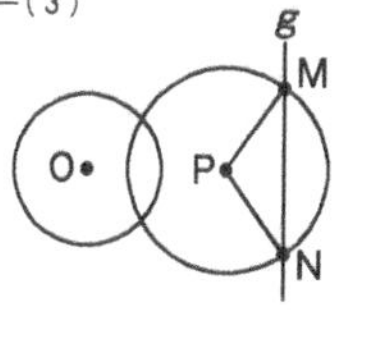

(4) radius　(5) $\overline{PN}$

(6) isosceles triangle

4 (1) (2)

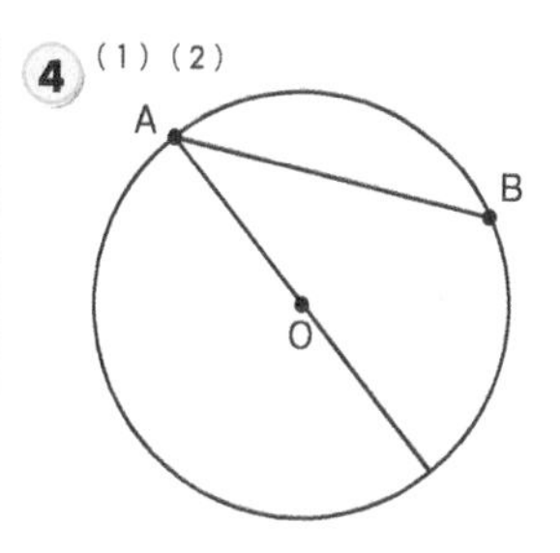

34 Construction 1 (Using Compass & Ruler) pp68,69

1 (1) (2)

(3) $\overline{CD} \perp \overline{PQ}$

(4) $\overline{QC}$

(5) $\overline{QD}$

(6) $\overline{QM}$

2 (1)–(3)

(4) longer than 2 cm

3

(1) 1 cm

(2) 5 cm

4 (1) (2)

(3) equilateral triangle

(4) $\overline{MN} \perp \overline{AB}$

(5) $\overline{MP}=\overline{NP}$, $\overline{AP}=\overline{BP}$

35 Construction 2 (Perpendicular) pp70,71

1 (1)–(3)

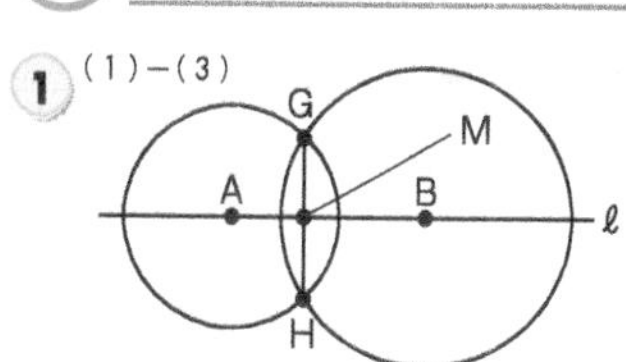

(4) $\overline{GH} \perp \overline{AB}$

(5) $\overline{GM}=\overline{HM}$

2 (1)–(5)

(6) $\overline{PQ} \perp \overline{AB}$

3 (1)–(4)

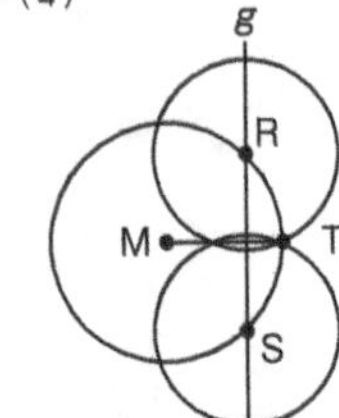

4 (1)–(3)

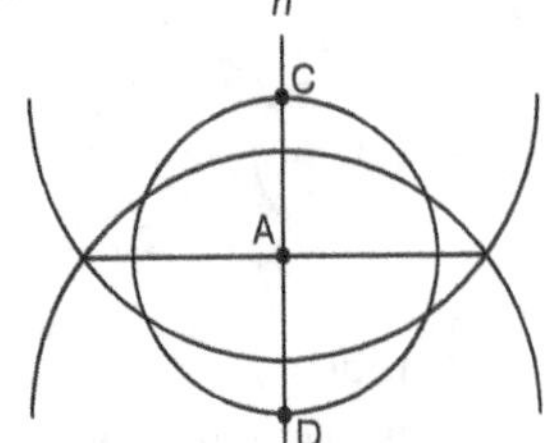

5

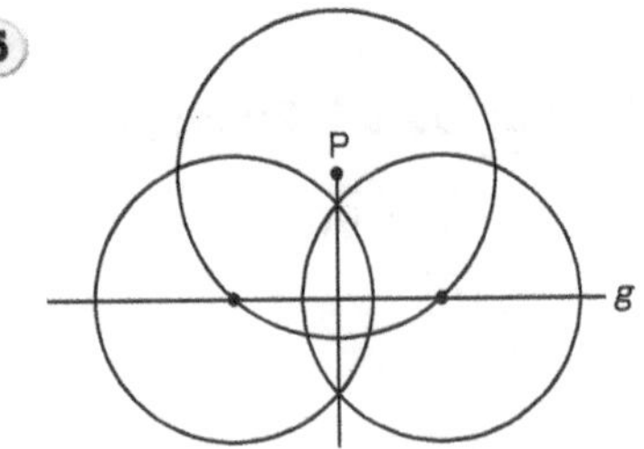

36 Construction 3 (Distance of Point & Line) pp72,73

1 (1) (2)

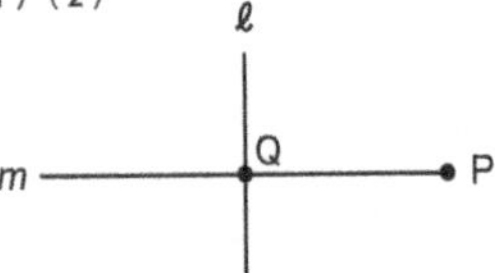

(3) 2 cm

2 (1) (2) (4)

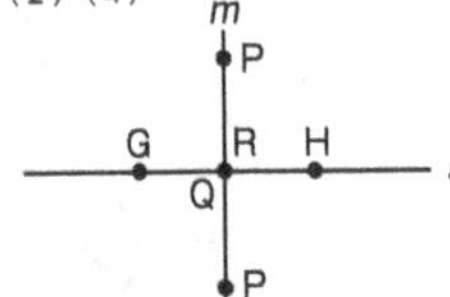

(3) ℓ ⊥ m

(5) 2

3 (1)–(6)

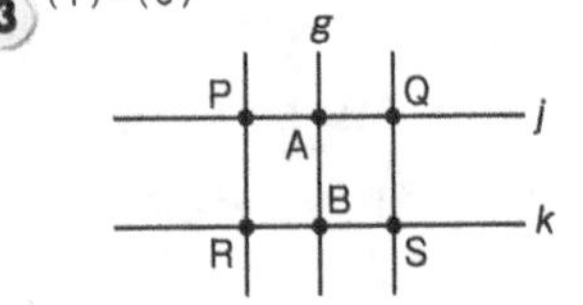

(7) 2

4

ℓ

37 Construction 4 (Perpendicular Bisector) pp74,75

1

A B

(1) $\overline{AP} = \overline{BP}$, $\overline{AP'} = \overline{BP'}$

(2) $\overline{AQ} = \overline{BQ}$, $\overline{AQ'} = \overline{BQ'}$

(3) $\overline{AR} = \overline{BR}$, $\overline{AR'} = \overline{BR'}$

(4) $\overline{AS} = \overline{BS}$, $\overline{AS'} = \overline{BS'}$

(5) same

(6) $\overline{AB} \perp \overline{SS'}$

(7) $\overline{AM} = \overline{BM}$

2

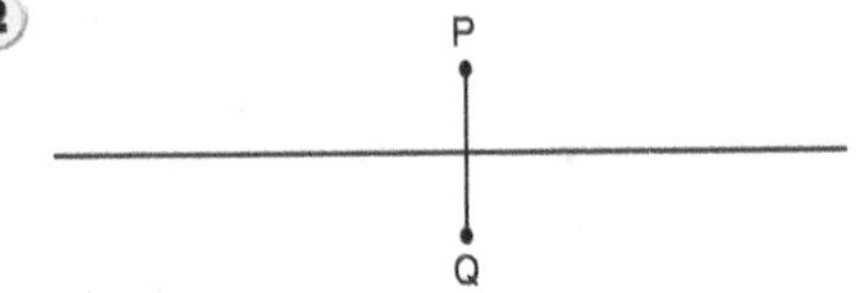

3 (1) (2)

B C A O D

(3) point O

4 (1)–(3)

A B C P

38 Construction 5 (Angle Bisector) pp76,77

1 (1)–(4)

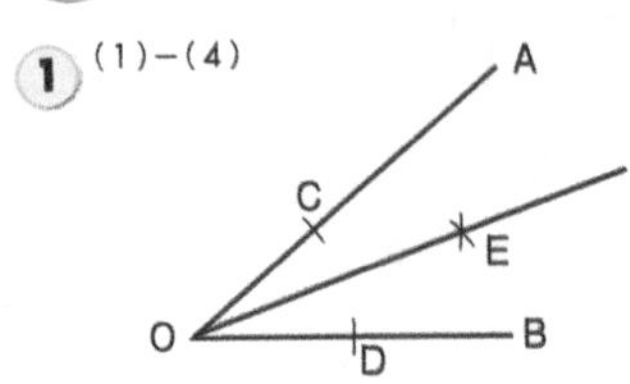

2 (1)

X O Y

(2)

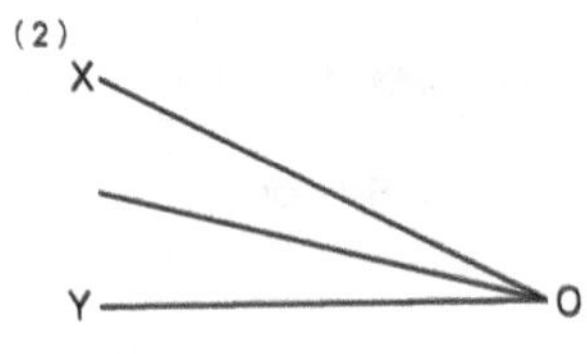

(3)

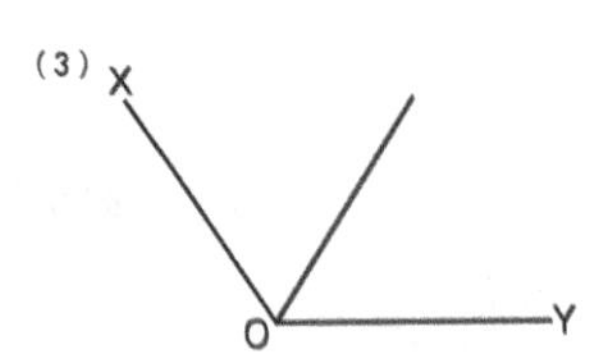

(4)

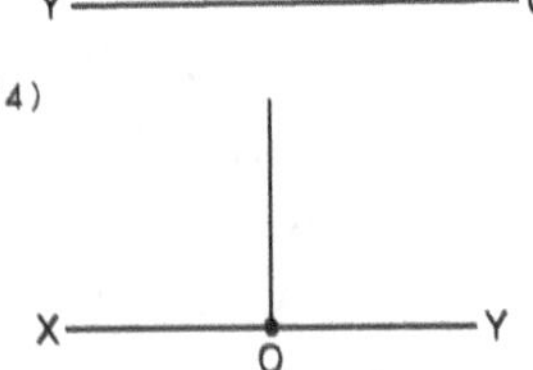

3 (1)

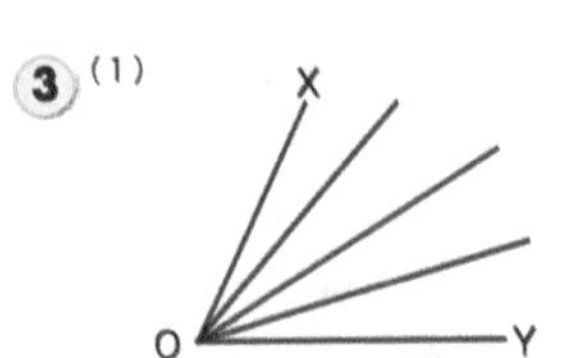

(2)

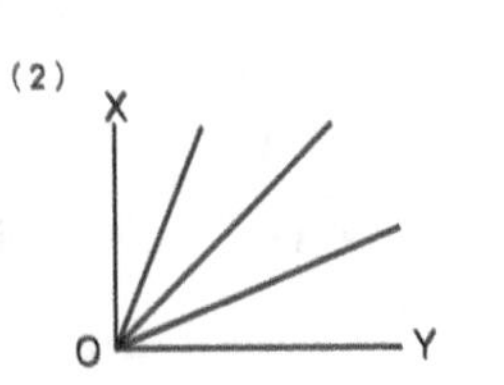

4 (1)

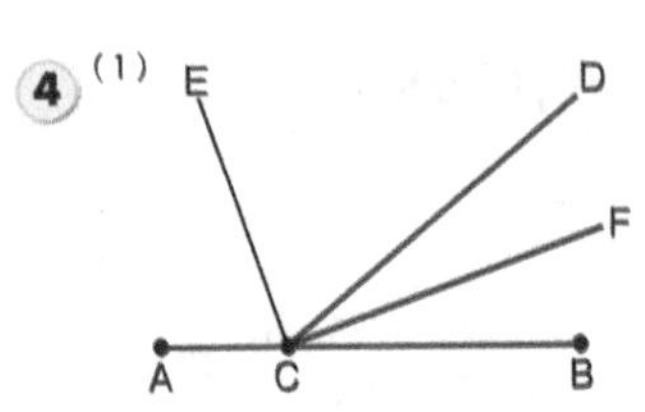

(2) 90°

(3) 90°

5 (1) (2)

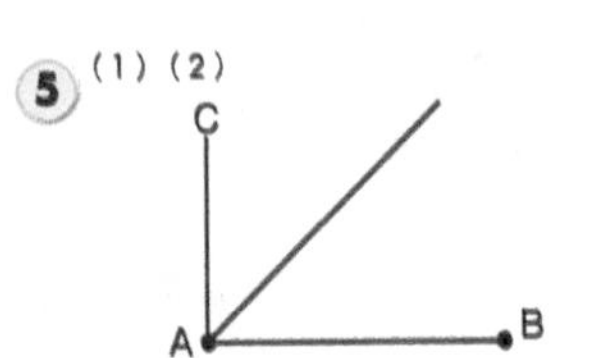

39 Construction 6 (Application 1) pp78,79

1 (1) (2)

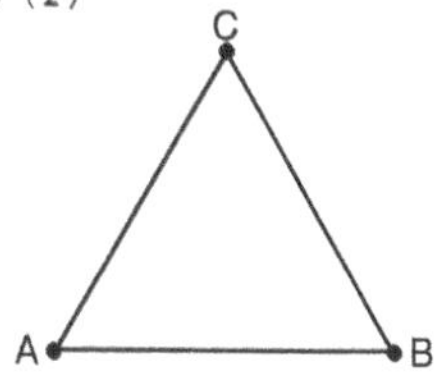

(3) AB = BC = CA

(4) ∠ABC = ∠BCA
= ∠CAB
= 60°

2 (1)–(4)

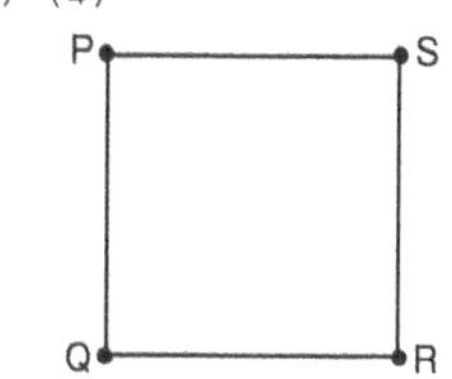

3 (1)–(3)

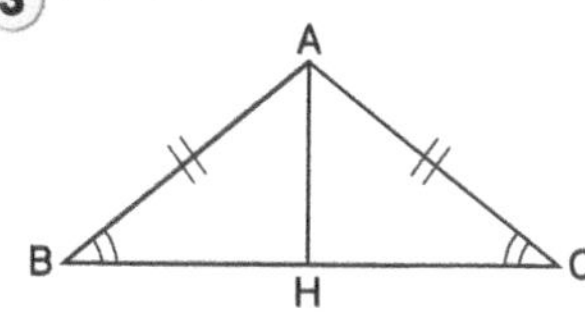

4 (1) (2)

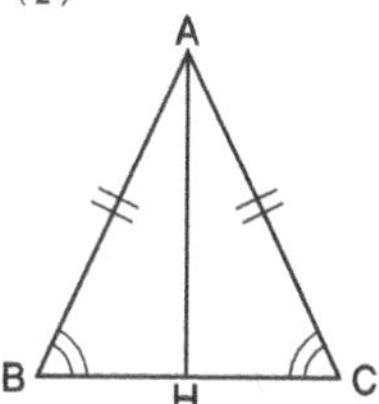

5

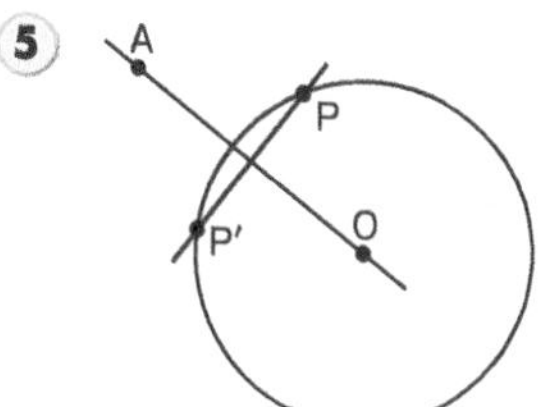

40 Construction 7 (Application 2) pp80,81

1 (1) (2) (4) (8)

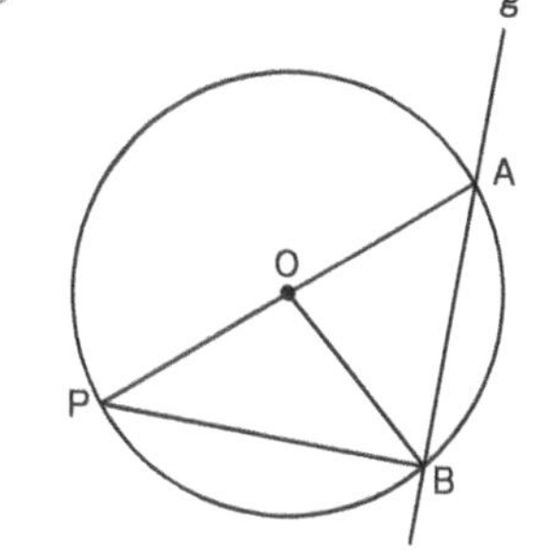

(3) isosceles triangle

(5) diameter

(6) 4 cm

(7) $\overline{AP}$

(9) $\overline{PB} \perp \overline{AB}$

(10) right triangle

2

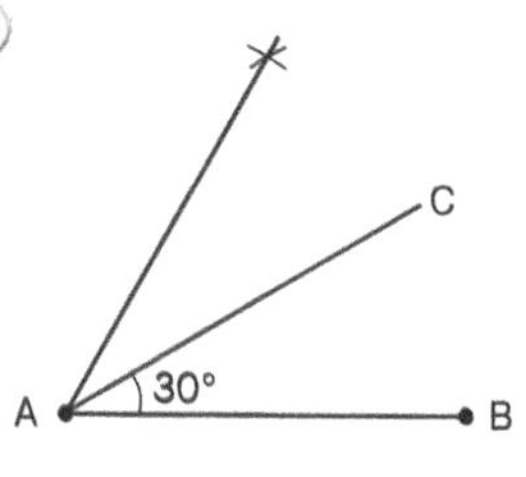

3 (1)–(4)

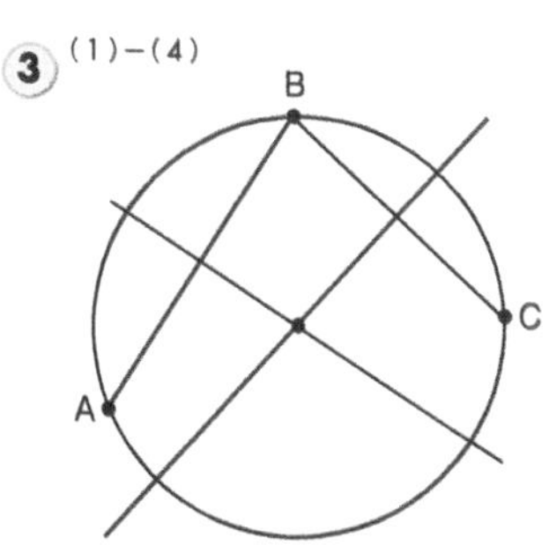

4 (1)–(3)

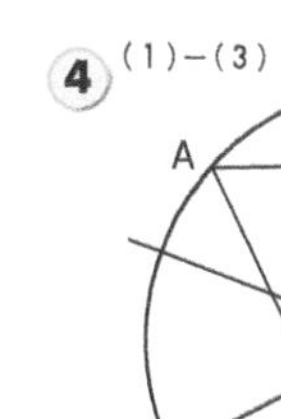

5

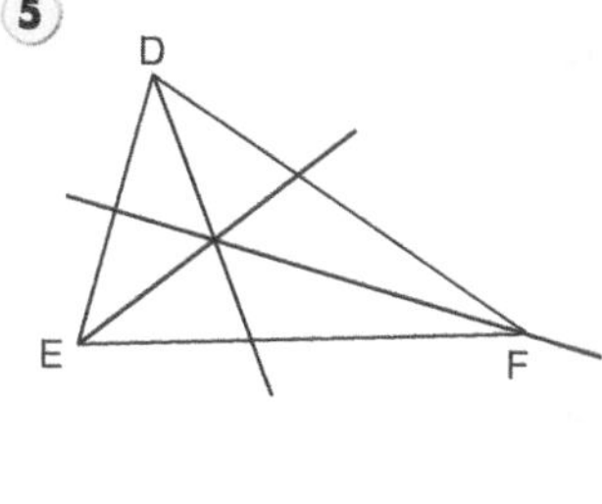

41 Review 1 pp82,83

1 (1) ∠AOB (2) $\overset{\frown}{CD}$ (3) $\overset{\frown}{CD} = \overset{\frown}{DE}$

2 (1) $(20+10\pi)$ cm (2) 50π cm²

3 (1) Perimeter: $\left(16+\frac{8}{3}\pi\right)$ cm

Area: $\frac{32}{3}\pi$ cm²

(2) Perimeter: $(6+4\pi)$ cm

Area: 6π cm²

4 (1) 4 cm (2) 8 cm (3) 45° (4) 24 cm

5 (1) 64π cm² (2) $\frac{1}{8}$ (3) 8π cm²

(4) $\frac{3}{4}$ (5) $\frac{a}{360}$

42 Review 2 pp84,85

1 (1) 30.5 cm² (2) 175 cm²

2 (1) 28 cm² (2) 16 cm² (3) 51 cm²

3 (1) 80 cm² (2) 150 cm² (3) 75 cm²

4 (1) $\frac{13}{2}x$ cm² (2) $\frac{21}{2}x$ cm²

5 (1) Perimeter: 14π cm

Area: $(48+9\pi)$ cm²

(2) Perimeter: $\left(10+\frac{15}{2}\pi\right)$ cm

Area: $\left(25+\frac{75}{4}\pi\right)$ cm²

6 (1) 10 cm (2) 6 cm (3) 4 cm (4) GH

(5) AN (6) JGDM

43 Review 3

pp 86, 87

1 (1) $364\ cm^3$ (2) $94\ cm^3$

2 (1) $158\ cm^2$ (2) $430\ cm^2$

3 (1) midpoint (2) $\overline{AB}$ ($\overline{AP}$), $\overline{DP}$

(3) $\overline{QO}$, $\overline{SO}$

4 (1)

P Q ℓ

M N m

(2) P, $\overline{QN}$, distance

5 (1)

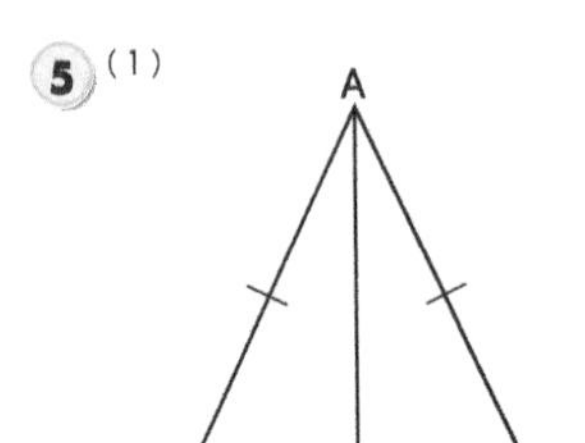

(2) ∠ACB

6 (1)

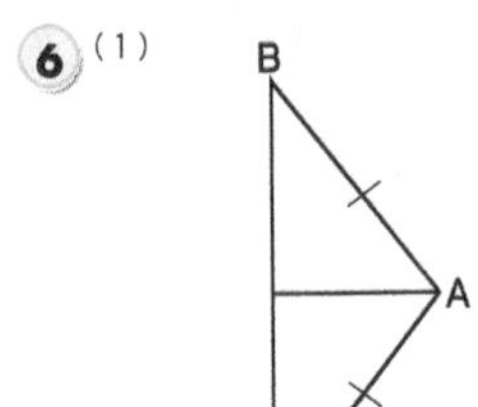

(2) midpoint of base BC

How to Solve

P.45 **22**–**3** (5)

(area of the shaded region) = (area of the sector 1 + area of the sector 2) – (area of the square)

Therefore, $\left(8 \times 8 \times \pi \times \frac{1}{4}\right) \times 2 - 64 = (32\pi - 64)\ cm^2$

(perimeter of the shaded region) = (length of the arc of the sector A + length of the arc of the sector B)

Therefore, $\left(16 \times \pi \times \frac{1}{4}\right) \times 2 = 8\pi\ cm$

P.45 **22**–**3** (6)

(area of the shaded region) = (area of the rectangle) – (area of the semicircle)

Therefore, $(12 \times 5) - (5 \times 5 \times \pi) \times \frac{1}{2} = \left(60 - \frac{25}{2}\pi\right) cm^2$

(perimeter of the shaded region) = (perimeter of the semicircle) + (12 + 12 – 5 – 5)

Therefore, $(12 + 12 - 5 - 5) + \left(2 \times 5 \times \pi \times \frac{1}{2}\right) = (14 + 5\pi)\ cm$

P.59 **29**–**3** (3)

(volume of solid A) = (volume of solid B) – (volume of solid C)

Therefore, $(3 + 2 + 4) \times 14 \times 4 - (2 \times 10 \times 4) = 424\ in.^3$

P.59 **29**–**4** (1)

(surface of solid A) = (surface of solid B) + (surface of solid C) – (overlapping face)

Therefore, $(24 \times 2 + 12 \times 2 + 8 \times 2) + (4 \times 2 + 4 \times 2 + 4 \times 2) - (4 \times 2) = 104\ in.^2$

P.85 **42**–**5** (2)

$(\text{perimeter}) = (5 + 5) + \left(2 \times 5 \times \pi \times \frac{3}{4}\right) = \left(10 + \frac{15}{2}\pi\right) cm$

$(\text{area}) = (5 \times 5) + \left(5 \times 5 \times \pi \times \frac{3}{4}\right) = \left(25 + \frac{75}{4}\pi\right) cm^2$

P.86 **43**–**2** (2)

(surface area) = $(64 \times 2 + 80 \times 2 + 80 \times 2) - (3 \times 3 \times 2) = 430\ cm^2$

KUMON MATH WORKBOOKS

Intro to Geometry

Workbook II

Table of Contents

KUMON

1 Line & Plane 1

Level

Score

Date / / Name

■ The Answer Key is on page 184.

1 Answer the questions about the rectangular prism below. 4 points per question

(1) Surface ABCD is a rectangle. What are the four edges that make up surface ABCD?

⟨Ans.⟩ , , ,

(2) Surface ABCD is called surface *a* and surface BCGF is called surface *b*. What are the four edges that make up surface *b*?

⟨Ans.⟩ , , ,

(3) Surface CDHG is called surface *c*. What are the four edges that make up surface *c*?

⟨Ans.⟩ , , ,

(4) Edge CG is perpendicular to surface *a*.
What are the other three edges perpendicular to surface *a*?

⟨Ans.⟩ CG , , ,

(5) Which edges are perpendicular to surface *b*?

⟨Ans.⟩ , , ,

(6) Which edges are perpendicular to surface *c*?

⟨Ans.⟩ , , ,

(7) Surface *a* and surface *b* share edge BC. What is the edge shared by surface *a* and surface *c*?

⟨Ans.⟩

(8) Which edge is shared by surface *b* and surface *c*?

⟨Ans.⟩

(9) Which surface shares edge AD with surface AEHD?

⟨Ans.⟩

(10) Which surface shares edge AB with surface *a*?

⟨Ans.⟩

2 **A rectangular prism is made of six rectangles. Let each of the surfaces that make up the rectangular prism be named as follows. Surface ABCD ⇒ *a*, surface BCGF ⇒ *b*, surface CDHG ⇒ *c*, surface DAEH ⇒ *d*, surface EABF ⇒ *e*, and surface FGHE ⇒ *f*. Answer the questions about the rectangular prism below.**

4 points per question

(1) Which surface is parallel to surface *a*?

〈Ans.〉

(2) Which surface is parallel to surface *b*?

〈Ans.〉

(3) Which surface is parallel to surface *c*?

〈Ans.〉

(4) Which edges are parallel to edge BC?

〈Ans.〉 , ,

(5) Which edges are parallel to edge BF?

〈Ans.〉 , ,

(6) Which edges are parallel to edge CD?

〈Ans.〉 , ,

(7) Which surfaces are perpendicular to surface *a*?

〈Ans.〉 , , ,

(8) Which surfaces are perpendicular to surface *b*?

〈Ans.〉 , , ,

(9) Which surfaces are perpendicular to surface *c*?

〈Ans.〉 , , ,

(10) Which edges intersect with surface *a*, but are not included in surface *a*?

〈Ans.〉 , , ,

(11) Which edges intersect with surface *c*, but are not included in surface *c*?

〈Ans.〉 , , ,

(12) Which edges do not intersect and are not included in surface *a*?

〈Ans.〉 , , ,

(13) Which edges do not intersect and are not included in surface *c*?

〈Ans.〉 , , ,

(14) Which edges are not parallel and not perpendicular to edge BF?

〈Ans.〉 , , ,

(15) Which edges are not parallel and not perpendicular to edge BC?

〈Ans.〉 , , ,

2 Line & Plane 2

Level ☆☆

Date / / Name

■ The Answer Key is on page 184.

1 Surface ABCD of the rectangular prism below is contained in large plane X. Similarly, surface EFGH is contained in large plane Y. Planes X and Y are spread out infinitely and never cross. This type of plane is called a parallel plane and is shown as X ∥ Y. Additionaly, a line segment on surface ABCD is similarly parallel to plane Y. For example, line segment AB parallel to plane Y is shown as AB ∥ Y. Use the information provided to answer the questions below.

4 points per question

(1) Plane X and plane Y are parallel planes, and edge AB is a line segment on plane X. Show the relationship between plane Y and line segment AB with the correct symbol.

〈Ans.〉

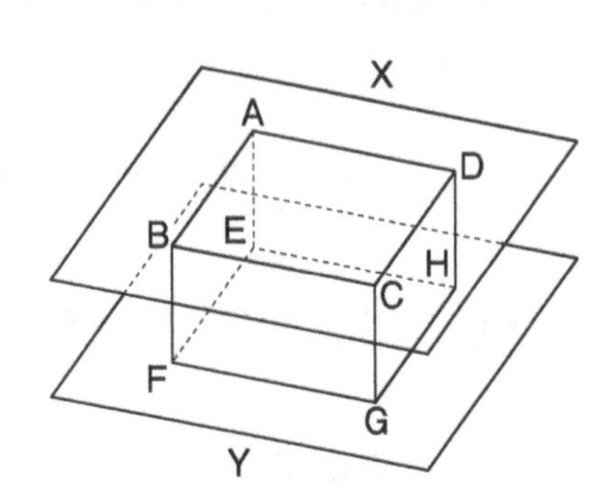

(2) Edge BC is also a line segment on plane X. Show the relationship between plane Y and line segment BC with the correct symbol.

〈Ans.〉

(3) Since line segment BF (edge BF) is perpendicular to surface ABCD, it is also perpendicular to plane X. This is shown as BF⊥X. Show the relationship between plane Y and line segment BF with the correct symbol.

〈Ans.〉

(4) Show the relationship between plane X and line segment DH with the correct symbol.

〈Ans.〉

(5) Show the relationship between plane X and line segment FG with the correct symbol.

〈Ans.〉

(6) Show the relationship between plane Y and line segment CD with the correct symbol.

〈Ans.〉

(7) Show the relationship between plane X and line segment AE with the correct symbol.

〈Ans.〉

(8) Let Z be the large surface containing surface BCGF. Show the relationship between plane Z and line segment AB with the correct symbol.

〈Ans.〉

(9) Show the relationship between plane Z and line segment EH with the correct symbol.

〈Ans.〉

(10) Show the relationship between plane Z and line segment GH with the correct symbol.

〈Ans.〉

2 **Surface ABCD of the rectangular prism below is contained in large plane X. Similarly, surface EFGH is contained in the large plane Y, and surface BCGF is contained in large plane Z. Use the information provided to answer the questions below.**

(1)-(8) 5 points per question (9)-(13) 4 points per question

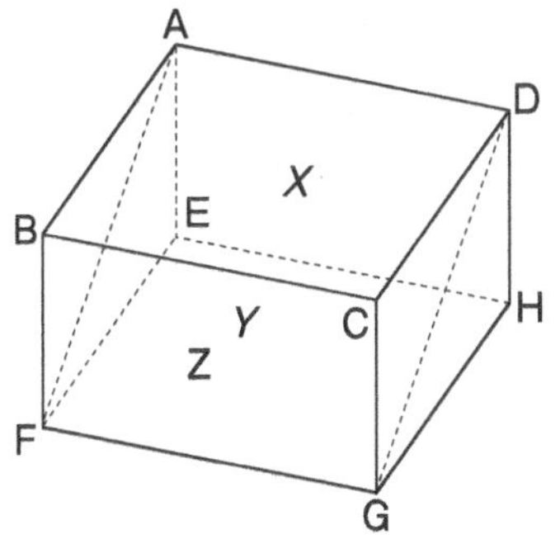

(1) The relationship between surface BCGF and surface ABCD is shown as BCGF⊥ABCD. Show the relationship between plane X and plane Z with the correct symbol.

〈Ans.〉

(2) Show the relationship between plane Y and plane Z with the correct symbol.

〈Ans.〉

(3) Show the relationship between surface AEHD and plane Y with the correct symbol. 〈Ans.〉

(4) Surface BCGF and surface ABCD share edge BC. Which line segment is formed by the intersection of plane X and plane Z? 〈Ans.〉

(5) Which line segment is formed by the intersection of plane Y and plane Z? 〈Ans.〉

(6) Which line segment is formed by the intersection of surface CDHG and plane X? 〈Ans.〉

(7) The surface including rectangle AFGD formed by connecting points A, F, D, and G is shown on plane W. Show the relationship between plane W and line segment BC with the correct symbol. 〈Ans.〉

(8) The line formed by the intersection of two planes is called the intersection line. What is the intersection line formed when plane W and plane X intersect with each other? 〈Ans.〉

(9) What is the intersection line formed by plane W and plane Y intersecting with each other? 〈Ans.〉

(10) A plane consists of two straight parallel lines. Which plane can be formed with line AB and line CD? 〈Ans.〉

(11) Which plane can be formed by line BF and line CG? 〈Ans.〉

(12) A plane consists of one intersection line. Which plane can be formed with line CD and line CG? 〈Ans.〉

(13) Straight lines that do not intersect each other and have a non-parallel relationship are skew lines. Which line segments are in a skewed position with line segment AD?

〈Ans.〉 , , ,

3 Line & Plane 3

Date / / Name

■ The Answer Key is on page 184.

Don't forget!

● The place where lines and planes intersect or are parallel, is called the **space**.

● The relationship between lines and planes in the space:

- plane and plane ⇒ Either parallel or intersect
- line and plane ⇒ 1. Line is included in plane.
  2. parallel
  3. intersect
- line and line ⇒ 1. parallel ... when one plane contains two straight lines.
  2. intersect ... when one plane contains two straight lines.
  3. skewed position ... when there is not one plane containing two straight lines.

1 **There are two parallel planes X and Y, where straight line ℓ is on plane X. Use the information provided to answer the questions below.** 6 points per question

(1) Show the relationship between line ℓ and plane Y with the correct symbol. 〈Ans.〉

(2) Does a line intersecting with plane X always intersect or never intersect with plane Y? 〈Ans.〉

(3) Show the relationship between line m perpendicular to plane X and plane Y with the correct symbol. 〈Ans.〉

X

Y

(4) Does a line intersecting line ℓ always intersect plane Y or not? p.192 〈Ans.〉

2 **The two parallel planes X and Y are perpendicular to line ℓ. Then, the intersection of plane X and line ℓ is point P, and the intersection of plane Y and line ℓ is point Q. Use the information provided to answer the questions below.** 7 points per question

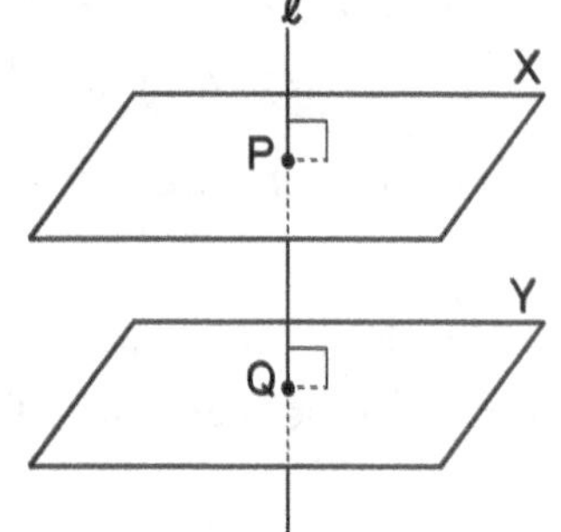

(1) Show the relationship between line m parallel to line ℓ and plane X with the correct symbol. 〈Ans.〉

(2) Show the relationship between line n on plane X passing through point P and line ℓ with the correct symbol. 〈Ans.〉

(3) Show the relationship between line n and plane Y with the correct symbol. 〈Ans.〉

(4) Is a plane passing through a point other than point P and point Q, and perpendicular to line ℓ, always parallel or never parallel to plane X? 〈Ans.〉

3 **In the following figure, the relationship between line ℓ and plane X is $\ell \perp X$, and the relationship between line ℓ and line n is $\ell \perp n$. Use the information provided to answer the questions below.**

6 points per question

(1) Show the relationship between line n and plane X with the correct symbol.

〈Ans.〉

(2) Line m is on plane X and crosses the intersection point of line ℓ and plane X. Show the relationship between line ℓ and line m using the correct symbol.

〈Ans.〉

(3) Let plane Y be the plane which includes line n, and let line k be the intersection line of plane Y and plane X. Show the relationship between line n and line k using the correct symbol.

p.192

〈Ans.〉

ℓ
n
P
O
X

4 **Plane X and plane Y intersect, and the intersection line is line ℓ. Use the information provided to answer the questions below.**

5 points per question

(1) Is line m, on plane X parallel to line ℓ, always or never parallel to plane Y?

〈Ans.〉

(2) Let the intersection line of plane Z and plane Y, both of which are parallel to plane X, be line n. Based on the information given, are lines n and ℓ always or never parallel?

〈Ans.〉

(3) Is a line on plane Z always or never parallel to plane X?

〈Ans.〉

Y
n
Z
ℓ
X

(4) Will a line on plane Z, except line n, be in a skewed position with line ℓ always or not?

〈Ans.〉

(5) Line k is perpendicular to line ℓ on plane Y. Show the relationship between line k and line n with the correct symbol.

p.192

〈Ans.〉

(6) Is the plane perpendicular to line ℓ, always or never perpendicular to line m?

〈Ans.〉

4 Line & Plane 4

Date / / Name

■ The Answer Key is on page 184.

1 Answer the questions about the rectangular prism in the figure on the right.

2 points per question

(1) If there are planes that contain the following two lines, write ○. If not, write ×.

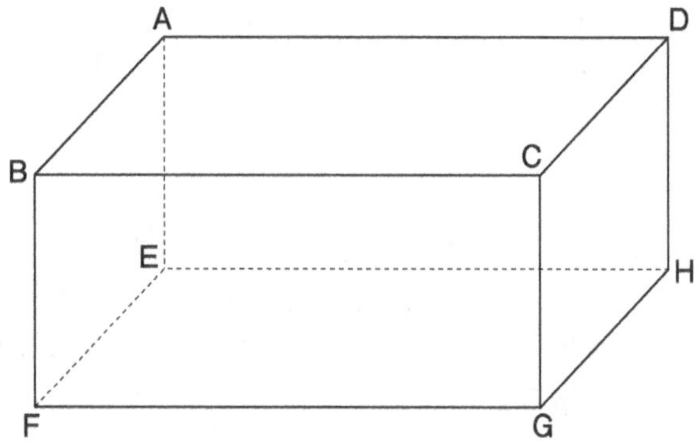

1 Edge BC and edge FG

〈Ans.〉

2 Edge AB and edge BC

〈Ans.〉

3 Edge BF and edge DH

〈Ans.〉

4 Edge BC and edge GH

〈Ans.〉

5 Edge AD and edge BF

〈Ans.〉

6 Diagonal BD and diagonal FH

〈Ans.〉

7 Diagonal BD and edge BF

〈Ans.〉

(2) Which of the following two line relationships is in a "parallel", "intersect", or "skewed position"?

1 Edge BC and edge FG

〈Ans.〉

2 Edge AB and edge BC

〈Ans.〉

3 Edge BF and edge DH

〈Ans.〉

4 Edge BC and edge GH

〈Ans.〉

5 Edge AD and edge BF

〈Ans.〉

6 Diagonal BD and diagonal FH

〈Ans.〉

7 Diagonal BD and edge BF

〈Ans.〉

2 Answer the questions about the triangular prism in the figure on the right.

3 points per question

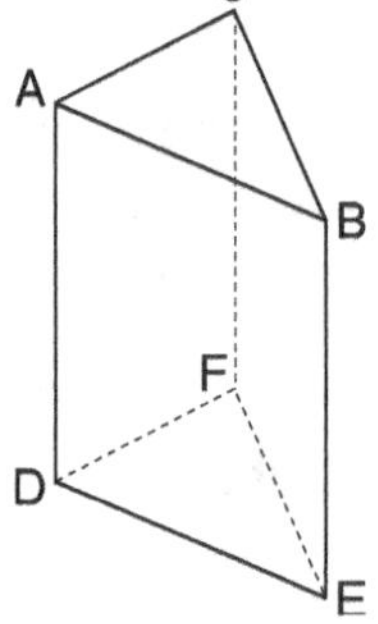

(1) Find the four lines that intersect line AB.

〈Ans.〉 , , ,

(2) Find the line parallel to line AB.

〈Ans.〉

(3) Find the three lines in a skewed position with line AB.

〈Ans.〉 , ,

3 **Answer the questions about the rectangular prism in the figure on the right.** 4 points per question

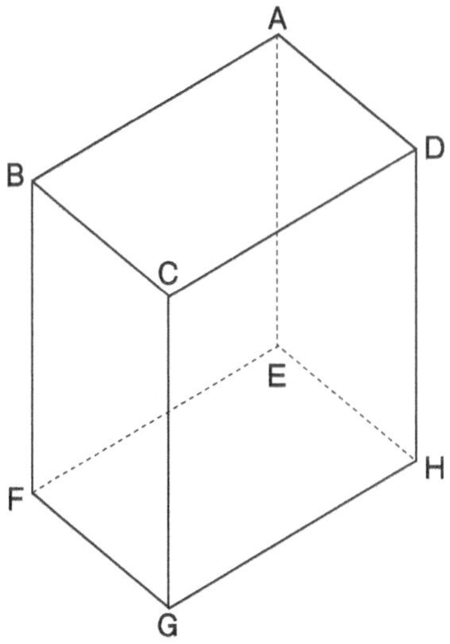

(1) Find all edges parallel to edge AB.

〈Ans.〉 ______ , ______ , ______

(2) Find all edges perpendicular to edge AB.

〈Ans.〉 ______ , ______ , ______ , ______

(3) Find all edges in a skewed position with edge AB.

〈Ans.〉 ______ , ______ , ______ , ______

(4) Find all edges parallel to surface ABFE.

〈Ans.〉 ______ , ______ , ______ , ______

(5) Find all surfaces parallel to surface ABFE.

〈Ans.〉 ______

(6) Find all edges perpendicular to surface ABCD.

〈Ans.〉 ______ , ______ , ______ , ______

(7) Find all edges perpendicular to surface BFGC.

〈Ans.〉 ______ , ______ , ______ , ______

4 **Answer the questions about the triangular prism in the figure on the right.** 7 points per question

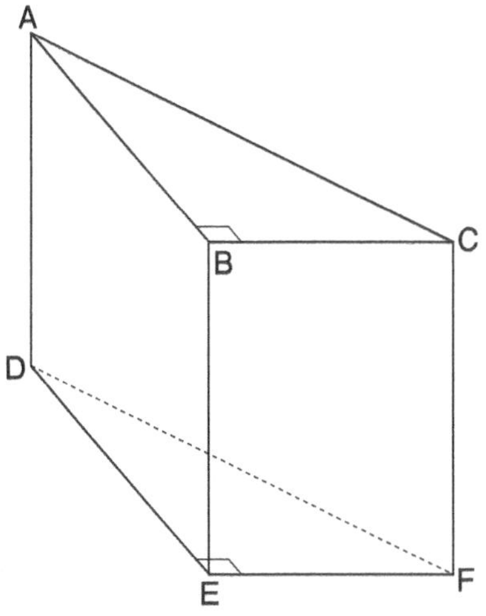

(1) Find all edges in a skewed position with edge AC.

〈Ans.〉 ______ , ______ , ______

(2) Find all edges parallel to surface BEFC.

〈Ans.〉 ______

(3) Find all edges perpendicular to surface ABC.

〈Ans.〉 ______ , ______ , ______

(4) Find all surfaces parallel to surface ABC.

〈Ans.〉 ______

(5) Find all edges perpendicular to surface ADEB.

〈Ans.〉 ______ , ______

5 Distance in Space

Level ☆☆

Date / /

Name

Score /100

■ The Answer Key is on page 184.

Don't forget!

● In a rectangular prism, three sets of opposing faces are parallel and three straight lines intersect perpendicularly at eight vertices.

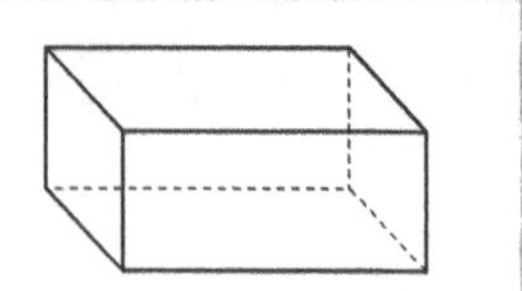

1 Answer the questions about the rectangular prism in the figure on the right. 6 points per question

(1) Find the distance between line AD and line BC.

〈Ans.〉

(2) Find the distance between point F and surface ABCD.

〈Ans.〉

(3) Find the distance between line FG and surface ABCD.

〈Ans.〉

(4) Find the distance between line EF and surface ABCD.

〈Ans.〉

(5) Find the distance between surface ABCD and surface EFGH.

〈Ans.〉

(6) Find the distance between surface AEFB and surface DHGC.

〈Ans.〉

(7) Find the intersection line of plane BFGC and plane EFGH.

〈Ans.〉

(8) Find the intersection line of plane AEFB and plane BFGC.

〈Ans.〉

Don't forget!

● The distance between two parallel planes in space is the same distance between two intersection points of a perpendicular line intersecting both planes.

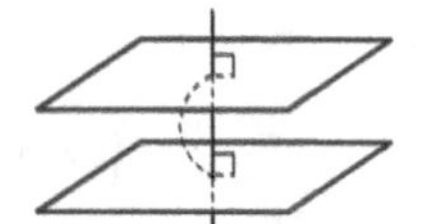

● The distance between the plane in space and the line parallel to that plane is the distance between the two intersection points with the line intersecting the plane perpendicularly to the plane.

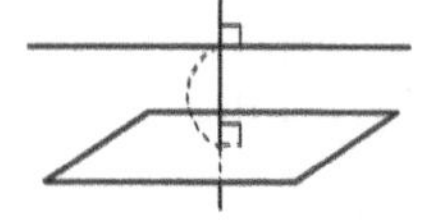

● The distance between the plane and the point in space is the distance between the intersection point with a line intersecting perpendicularly to the plane passing through that point.

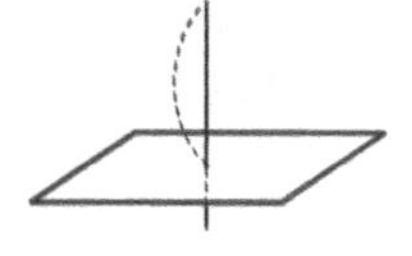

2 Answer the questions about the triangular prism in the figure on the right.

6 points per question

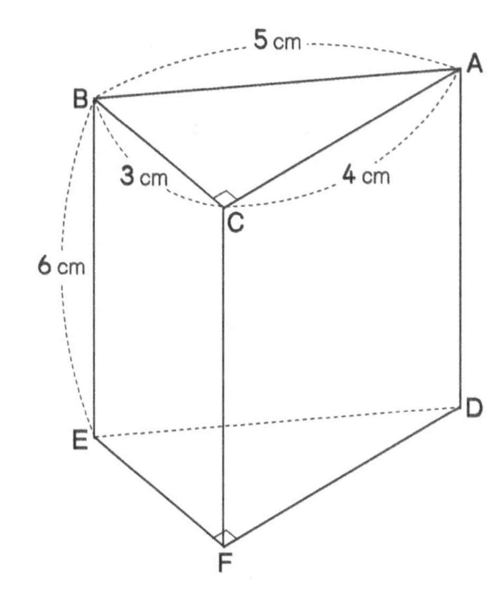

(1) All side shapes are rectangular. What is the distance between line AD and line BE?

〈Ans.〉

(2) What is the distance between point A and surface BCFE?

〈Ans.〉

(3) What is the distance between line BE and surface ADFC?

〈Ans.〉

(4) What is the distance between surface ABC and surface DEF?

〈Ans.〉

(5) Find all the lines in a skewed position with line AD.

〈Ans.〉 ,

(6) Find all the lines in a skewed position with line EF.

〈Ans.〉 , ,

3 Plane X and plane Y intersect perpendicularly, and point P on plane Y is 5 centimeters away from plane X. Use the information provided to answer the questions below.

8 points per question

(1) When drawing line m in contact with plane Y parallel to plane X passing through point P, what is the distance between line m and plane X?

〈Ans.〉

(2) Draw line n perpendicular to plane Y passing through point P. What is the distance between the plane formed by lines n and m and plane X?

p.192

〈Ans.〉

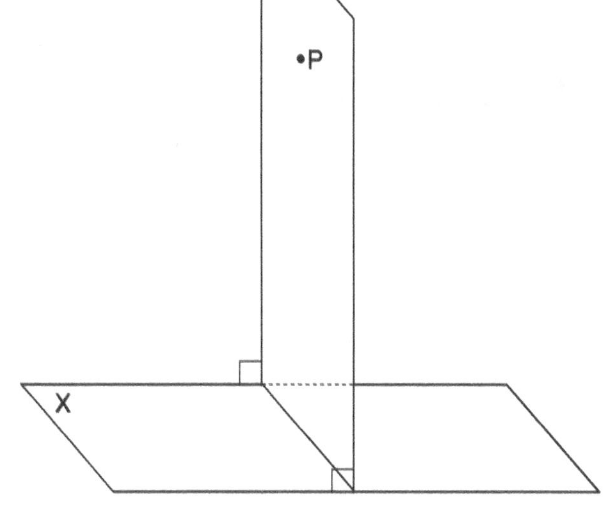

6 Solids 1

Level ☆☆

Date / /

Name

Score /100

■ The Answer Key is on page 184.

1 **The following solids are prisms with a regular polygon base. Write the name of the solids.** 3 points per question

(1) regular 〈Ans.〉

(2) regular 〈Ans.〉

(3) regular 〈Ans.〉

(4) regular 〈Ans.〉

Don't forget!

● The solid whose base is a regular polygon is called a **regular prism**. If the base is an equilateral triangle, it is a regular triangular prism. If it has a regular hexagon base, it is called a regular hexagonal prism. The regular quadrangular prism is a rectangular prism with a square base.

2 **Answer the following questions based on the regular triangular prism on the right.** 3 points per question

(1) What shape is the base?

〈Ans.〉

(2) How many bases are there?

〈Ans.〉

(3) What shape is the side?

〈Ans.〉

(4) How many sides are there?

〈Ans.〉

3 **Answer the following questions based on the regular pentagonal prism on the right.** 4 points per question

(1) What shape is the base?

〈Ans.〉

(2) How many bases are there?

〈Ans.〉

(3) What shape is the side?

〈Ans.〉

(4) How many sides are there?

〈Ans.〉

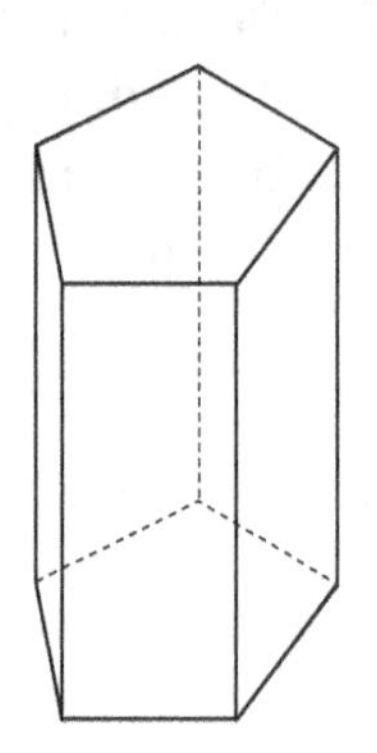

4 **Answer the following questions based on a regular hexagonal prism on the right.** 4 points per question

(1) What shape is the base?

⟨Ans.⟩ ____________

(2) How many bases are there?

⟨Ans.⟩ ____________

(3) What shape is the side?

⟨Ans.⟩ ____________

(4) How many sides are there?

⟨Ans.⟩ ____________

Don't forget!

●The shape of the sides of a regular prism are all rectangles.

5 **The figure on the right is a regular pentagonal prism whose base is a regular pentagon. Use the information given to answer the questions below.** 4 points per question

(1) Find all edges parallel to surface BGHC.

⟨Ans.⟩ ____ , ____ , ____

(2) Find all surfaces perpendicular to edge CH.

⟨Ans.⟩ ____ , ____

(3) Which surfaces are parallel to each other?

⟨Ans.⟩ ____ , ____

(4) Which surface is parallel to edge BC?

⟨Ans.⟩ ____________

(5) Find all edges parallel to edge CH.

⟨Ans.⟩ ____ , ____ , ____ , ____

(6) How many edges are in a skewed position with edge BC?

⟨Ans.⟩ ____________

6 **The three sets of sides facing each other are parallel in a regular hexagon. Answer the questions below about a regular hexagonal prism on the right, whose base is a regular hexagon.** 4 points per question

(1) Find all edges parallel to surface BCIH.

⟨Ans.⟩ ____ , ____ , ____ , ____ , ____ , ____

(2) Find all surfaces perpendicular to edge DJ.

⟨Ans.⟩ ____ , ____

(3) How many parallel surfaces are there?

⟨Ans.⟩ ____________

(4) Find all surfaces parallel to edge BC.

⟨Ans.⟩ ____ , ____

(5) Find all edges in a skewed position with edge BC. ⟨Ans.⟩ ____ , ____ , ____ , ____ , ____ , ____ , ____ , ____

7 Solids 2

Level ☆☆

Date / /

Name

Score /100

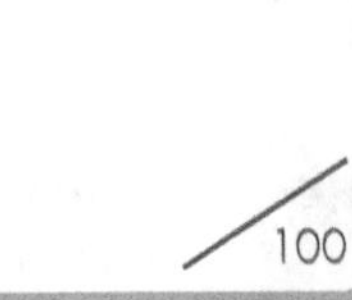

■ The Answer Key is on page 184.

Don't forget!

● A **pyramid** is a solid whose base is polygonal and sides are all triangles.

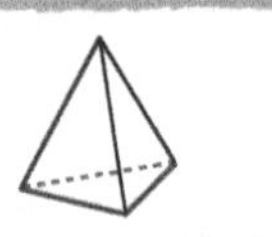

1 **Mark ○ for the pyramids and × for the non-pyramids from the following solids.** 4 points per question

(1) [ ] 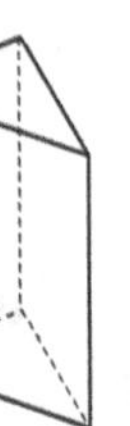

(2) [ ] 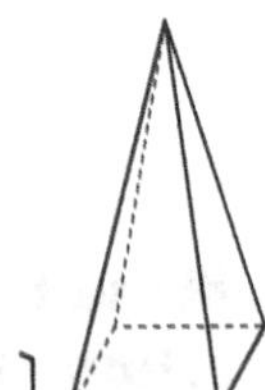

(3) [ ] 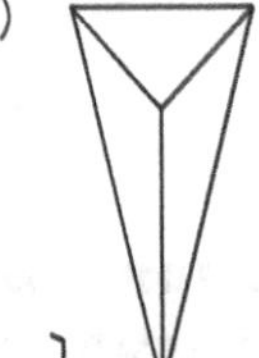

(4) [ ] 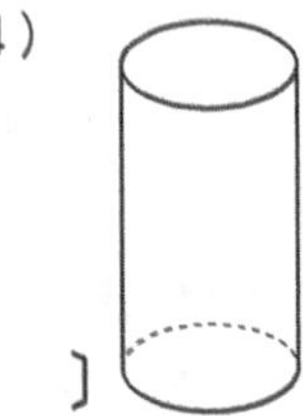

(5) [ ] 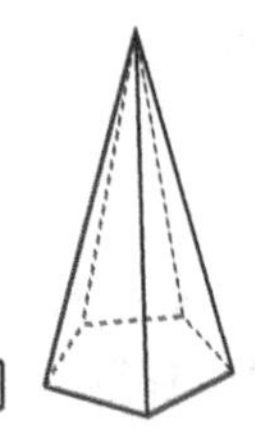

(6) [ ] 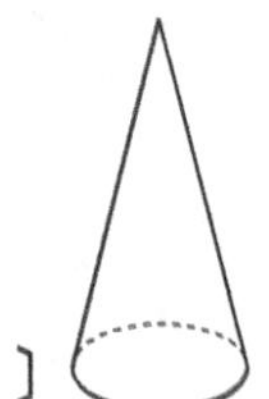

Don't forget!

● A **regular pyramid** is the pyramid whose base is a regular polygon and whose sides are all congruent isosceles triangles.

2 **The figure on the right is a square pyramid. Using the information given, answer the questions below.** 4 points per question

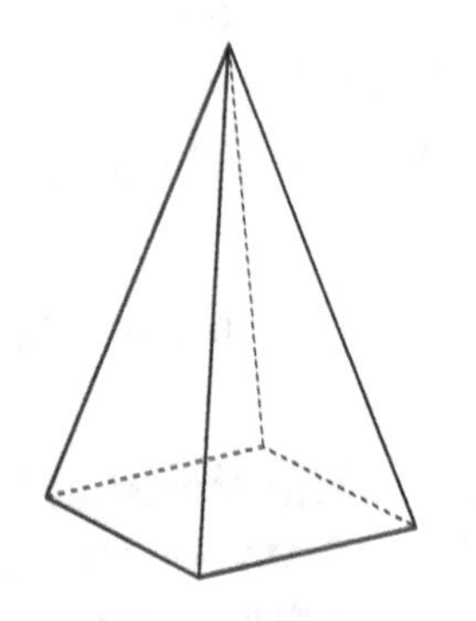

(1) What shape is the base?

⟨Ans.⟩ ______

(2) How many bases are there?

⟨Ans.⟩ ______

(3) What shape is the side?

⟨Ans.⟩ ______

(4) How many sides are there?

⟨Ans.⟩ ______

3 **The figure on the right is a regular hexagonal pyramid. Using the information given, answer the questions below.** 5 points per question

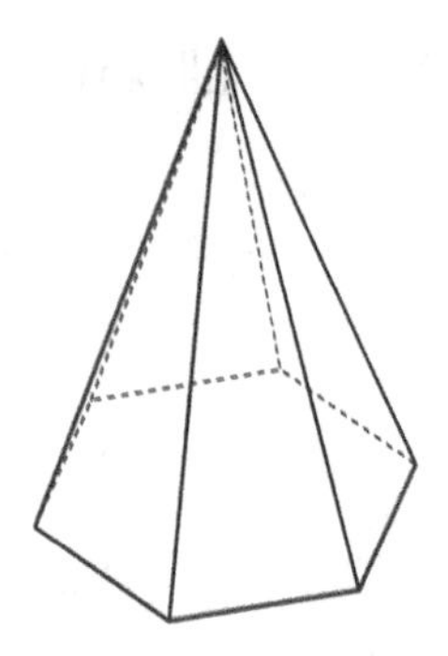

(1) What shape is the base?

⟨Ans.⟩ ______

(2) How many bases are there?

⟨Ans.⟩ ______

(3) What shape is the side?

⟨Ans.⟩ ______

(4) How many sides are there?

⟨Ans.⟩ ______

4 **The figure on the right is a square pyramid.**
Using the information given, answer the questions below.

4 points per question

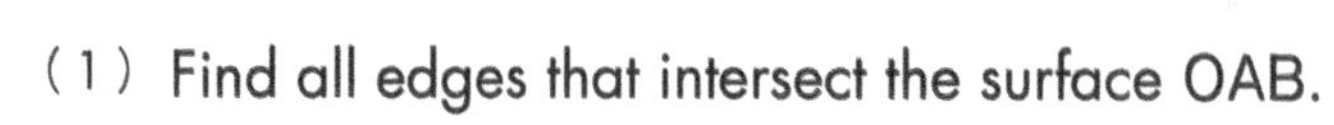

(1) Find all edges that intersect the surface OAB.

〈Ans.〉 , , ,

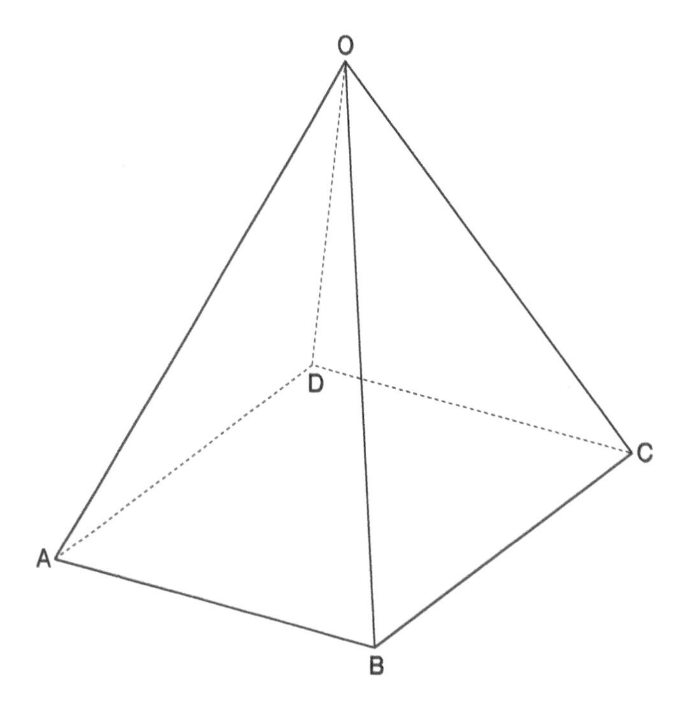

(2) Find all edges parallel to the surface OAB.

〈Ans.〉

(3) Find all edges in a skewed position with edge AB.

〈Ans.〉 ,

(4) Find all edges in a skewed position with edge OB.

〈Ans.〉 ,

(5) Find all edges contained in the surface OAB.

〈Ans.〉 , ,

5 **The pyramid on the right is not a square pyramid. Edge OD, edge AD, and edge CD, are perpendicular to each other. Then edge AD and edge BC are parallel. Using the information given, answer the questions below.**

4 points per question

(1) Find all surfaces perpendicular to surface ODC.

〈Ans.〉 ,

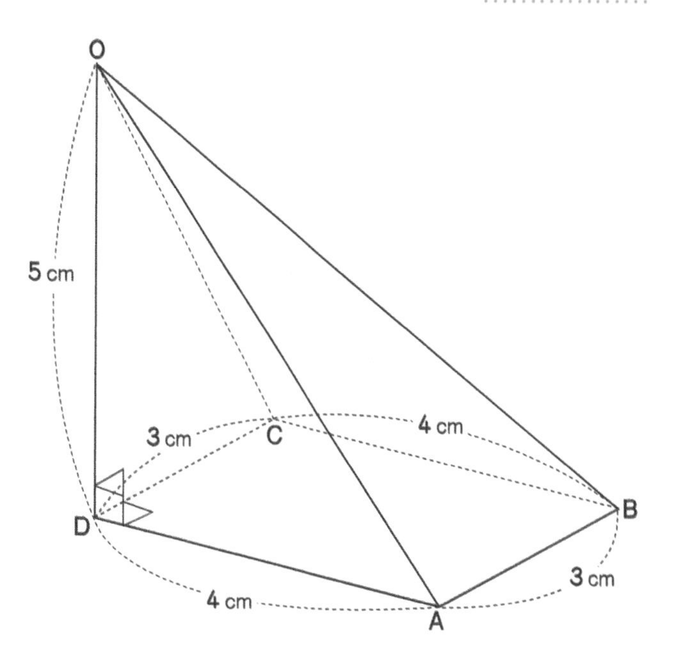

(2) What is the distance between line BC and surface ODA?

〈Ans.〉

(3) What is the distance between vertex A and surface ODC?

〈Ans.〉

(4) What is the distance between the plane passing through vertex O and perpendicular to line OD and surface ABCD?

〈Ans.〉

(5) Find all lines in a skewed position with line OD.

〈Ans.〉 ,

8 Net 1

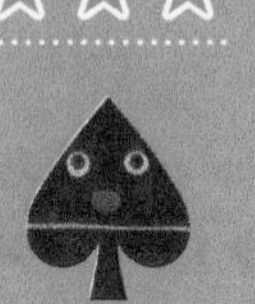

■ The Answer Key is on page 185.

1 **The figures on the right show the net and the built shape of a rectangular prism. Answer the questions below.**

4 points per question

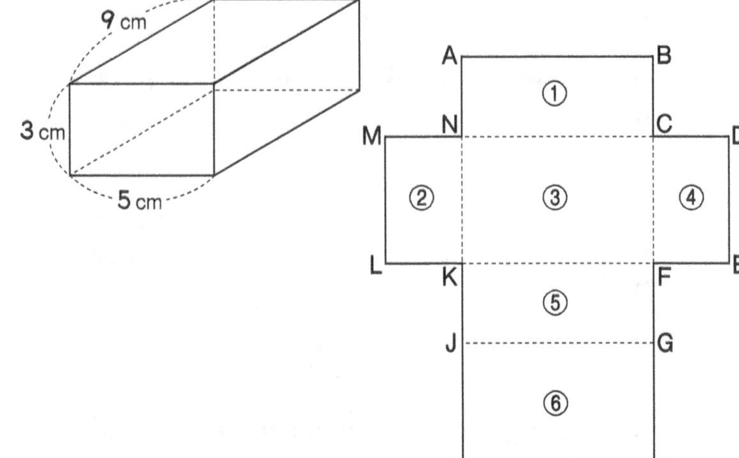

(1) What is the length of edge AB?

〈Ans.〉

(2) When assembling the net, which edge overlaps with edge AB?

〈Ans.〉

(3) When assembling the net, which edge overlaps with edge DE?

〈Ans.〉

(4) When assembling the net, which vertex overlaps with vertex E?

〈Ans.〉

(5) When assembling the net, which vertex overlaps with vertex A?

〈Ans.〉 ,

(6) When assembling the net, which surface is parallel to surface MLKN? Answer with the correct number.

〈Ans.〉

(7) When assembling the net, which surface is parallel to surface KJGF? Answer with the correct number.

〈Ans.〉

(8) When assembling the net, find all surfaces perpendicular to surface ABCN. Answer with the correct number.

〈Ans.〉 , , ,

(9) What is the distance between surface ABCN and surface KFGJ?

〈Ans.〉

(10) What is the distance between line CD and surface MNKL?

〈Ans.〉

2 **The figure on the right shows the net of a rectangular prism. Answer the questions below about the rectangular prism by assembling this net.** 4 points per question

(1) Which surface is parallel to surface KLMN? Answer with the correct number. 〈Ans.〉

(2) Which edge overlaps with edge AB? 〈Ans.〉

(3) Find all surfaces perpendicular to edge AB. Answer with the correct number. 〈Ans.〉 ,

(4) What is the relative position between edge AB and edge GD? 〈Ans.〉

(5) What is the relative position between edge AB and edge HC? 〈Ans.〉

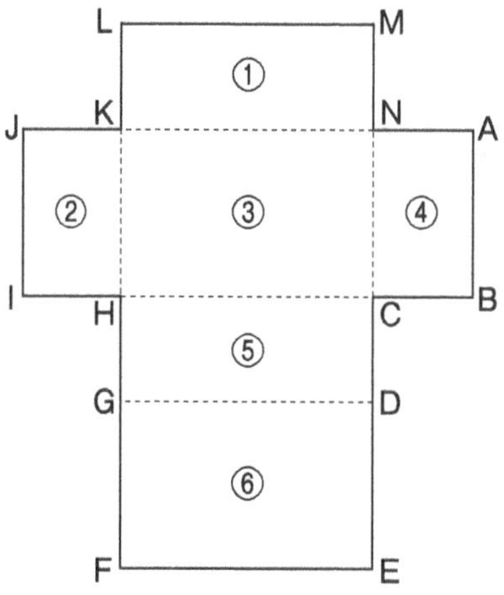

L M ① J K N A ② ③ ④ I H C B ⑤ G D ⑥ F E

3 **The figure on the right shows the net of a triangular prism whose base is a right triangle. Answer the questions below about a triangular prism by assembling this net.** 5 points per question

(1) Draw the built shape of a triangular prism using the net on the right.

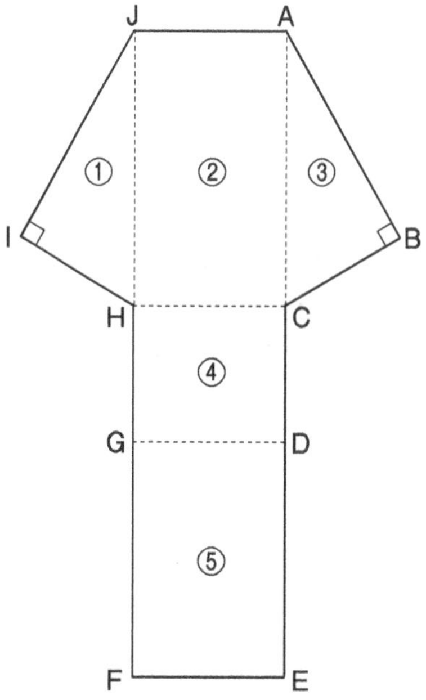

J A ① ② ③ I B H C ④ G D ⑤ F E

(2) Which surfaces are parallel to each other? Answer with the correct number. 〈Ans.〉 ,

(3) Which vertex overlaps with vertex F? 〈Ans.〉

(4) Which surface is perpendicular to edge AB? Answer with the correct number. 〈Ans.〉

(5) Which edge overlaps with edge AB? 〈Ans.〉

(6) Which edges are perpendicular to surface GDEF? 〈Ans.〉 ,

(7) Which edge is parallel to surface GDEF? 〈Ans.〉

(8) Which surface is perpendicular to edge BC? Answer with the correct number. 〈Ans.〉

9 Net 2

Level ★★★

Date / /

Name

Score /100

■ The Answer Key is on page 185.

1 **Answer the questions about a rectangular prism in the figure on the right.**

5 points per question

(1) Excluding vertices B, D, and F, what is the vertex on plane X made by diagonal BD and vertex F?

〈Ans.〉

(2) Find all edges parallel to plane X.

〈Ans.〉 ,

(3) Find all edges perpendicular to line BD.

〈Ans.〉 ,

(4) Find all lines parallel to line BD.

〈Ans.〉

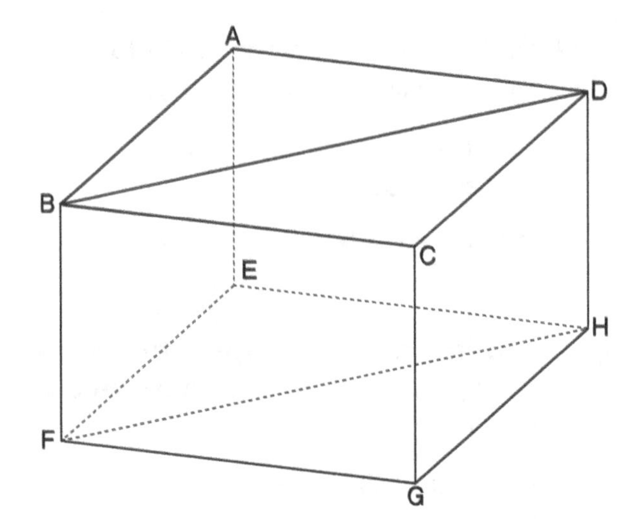

2 **The figure on the right is the net of a rectangular prism. A diagonal line is drawn on surface ABKN. Answer the questions below about a rectangular prism by assembling this net.**

4 points per question

(1) Which surface is parallel to surface KLMN? Answer with the correct number.

〈Ans.〉

(2) Which edge overlaps with edge AB?

〈Ans.〉

(3) Find all surfaces perpendicular to edge AB. Answer with the correct number.

〈Ans.〉 ,

(4) What is the relative position between edge AB and edge CJ?

〈Ans.〉

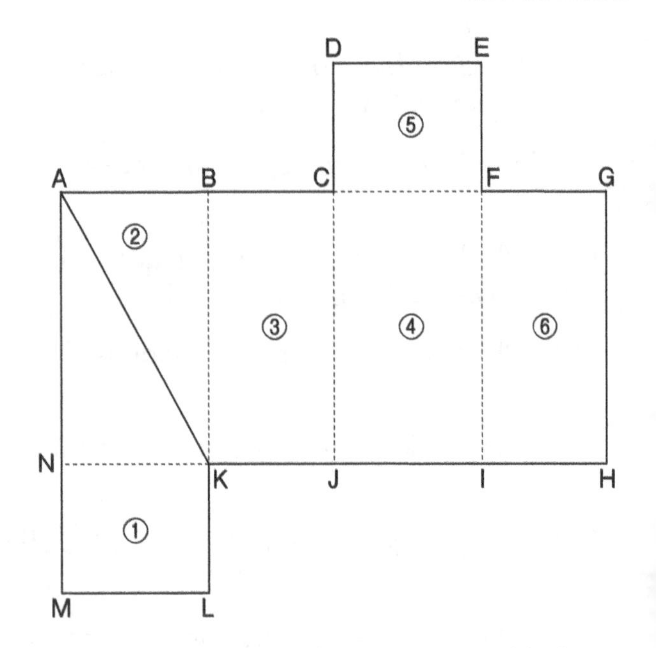

(5) In the net above, draw a diagonal which is parallel to diagonal AK when the net is assembled.

3 **The figure on the right is the net of a cube. Answer the questions below about the cube by assembling the net.** 4 points per question

(1) Find all surfaces parallel to edge AB.

⟨Ans.⟩ ______ , ______

(2) Which two surfaces intersect to form edge AB?

⟨Ans.⟩ ______ , ______

(3) Which surface is parallel to surface a?

⟨Ans.⟩ ______

(4) Find all surfaces perpendicular to edge AB.

⟨Ans.⟩ ______ , ______

(5) A die is a cube. A die is designed so that the sum of the numbers on opposite sides equals 7. When b is 1, what side is 6?

⟨Ans.⟩ ______

(6) When a is 3, what side is 4?

⟨Ans.⟩ ______

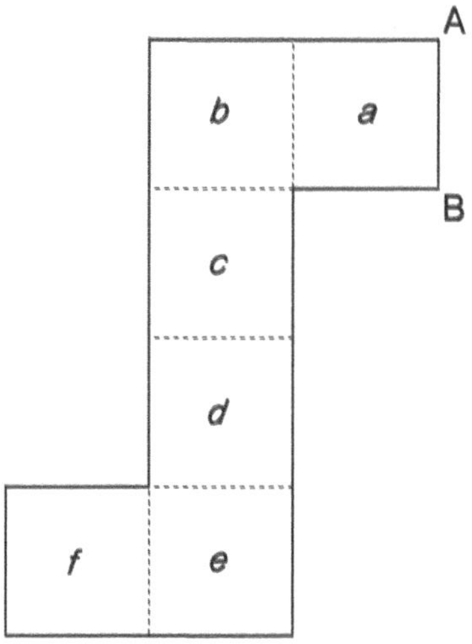

A
b a
B
c
d
f e

4 **Using the correct symbol, show the relative position between the lines and the surfaces described below.** 4 points per question

(1) Plane X and plane Y are parallel. If so, what is the relationship between line ℓ on plane X and plane Y?

⟨Ans.⟩ ______

(2) Line ℓ and line m are parallel, and line ℓ is perpendicular to plane X. If so, what is the relationship between line m and plane X?

⟨Ans.⟩ ______

(3) Plane X and plane Y are parallel, and line ℓ is perpendicular to plane X. If so, what is the relationship between line ℓ and plane Y?

⟨Ans.⟩ ______

(4) Plane X and line ℓ are parallel. Let the intersection between plane Y, including line ℓ, and plane X be line m, what is the relationship between line ℓ and line m?

⟨Ans.⟩ ______

5 **Mark ○ for the correct sentence and × for the incorrect sentence describing the relative position between the lines and the planes in space below.** 5 points per question

(1) Two planes parallel to one line are always parallel. [ ] p.192

(2) Two planes perpendicular to one line are always parallel. [ ]

(3) Two lines parallel to one plane are always parallel. [ ] p.192

(4) Two lines perpendicular to one plane are always parallel. [ ]

10 Cross Section of a Solid

Level ★★★

Date / /   Name

Score /100

■ The Answer Key is on page 185.

1 The figure on the right is a square pyramid whose base is a square. Using the information given, answer the questions below. 7 points per question

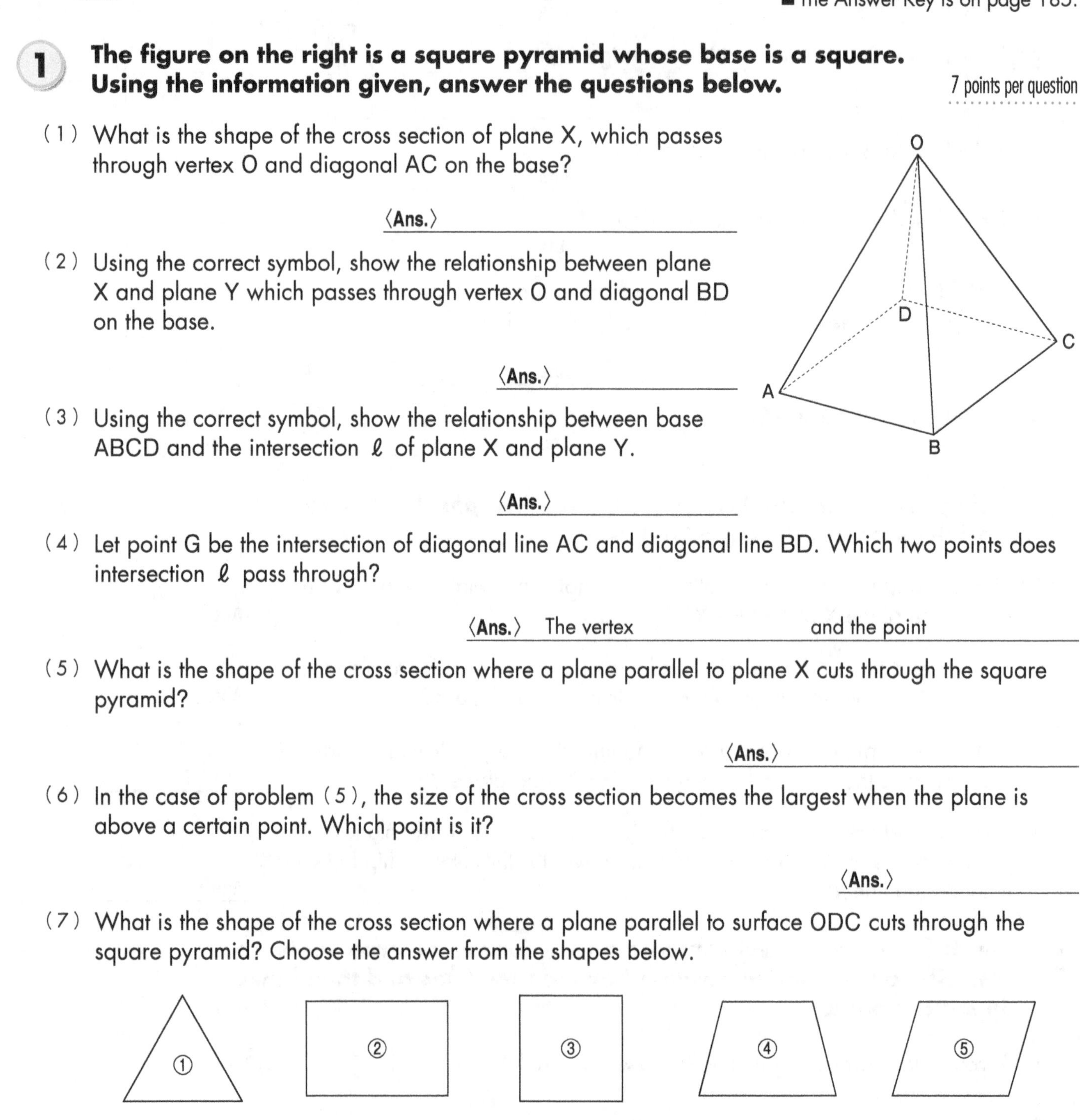

(1) What is the shape of the cross section of plane X, which passes through vertex O and diagonal AC on the base?

〈Ans.〉

(2) Using the correct symbol, show the relationship between plane X and plane Y which passes through vertex O and diagonal BD on the base.

〈Ans.〉

(3) Using the correct symbol, show the relationship between base ABCD and the intersection ℓ of plane X and plane Y.

〈Ans.〉

(4) Let point G be the intersection of diagonal line AC and diagonal line BD. Which two points does intersection ℓ pass through?

〈Ans.〉 The vertex   and the point

(5) What is the shape of the cross section where a plane parallel to plane X cuts through the square pyramid?

〈Ans.〉

(6) In the case of problem (5), the size of the cross section becomes the largest when the plane is above a certain point. Which point is it?

〈Ans.〉

(7) What is the shape of the cross section where a plane parallel to surface ODC cuts through the square pyramid? Choose the answer from the shapes below.

〈Ans.〉

(8) What is the shape of the cross section where a plane parallel to base ABCD cuts through the square pyramid?

〈Ans.〉

2 The figure on the right is a regular triangular pyramid whose base is an equilateral triangle. Using the information given, answer the questions below.

6 points per question

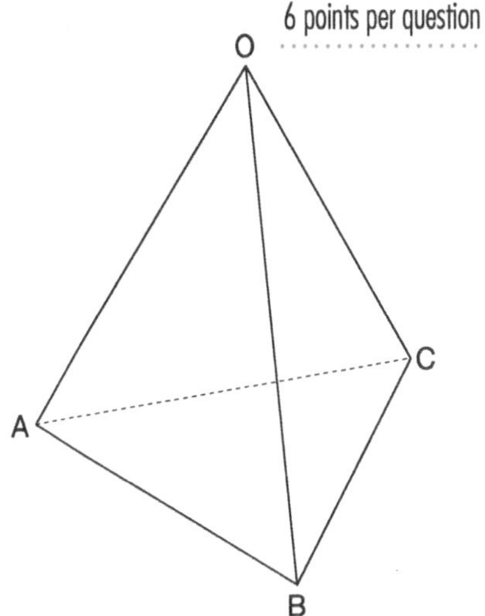

(1) What is the shape of the cross section where a plane parallel to base ABC cuts through the regular triangular pyramid?

〈Ans.〉

(2) What is the shape of the cross section where a plane parallel to surface OAB cuts through the regular triangular pyramid? p.192

〈Ans.〉

(3) What is the shape of the cross section where a plane parallel to edge AB and edge OC cuts through the regular triangular pyramid? Choose the answer from the shapes below. p.192

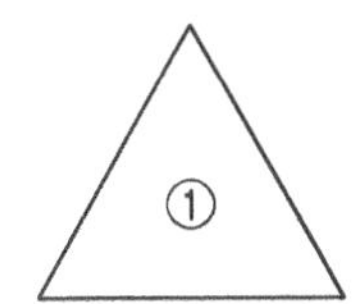

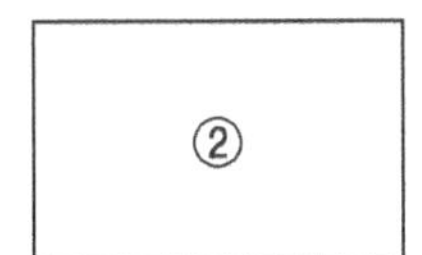

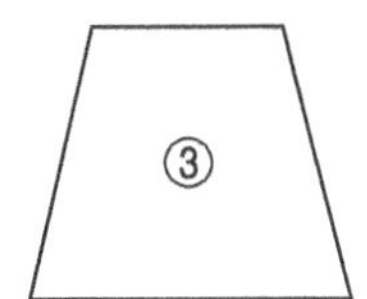

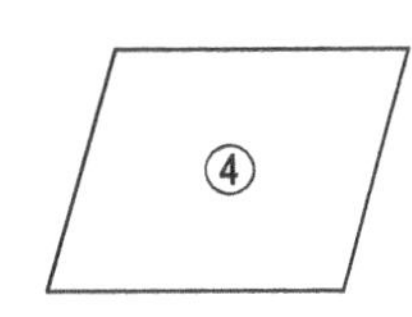

〈Ans.〉

(4) When the lengths of the six sides are equal, answer the name of the figure that can be formed by connecting the midpoints of sides OA, OB, CA, and CB. p.192

〈Ans.〉

3 The figure on the right is a cube. Answer the questions below.

10 points per question

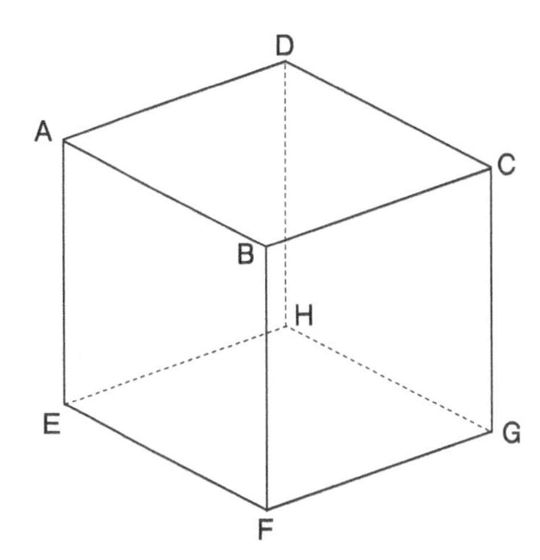

(1) What is the shape of the cross section where the plane formed by diagonal lines AC and EG cuts through the cube?

〈Ans.〉

(2) What is the shape of the cross section where the plane passing through the midpoint of four edges DC, BC, EH, and EF throughout the cube? p.192

〈Ans.〉

Solids 3 (Cylinder & Cone)

Level ★★

Date / /　Name

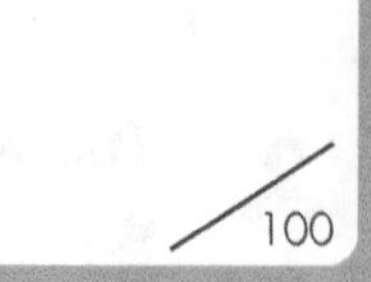

■ The Answer Key is on page 185.

Don't forget!

● A prism is made up of two polygon bases, and a solid whose bases are circular is a **cylinder**. A pyramid has the base of a polygon, and the solid whose base is circular is a **cone**. The sides of a cylinder and a cone are curved surfaces, not flat surfaces.

1 **Mark ○ for the cylinder and △ for the cone in the figures below.** 　4 points per question

(1) 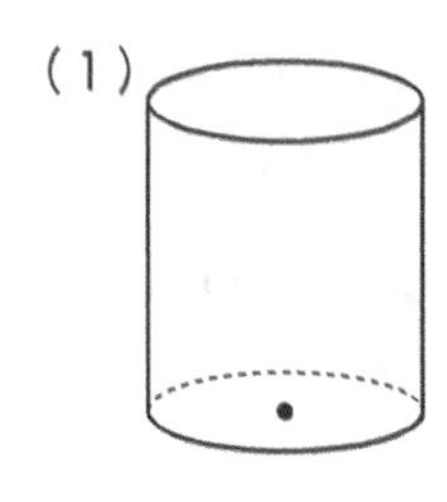
[　　]

(2) [　　]

(3) [　　]

(4) [　　]

(5) [　　]

(6) 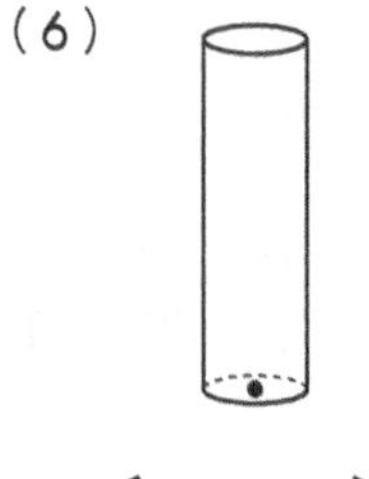
[　　]

2 **Connect each solid, cylinder or cone, to the correct net on the right.** 　4 points per question

(1) 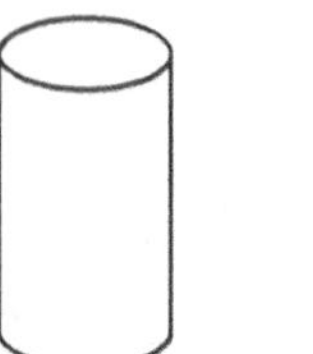 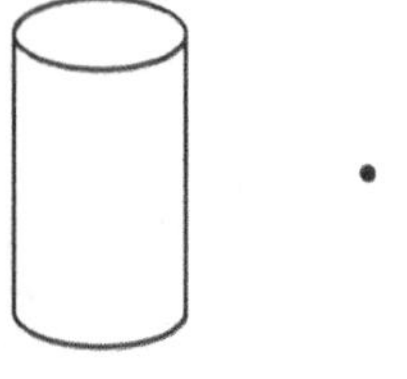 •　　• 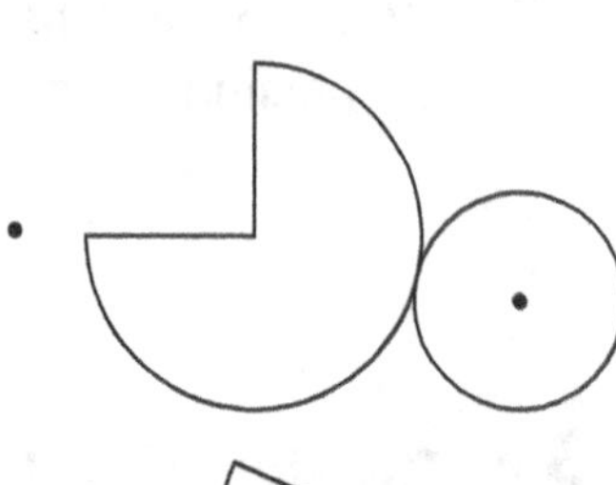

(2) 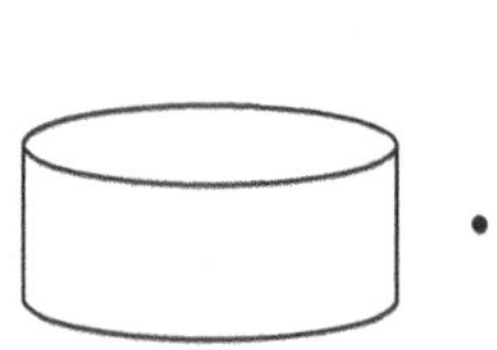 •　　• 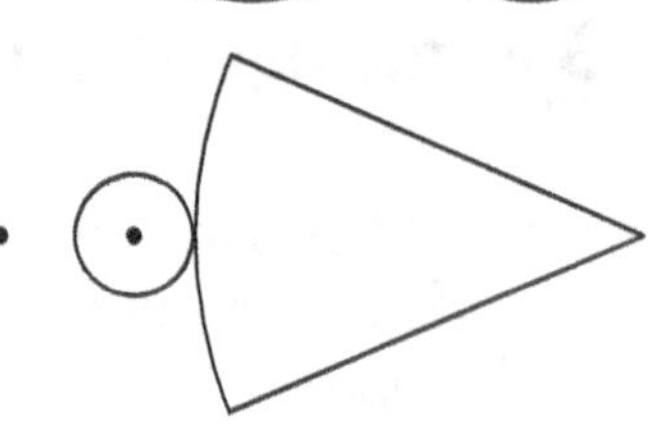

(3) 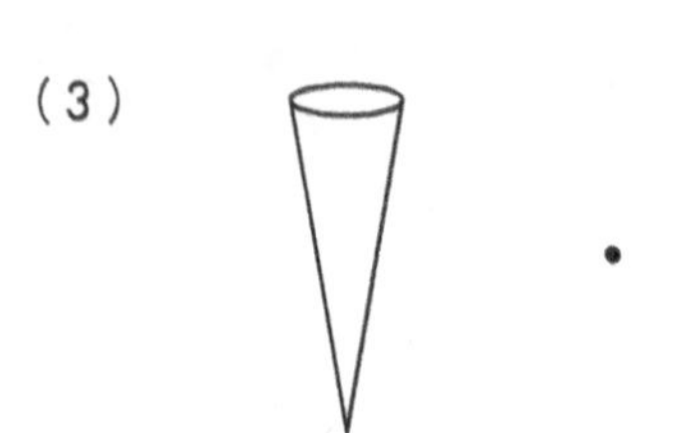 •　　• 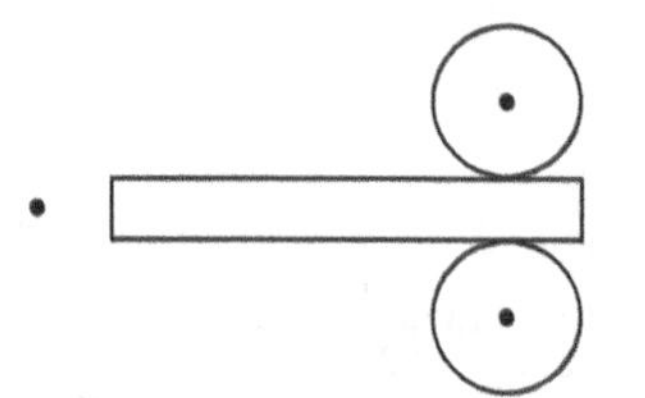

(4) 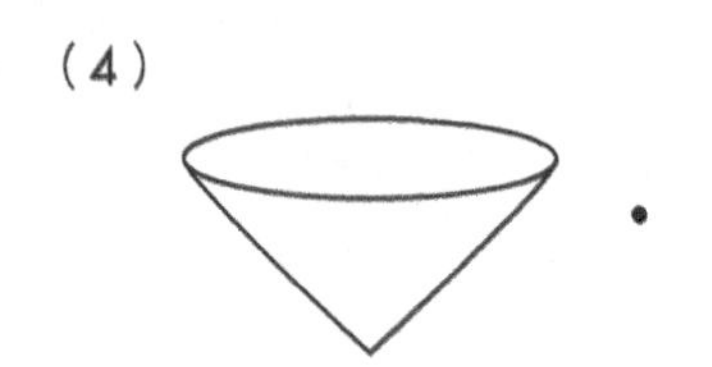 •　　• 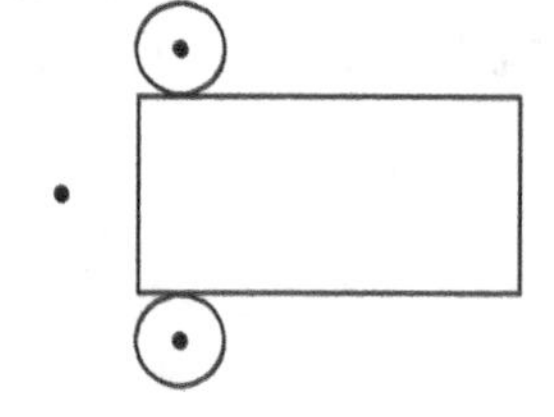

3 Answer the following questions about the cylinder on the right.

12 points per question

(1) Find the circumference of the base in terms of π.

〈Ans.〉 ________________

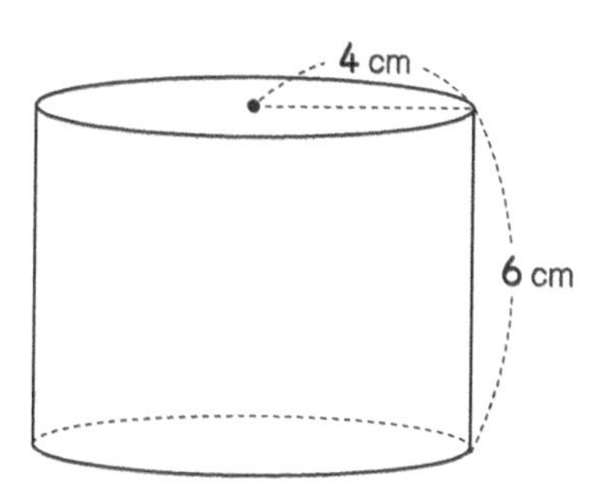

(2) What is the shape of the side surface in the net of this cylinder?

〈Ans.〉 ________________

(3) The figure on the right is the net of the cylinder above. Find the lengths of a and b.

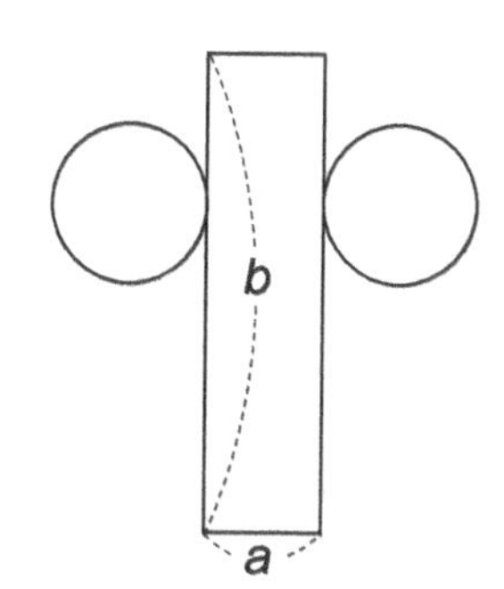

〈Ans.〉 a: ________________

〈Ans.〉 b: ________________

4 The figures on the right are cones A, B, and C with a base radius of 10 inches. Using the information provided, answer the questions below.

12 points per question

(1) Write the lengths of a, b, and c of the sides of each cone in descending order.

〈Ans.〉 ________ , ________ , ________

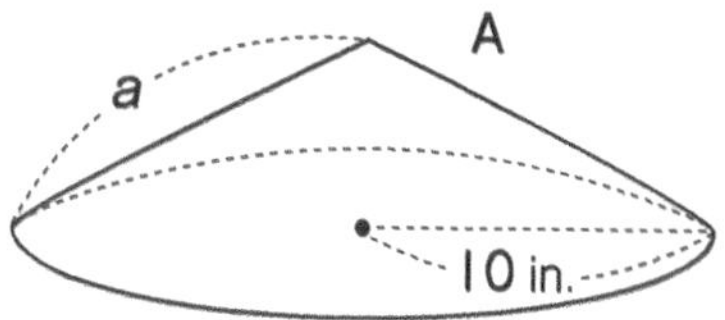

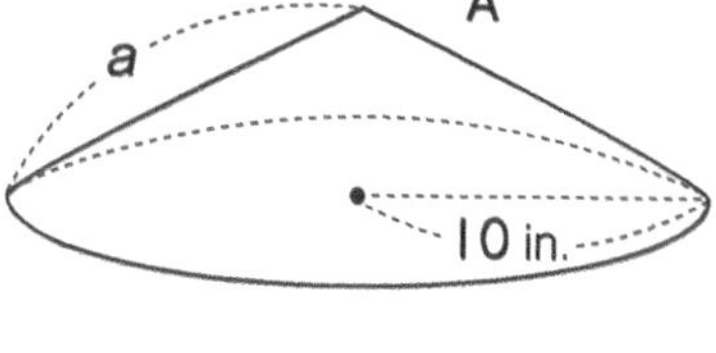

(2) Write A, B, or C in the spaces below to match each net of the side to its corresponding cone.

① 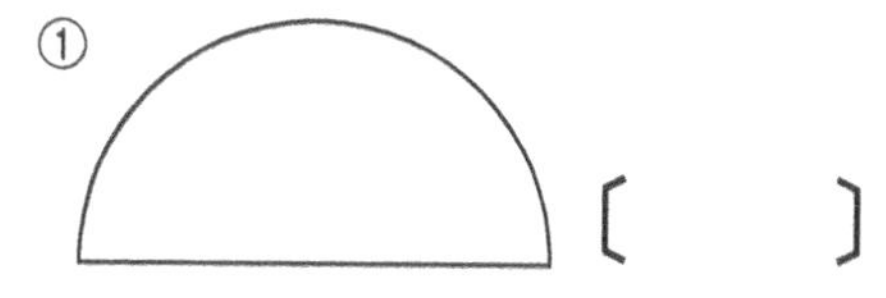 [   ]

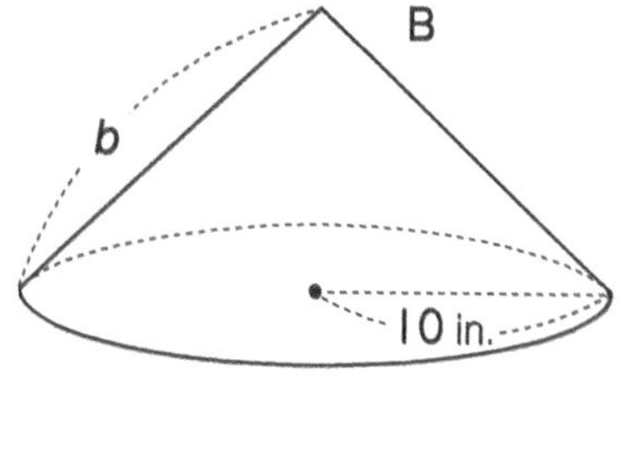

② 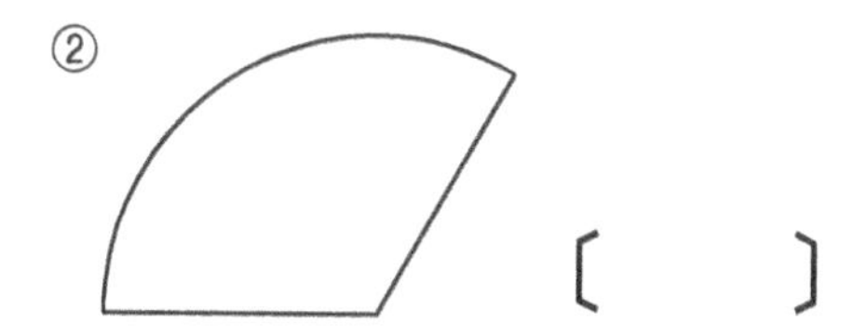 [   ]

③ 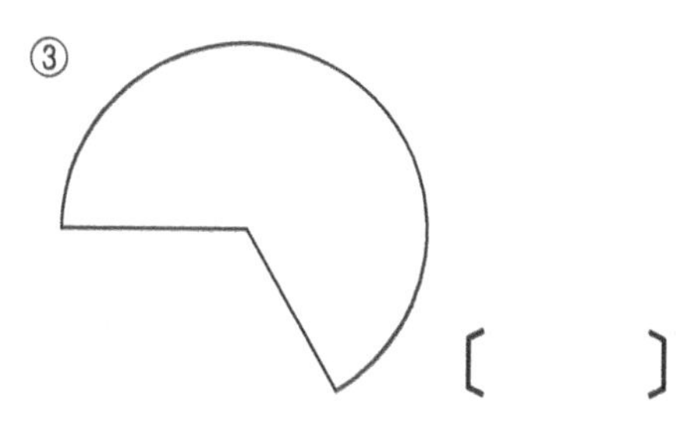 [   ]

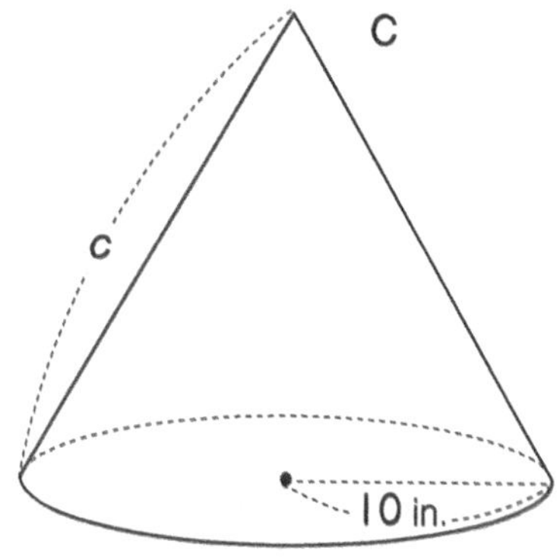

12 Projection of a Solid

Level ★★

Date / / Name

Score /100

■ The Answer Key is on page 185.

1 **Draw figures (*a*) as seen from the top of the built shapes below. Next, draw figures (*b*) as seen from the side from the direction of the arrow.** 10 points per question

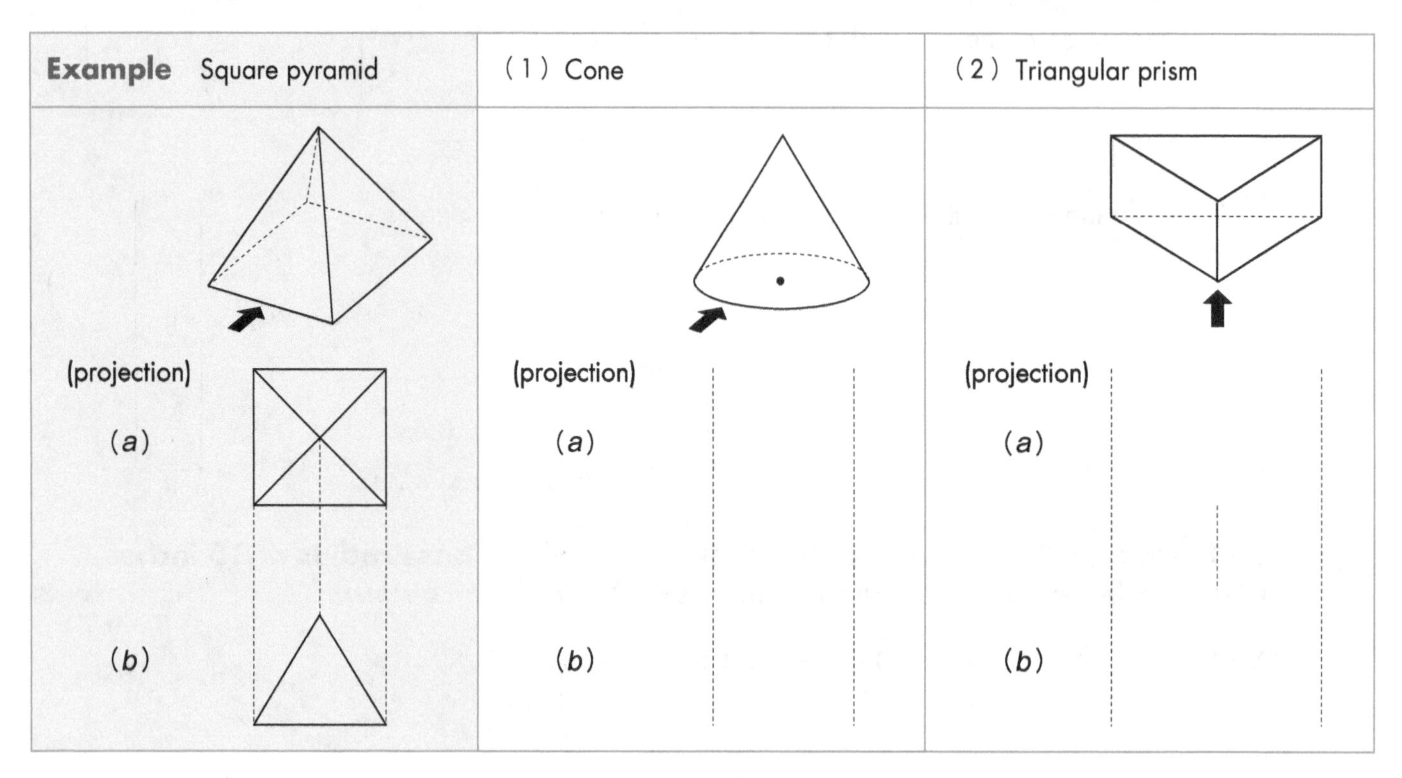

(3) Cylinder	(4) Regular triangular pyramid	(5) Cube
(projection) (*a*) (*b*)	(projection) (*a*) (*b*)	(projection) (*a*) (*b*)

Don't forget!

●The shape of a solid viewed directly above or from the side is called the **projection**.

2 **Draw the built shapes that are shown by the projections below, and write its name.**

10 points per question

Example

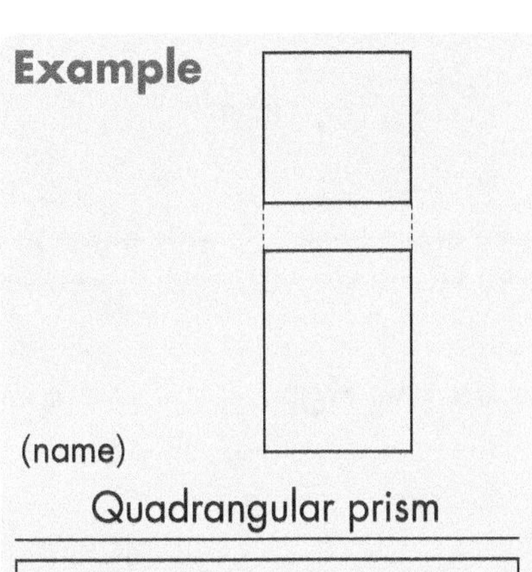

(name)

Quadrangular prism

(built shape)

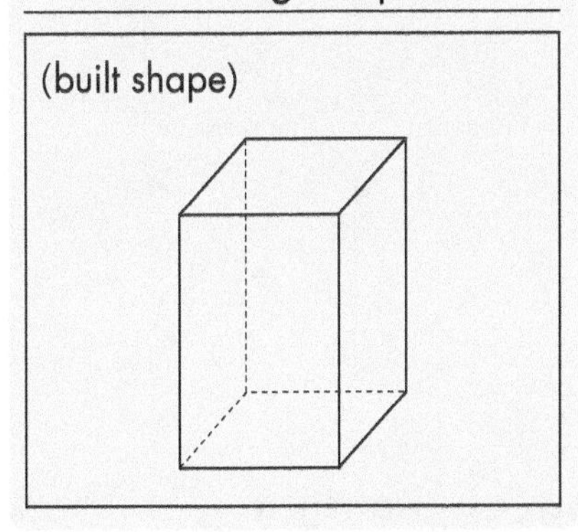

(1)

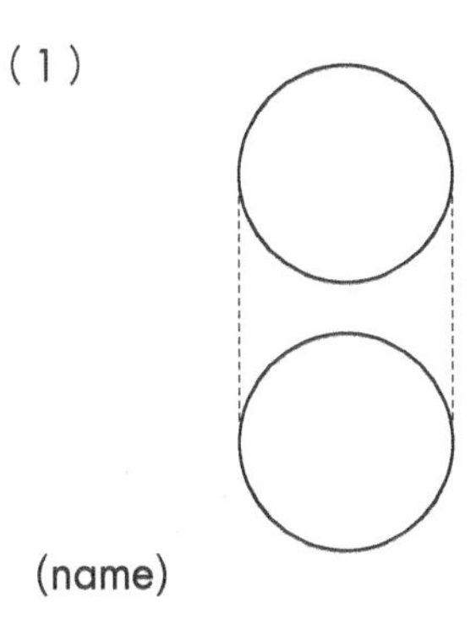

(name)

(built shape)

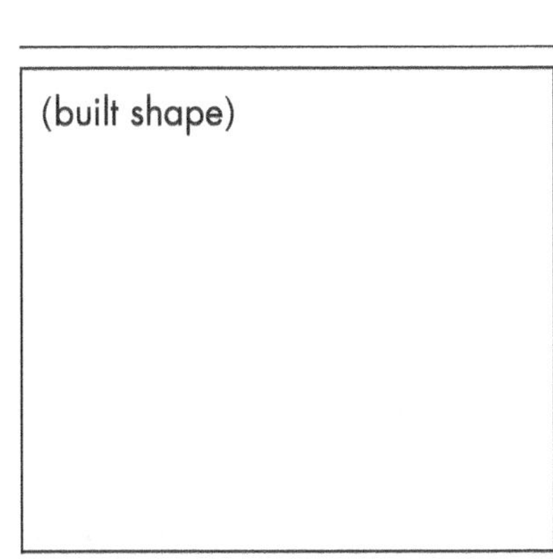

(2)

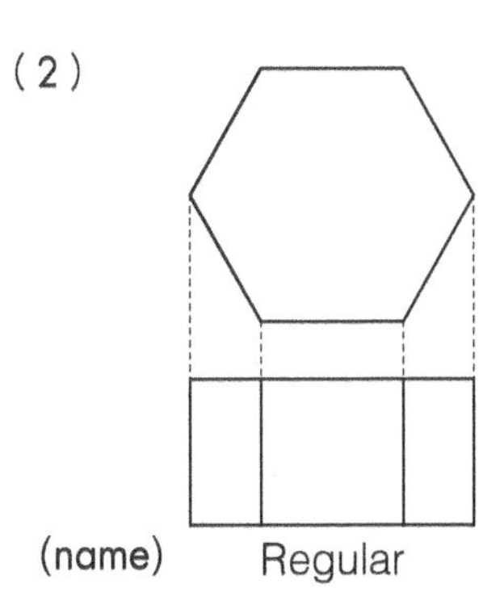

(name) Regular

(built shape)

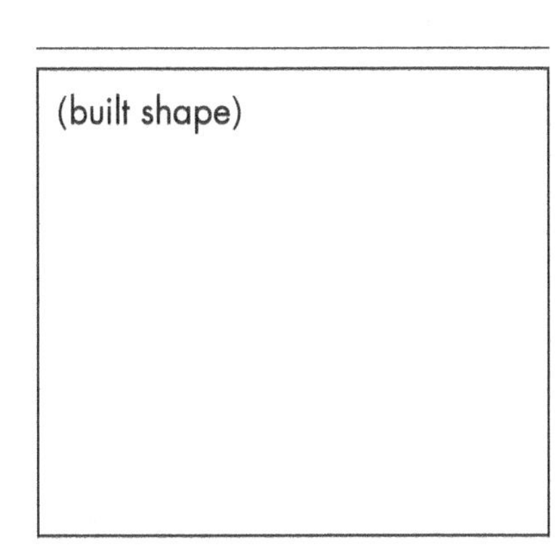

(3)

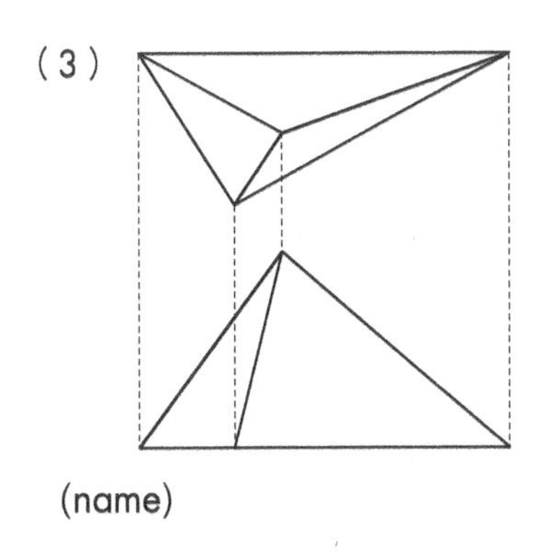

(name)

(built shape)

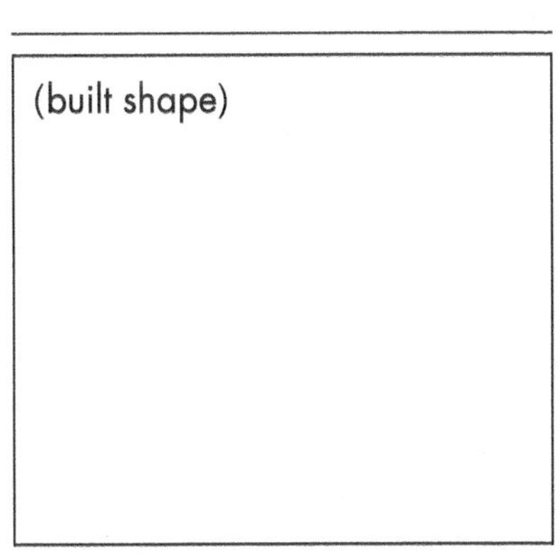

(4)

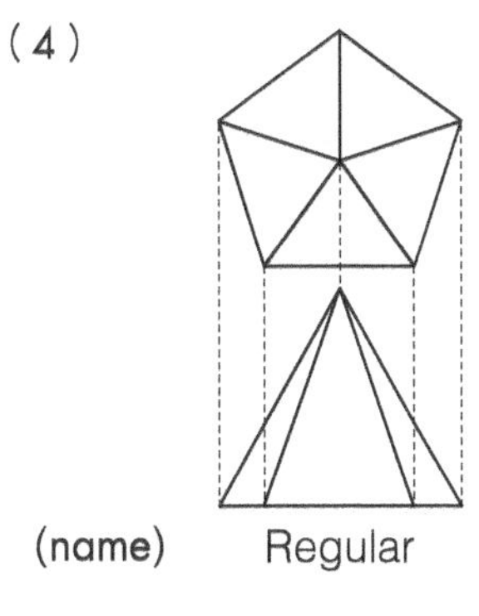

(name) Regular

(built shape)

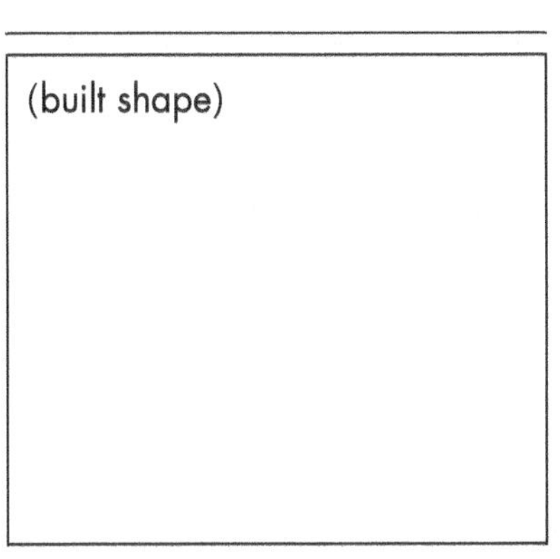

(5)

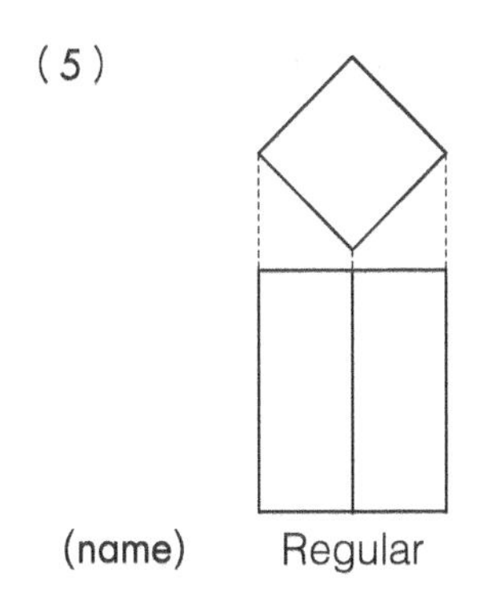

(name) Regular

(built shape)

13 Volume 1 (Prism & Cylinder)

Date / /

Name

■ The Answer Key is on page 186.

Don't forget!

● The volume of a prism or a cylinder can be found by multiplying the area of its base times its height.

Volume of prism or cylinder (V) = area of the base (B) × height (h)

$$V = B \times h$$

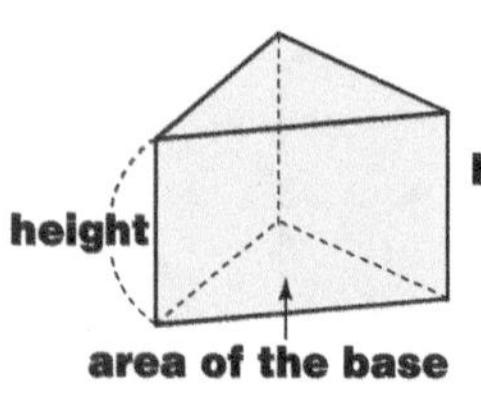

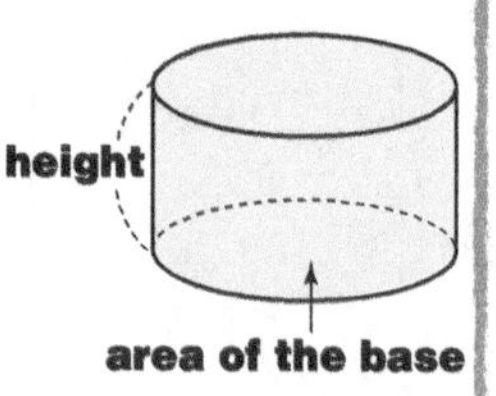

1 **Answer the questions below using the triangular prism on the right.** 7 points per question

(1) Find the area of the base.

〈Ans.〉

(2) Find the volume of the triangular prism.

〈Ans.〉

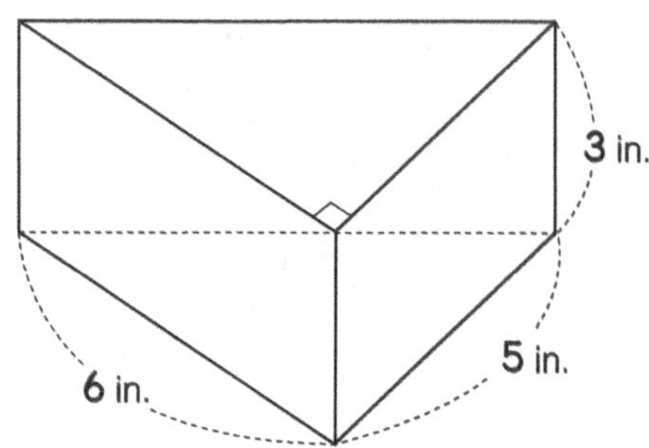

2 **Answer the questions below using the rectangular prism on the right.** 7 points per question

(1) Find the area of the base.

〈Ans.〉

(2) Find the volume of the rectangular prism.

〈Ans.〉

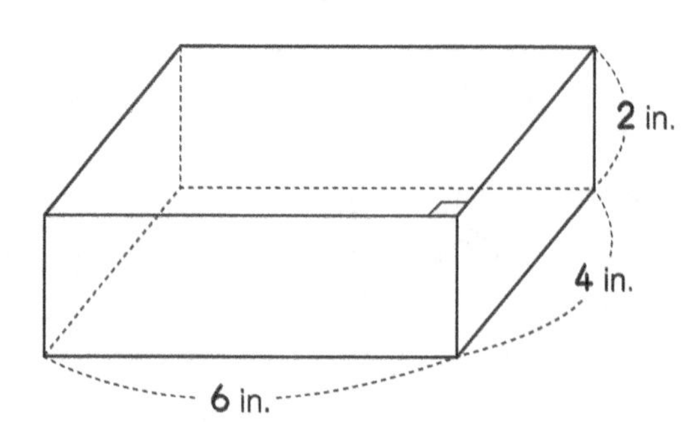

3 **Answer the questions below using the cylinder on the right.** 8 points per question

(1) Find the area of the base.
(Pi = π)

〈Ans.〉

(2) Find the volume of the cylinder.
(Pi = π)

〈Ans.〉

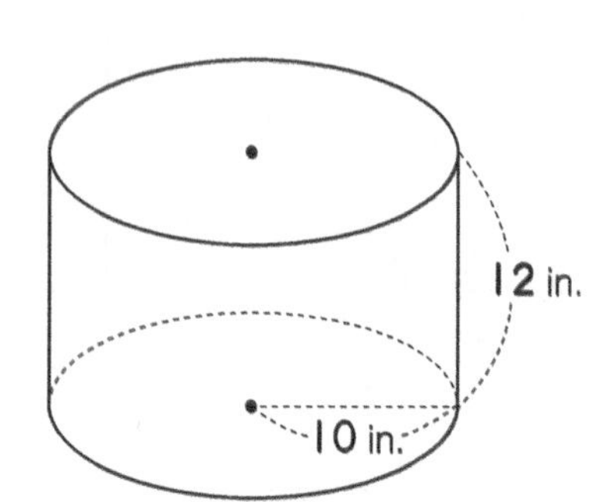

4 Find the volume of the solids below.

7 points per question

(1)

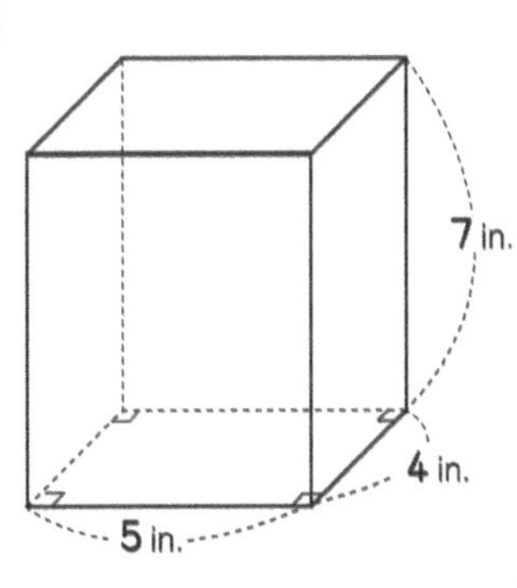

〈Ans.〉

(2)

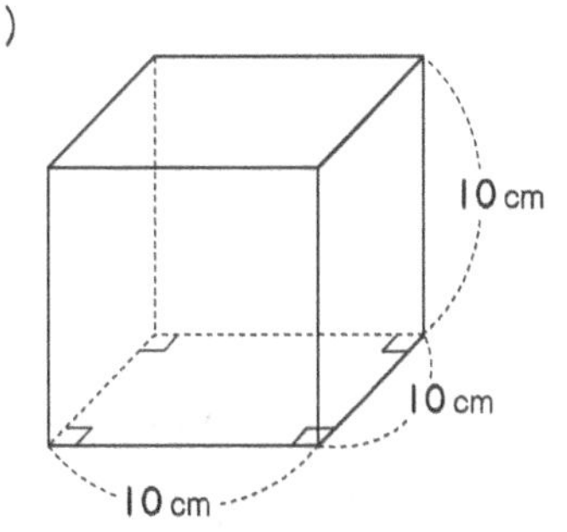

〈Ans.〉

(3)

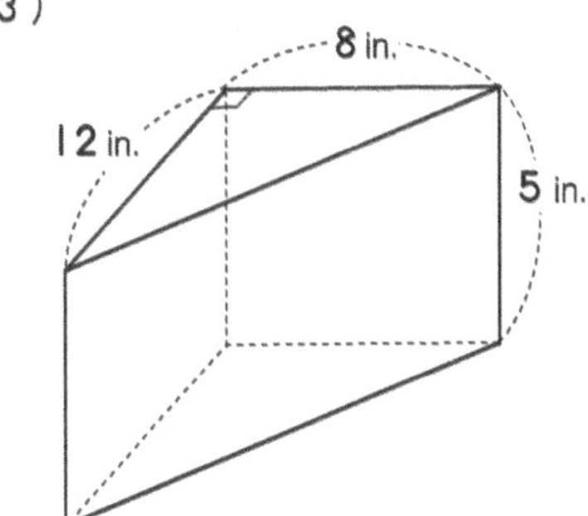

〈Ans.〉

(4)

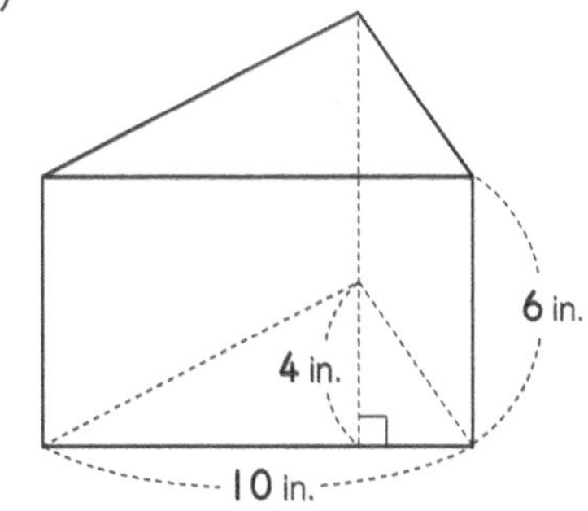

〈Ans.〉

(5)

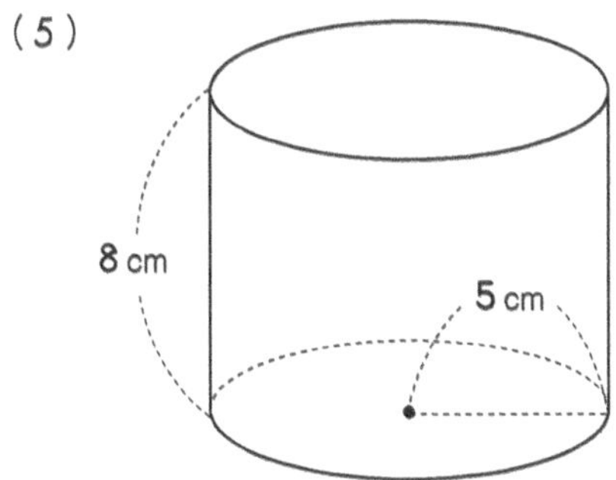

(Pi = π)

〈Ans.〉

(6)

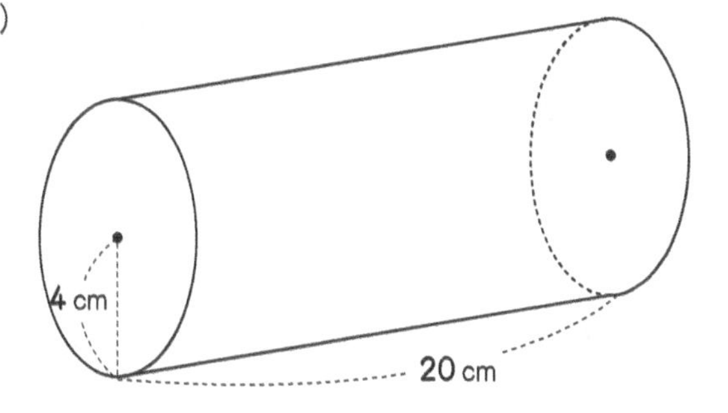

(Pi = π)

〈Ans.〉

(7)

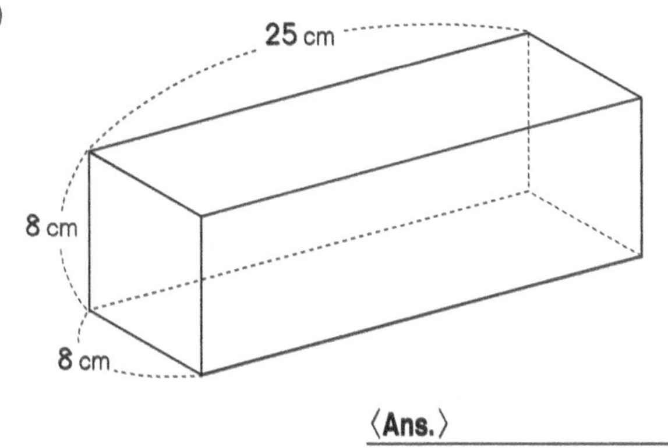

〈Ans.〉

(8)

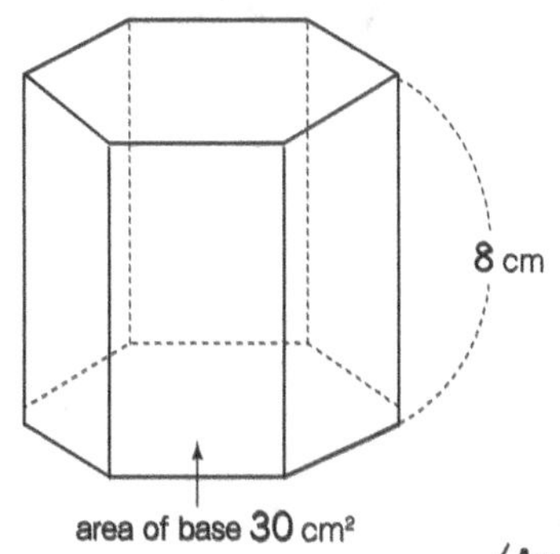

〈Ans.〉

14 Volume 2 (Pyramid & Cone)

Level ★★

Date / /

Name

■ The Answer Key is on page 186.

Don't forget!

● The volume of a pyramid or a cone can be found by multiplying the area of its base by its height, and then by multiplying the product by $\frac{1}{3}$.

Volume of pyramid or cone (V) = $\frac{1}{3}$ × area of the base (B) × height (h)

$$V = \frac{1}{3} \times B \times h$$

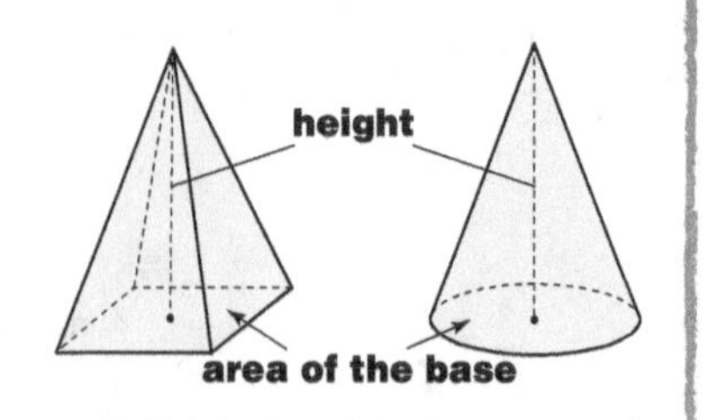

1 **Answer the questions below using the triangular pyramid on the right.** 7 points per question

(1) Find the area of the base.

⟨Ans.⟩ ______________

(2) Find the volume of the triangular pyramid.

⟨Ans.⟩ ______________

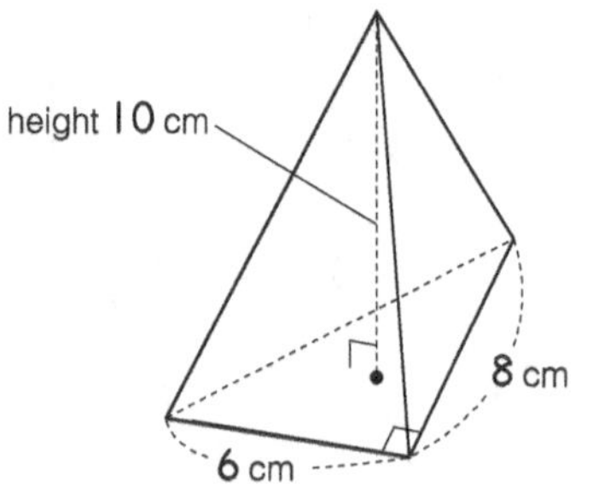

2 **Answer the questions below using the square pyramid on the right.** 7 points per question

(1) Find the area of the base.

⟨Ans.⟩ ______________

(2) Find the volume of the square pyramid.

⟨Ans.⟩ ______________

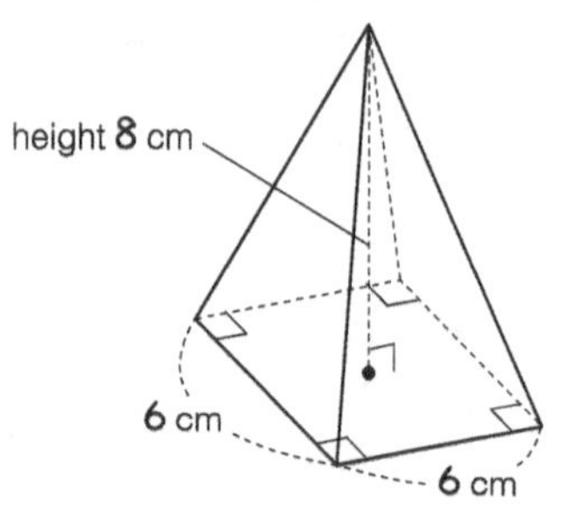

3 **Answer the questions below for a cone on the right.** 8 points per question

(1) Find the area of the base.
(Pi = π)

⟨Ans.⟩ ______________

(2) Find the volume of the cone.
(Pi = π)

⟨Ans.⟩ ______________

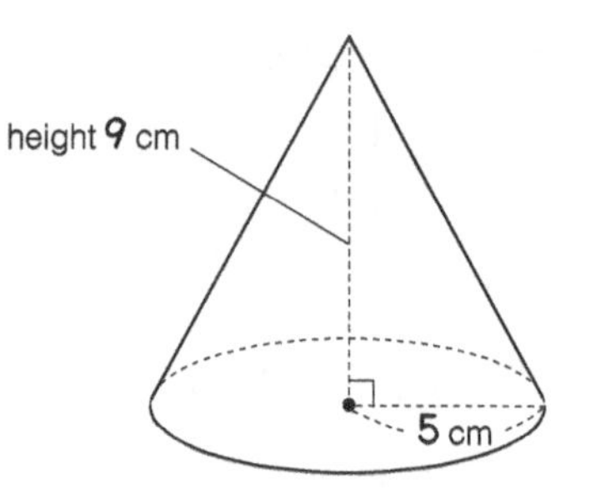

4 Find the volume of the solids below.

7 points per question

(1)

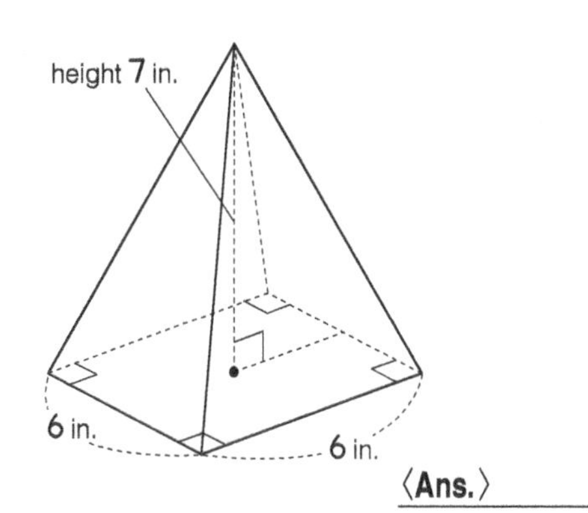

⟨Ans.⟩

(2)

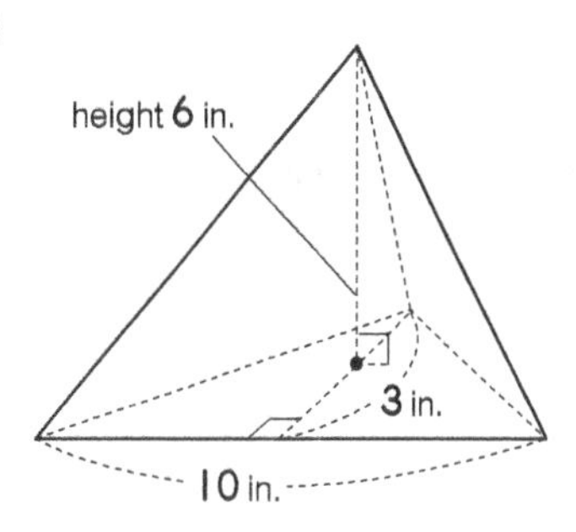

⟨Ans.⟩

(3)

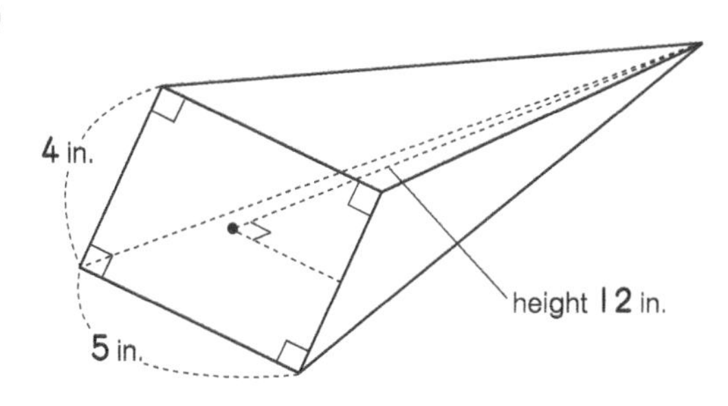

⟨Ans.⟩

(4)

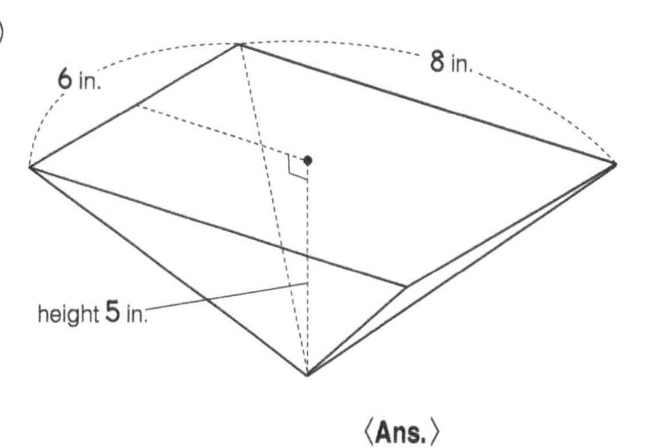

⟨Ans.⟩

(5)

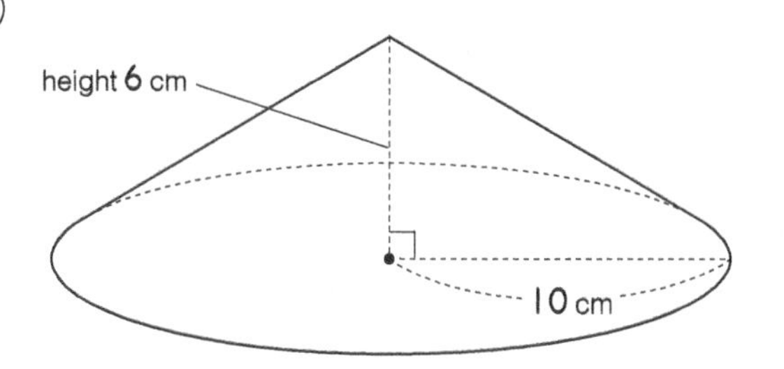

(Pi = π)

⟨Ans.⟩

(6)

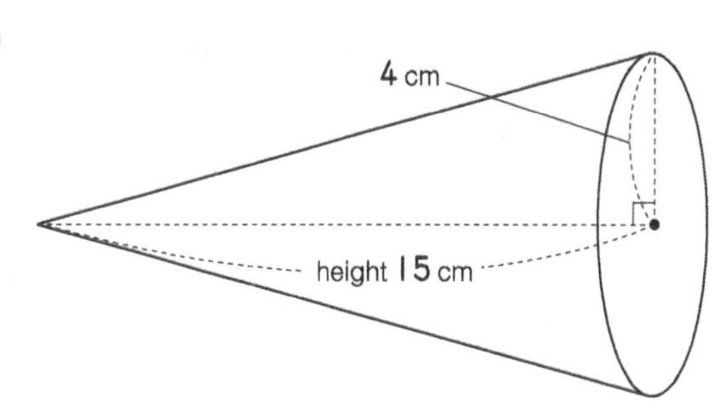

(Pi = π)

⟨Ans.⟩

(7)

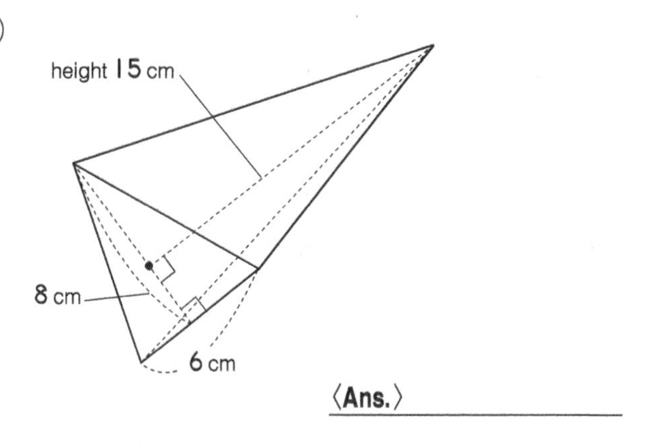

⟨Ans.⟩

(8)

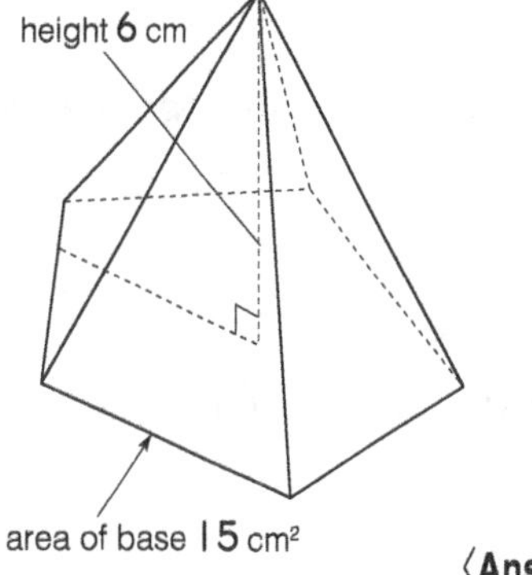

⟨Ans.⟩

15 Nets of Prisms & Cylinders

Level ★★

Date / /　　Name

■ The Answer Key is on page 186.

1 **The figures on the right are a triangular prism and its net. Answer the questions below.** 5 points per question

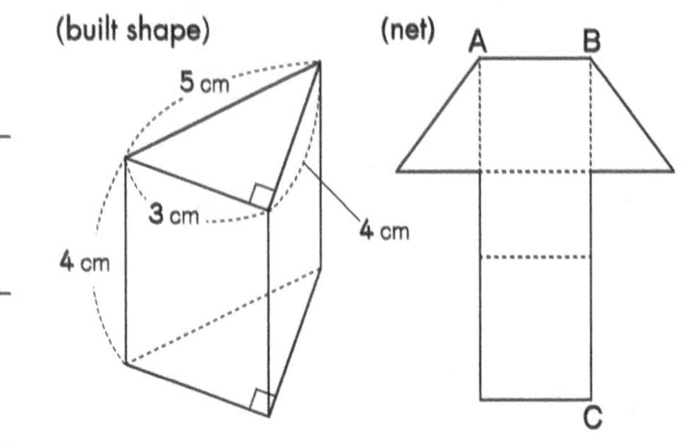

(1) Find the length of AB.

〈Ans.〉

(2) Find the area of the base.

〈Ans.〉

(3) Find the length of BC.

〈Ans.〉

(4) Find the lateral area of this triangular prism.

〈Ans.〉

2 **Using the figure of a rectangular prism with base 'a' and its net on the right, answer the questions below.** 6 points per question

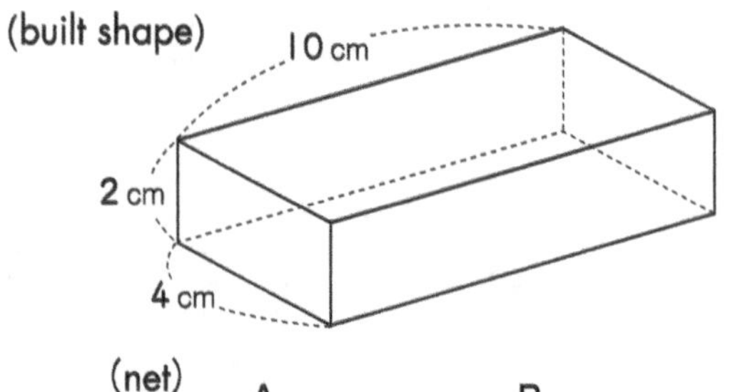

(net) A B, a, a, D C

(1) Find the length of AB.

〈Ans.〉

(2) Find the area of base '*a*'.

〈Ans.〉

(3) Find the length of BC.

〈Ans.〉

(4) Find the lateral area of this rectangular prism.  p.192

〈Ans.〉

(5) The sum of area of two bases and the lateral area is the surface area of a rectangular prism. Find the surface area of this rectangular prism.

〈Ans.〉

3 Using the figure of the cylinder and its net on the right, answer the questions below.

5 points per question

(1) Find the diameter of the circle of the base.

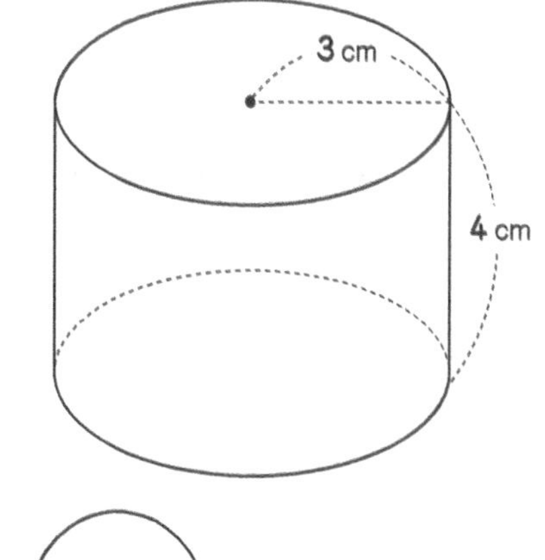

〈Ans.〉

(2) Find the circumference of the base.
(Pi = π)

〈Ans.〉

(3) Find the length of AB.
(Pi = π)

〈Ans.〉

(4) Find the length of BC.

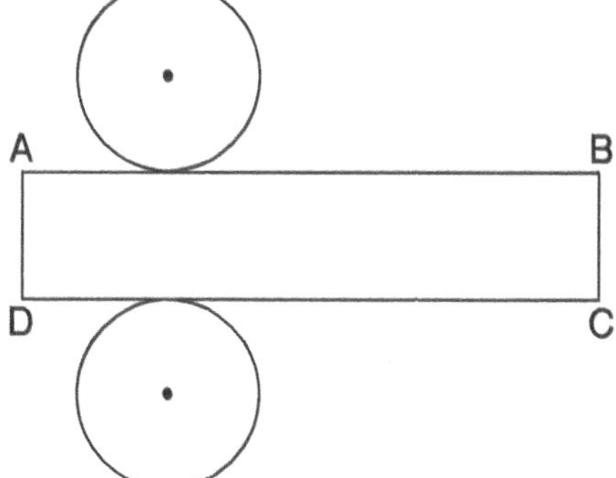

〈Ans.〉

(5) Find the area of the rectangle ABCD of the side.
(Pi = π)

〈Ans.〉

4 Using the figure of the cylinder and its net on the right, answer the questions below in terms of π.

5 points per question

(1) Find the area of the base.

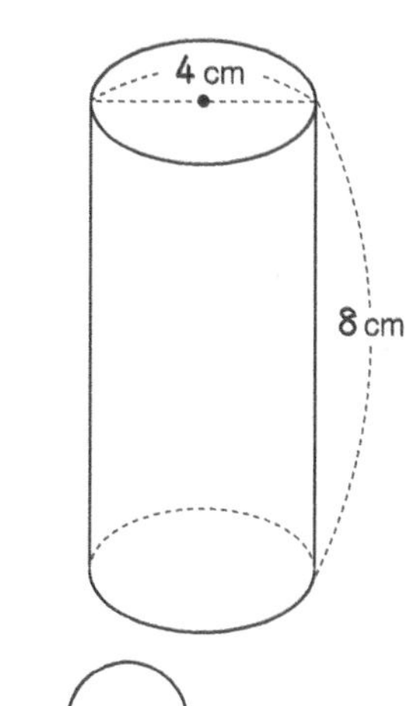

〈Ans.〉

(2) Find the circumference of the base.

〈Ans.〉

(3) Find the length of AB.

〈Ans.〉

(4) Find the area of rectangle ABCD of the side.

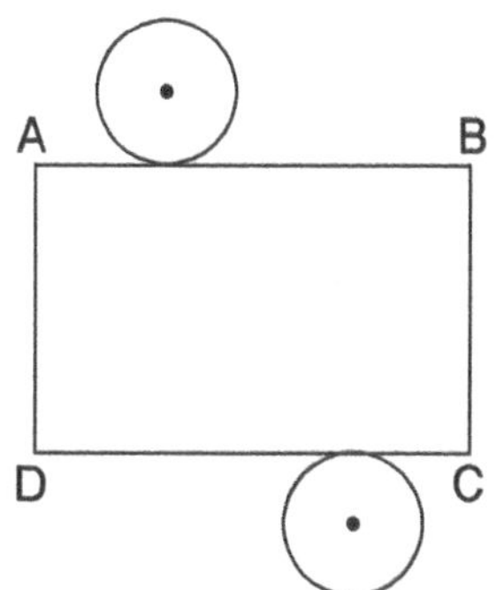

〈Ans.〉

(5) Find the surface area of this cylinder.

〈Ans.〉

Surface Area 1 (Prism & Cylinder)

Level ★★★

Date / /

Name

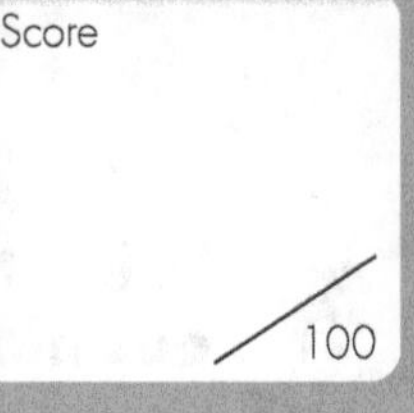

■ The Answer Key is on page 186.

1 **Assembling the nets below, create prisms and cylinders. Find the area of the base and the surface area of each solid below.**

8 points per question

(1)

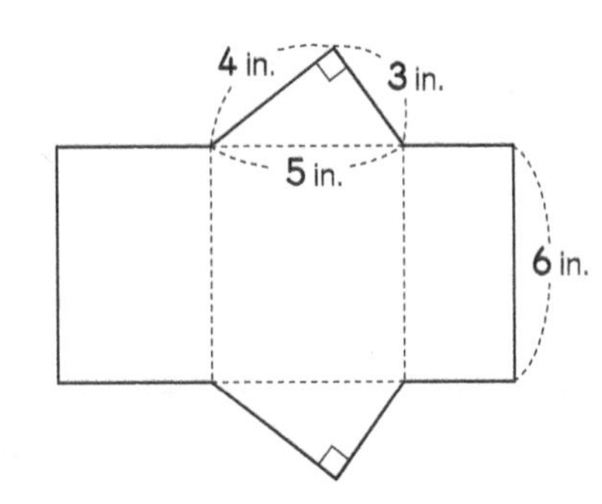

p.192

〈Ans.〉

area of the base:

surface area:

(2)

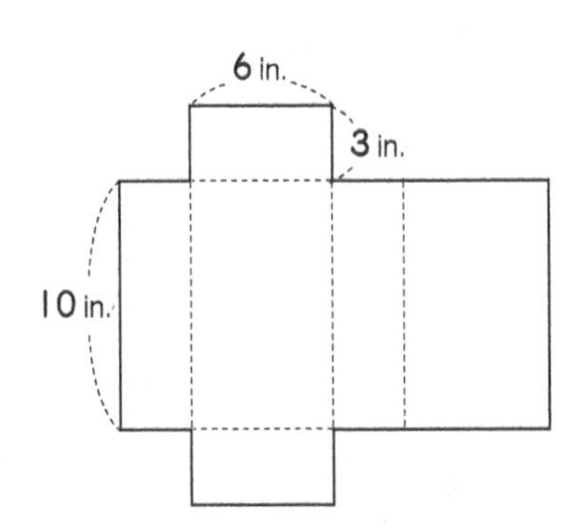

〈Ans.〉

area of the base:

surface area:

(3)

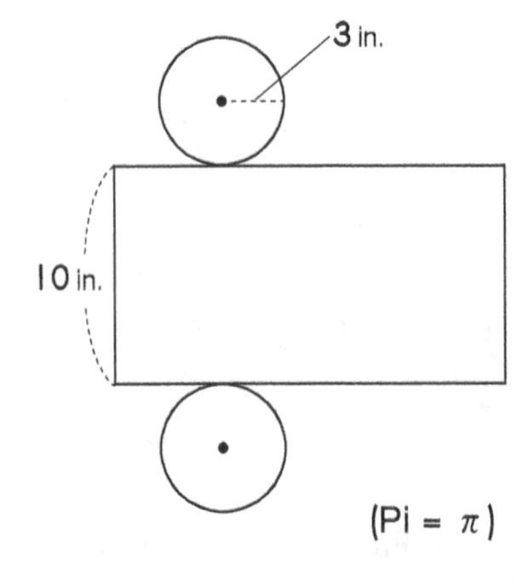

(Pi = π)

〈Ans.〉

area of the base:

surface area:

(4)

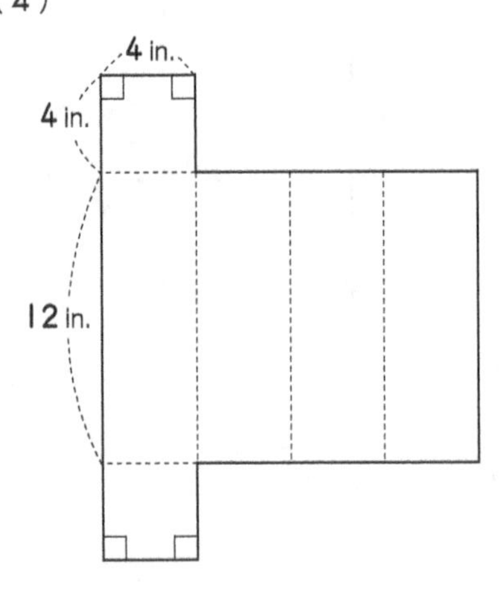

〈Ans.〉

area of the base:

surface area:

(5)

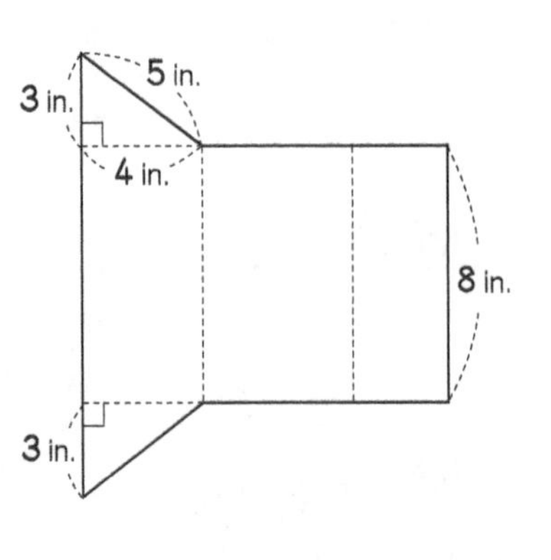

〈Ans.〉

area of the base:

surface area:

(6)

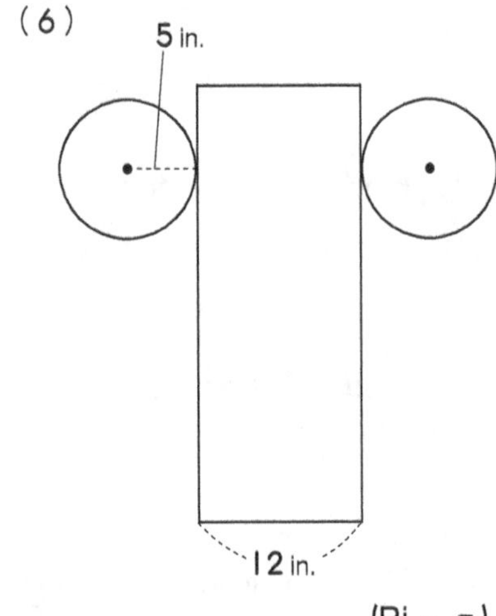

(Pi = π)

〈Ans.〉

area of the base:

surface area:

2 Find the lateral area and the surface area of each solid below.

(1)-(4) 8 points per question (5)(6) 10 points per question

(1)

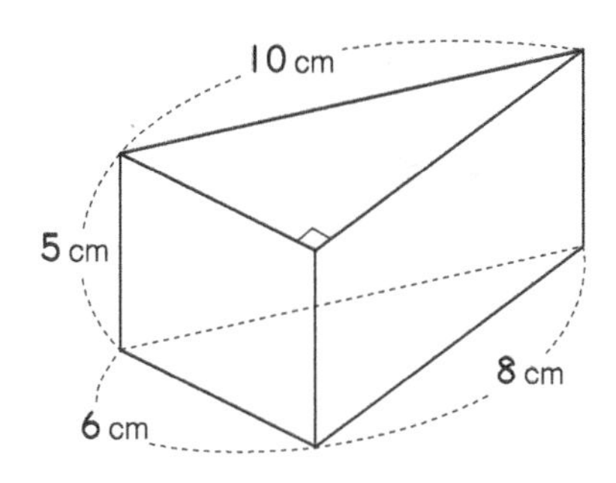

〈Ans.〉 lateral area: ______

surface area: ______

(2)

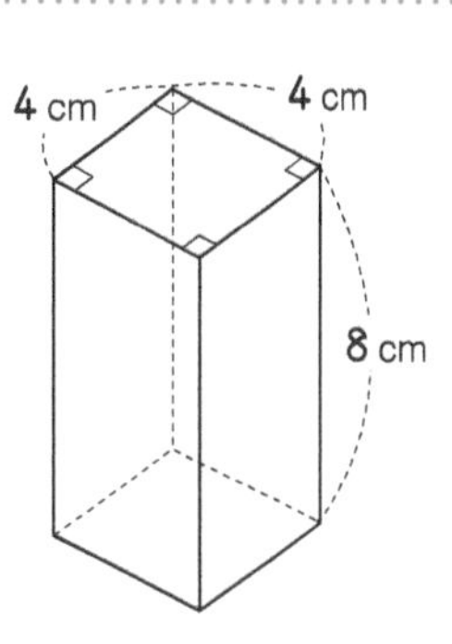

〈Ans.〉 lateral area: ______

surface area: ______

(3)

p.192

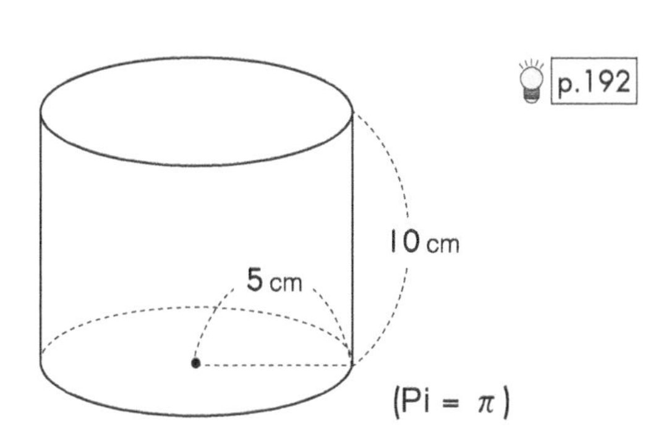

〈Ans.〉 lateral area: ______

surface area: ______

(4)

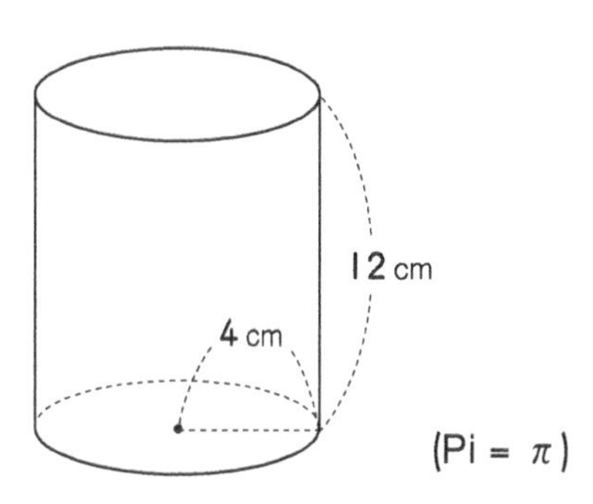

〈Ans.〉 lateral area: ______

surface area: ______

(5)

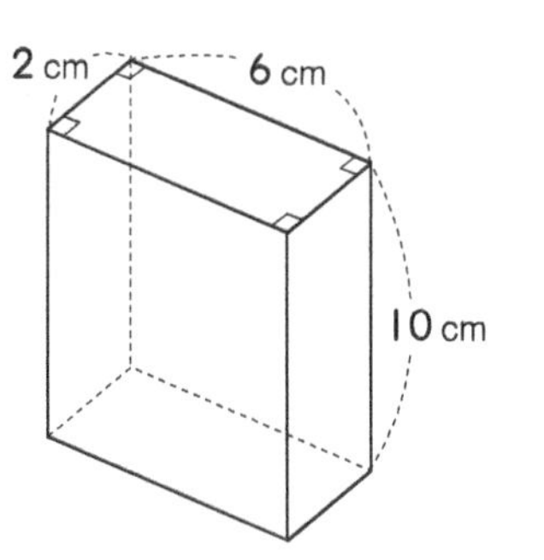

〈Ans.〉 lateral area: ______

surface area: ______

(6)

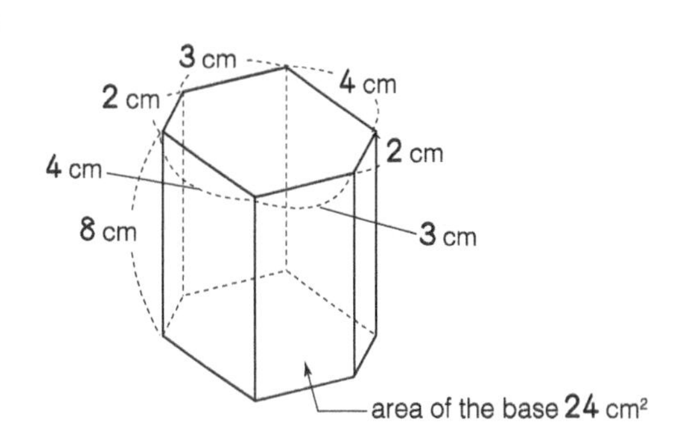

〈Ans.〉 lateral area: ______

surface area: ______

17 Net of a Pyramid

Level 

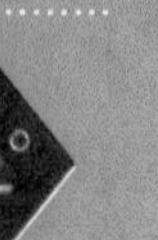

Date / /　Name

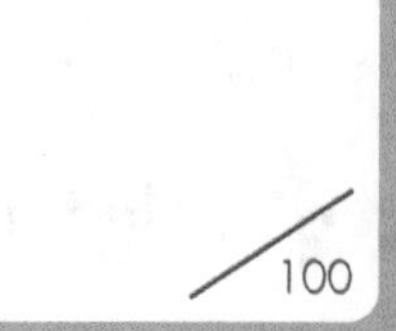

■ The Answer Key is on page 186.

1 **There are various nets below. Choose the nets that do not become pyramids when assembled.** 10 points for completion

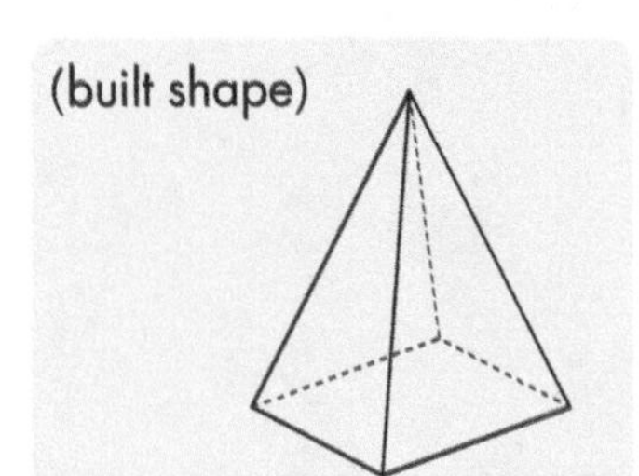

①
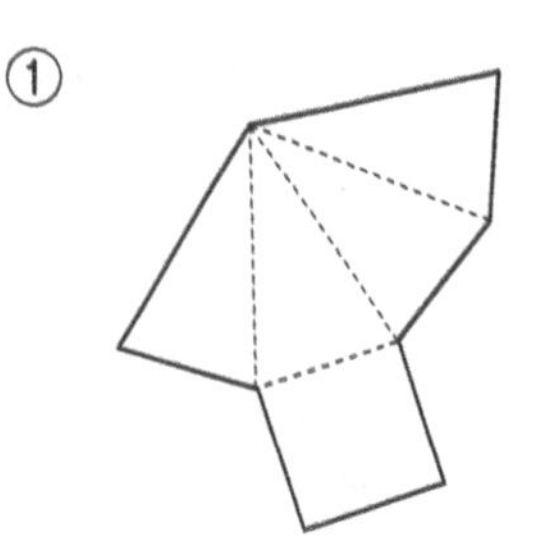

②
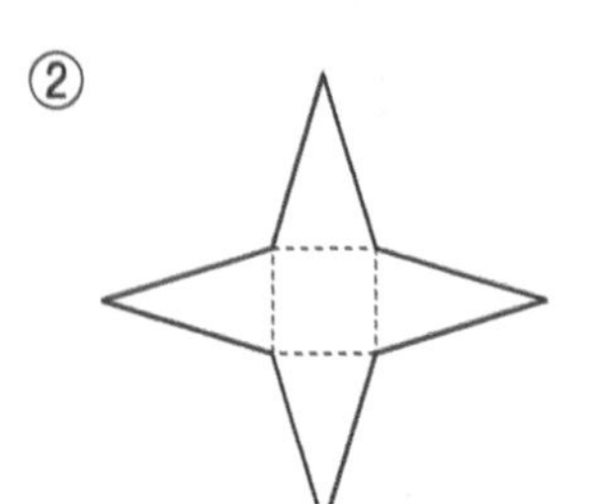

③
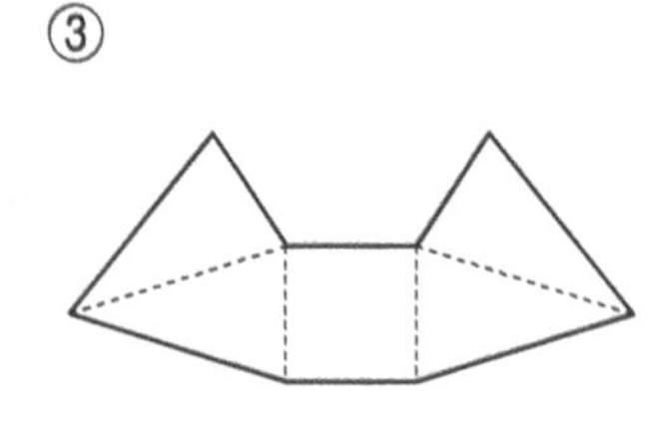

④
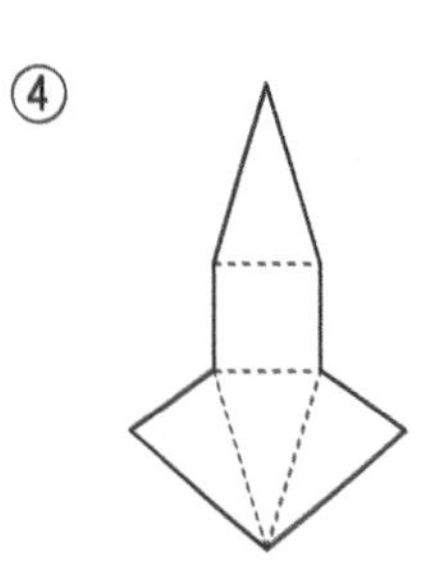

⑤
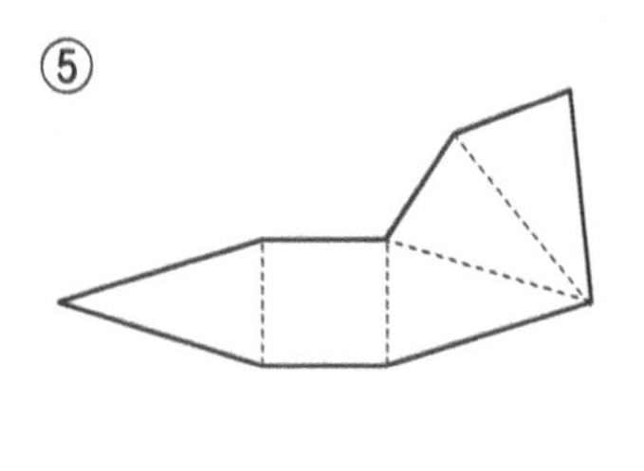

〈Ans.〉 ,

2 **The figure on the right is the net of a square pyramid. Answer the questions below.** 8 points per question

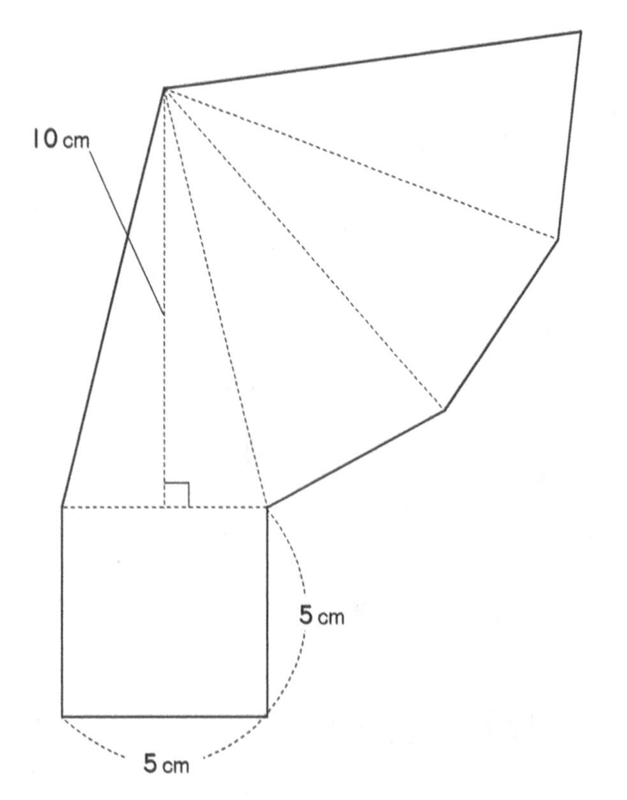

(1) What type of figure makes up a side of the square pyramid?

〈Ans.〉

(2) Find the area of a side.

〈Ans.〉

(3) Find the lateral area.

〈Ans.〉

(4) Find the area of the base.

〈Ans.〉

(5) Find the surface area.

〈Ans.〉

3 Find the lateral area and the surface area of each solid below.

9 points per question

(1)

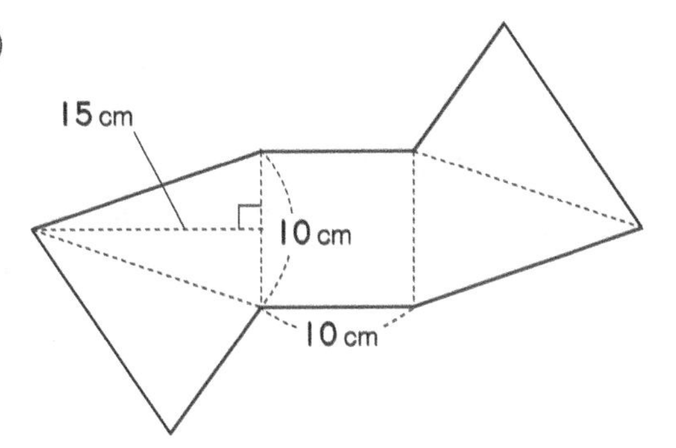

(2)

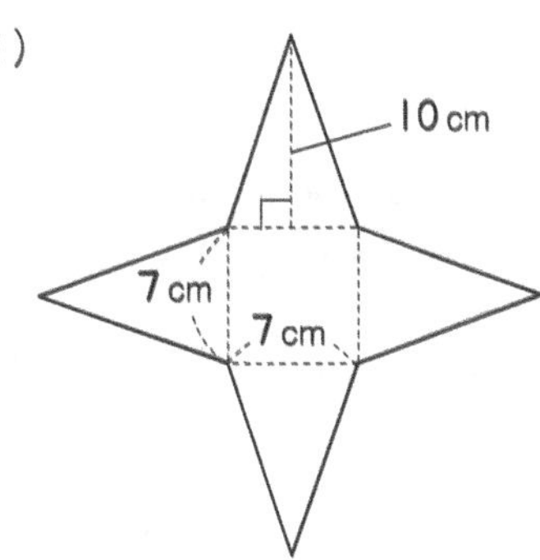

⟨Ans.⟩ lateral area:

surface area:

⟨Ans.⟩ lateral area:

surface area:

4 Find the surface area of each solid below.

8 points per question

(1)

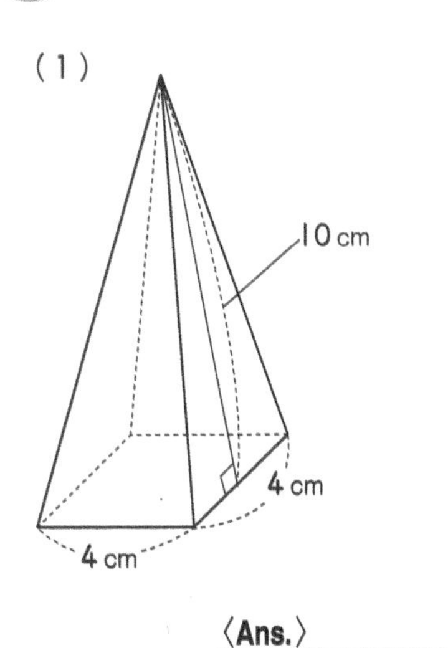

p.192

⟨Ans.⟩

(2)

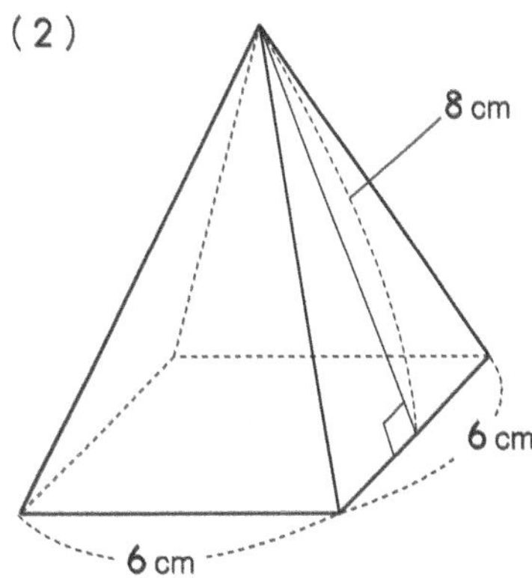

⟨Ans.⟩

(3)

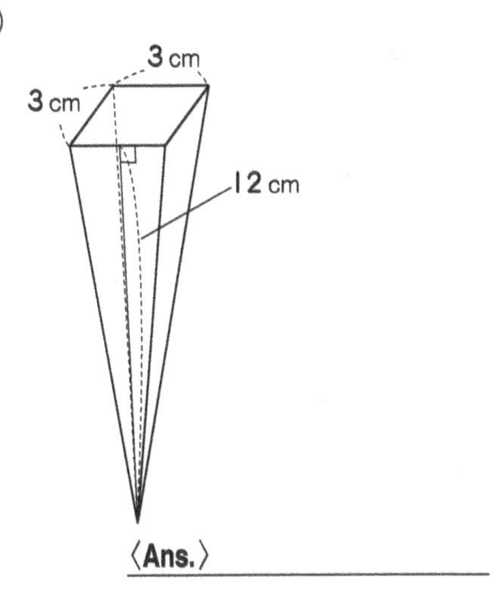

⟨Ans.⟩

(4)

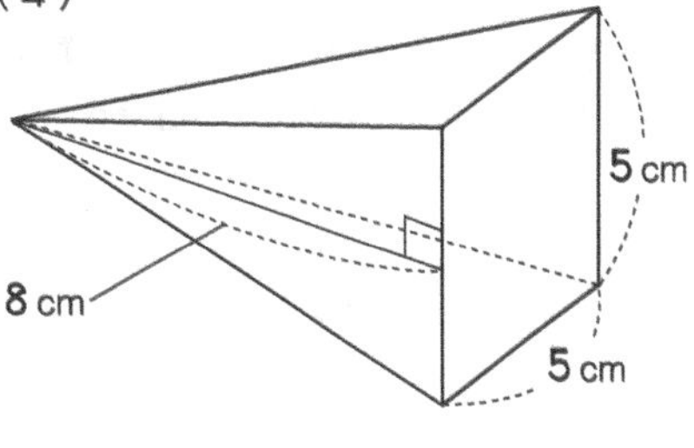

⟨Ans.⟩

18 Net of a Cone

Date / / Name

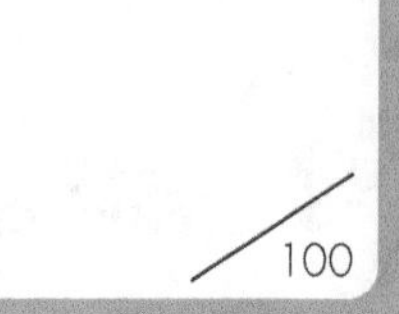

■ The Answer Key is on page 187.

1 The figures on the right are a cone and its net. The lateral area of a cone is a sector of the net. Using the information, given answer the questions below.

4 points per question

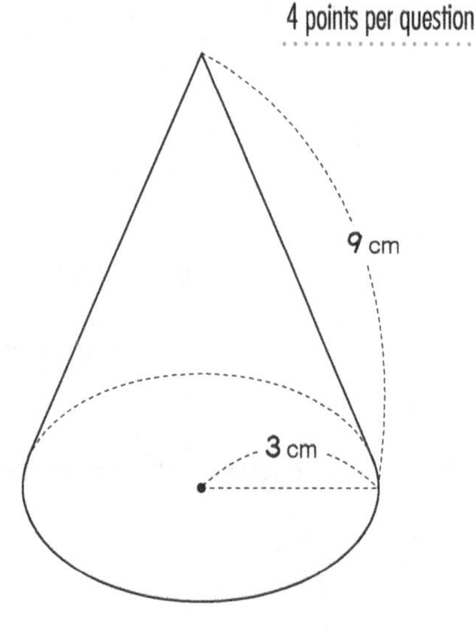

(1) Find the length of the circumference of the base.
(Pi = π)

〈Ans.〉

(2) What is the same length as arc AB?

〈Ans.〉

(3) Find the length of the circumferece of circle O.
(Pi = π)

〈Ans.〉

(4) What is the ratio of the length of arc AB to the length of the circumference of circle O? Show it in fractions.

〈Ans.〉

(5) What is the central angle of sector OAB?

〈Ans.〉

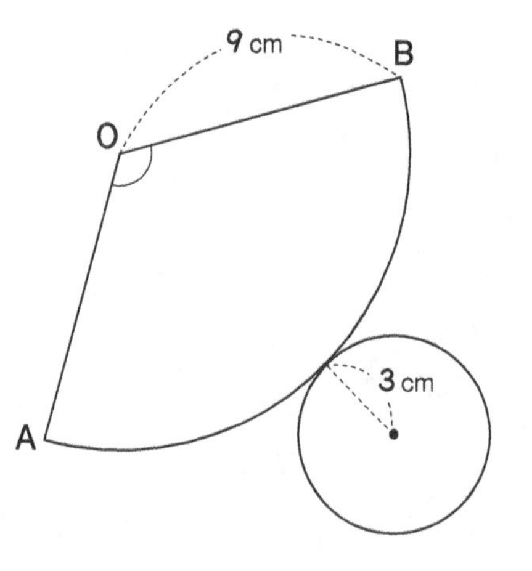

(6) Find the area of sector OAB.
(Pi = π)

〈Ans.〉

(7) Find the area of the base.
(Pi = π)

〈Ans.〉

(8) Find the surface area of the cone.
(Pi = π)

〈Ans.〉

2 Below are figures of cones and their nets. Find the measures of the central angle of each sector.

7 points per question

(1)

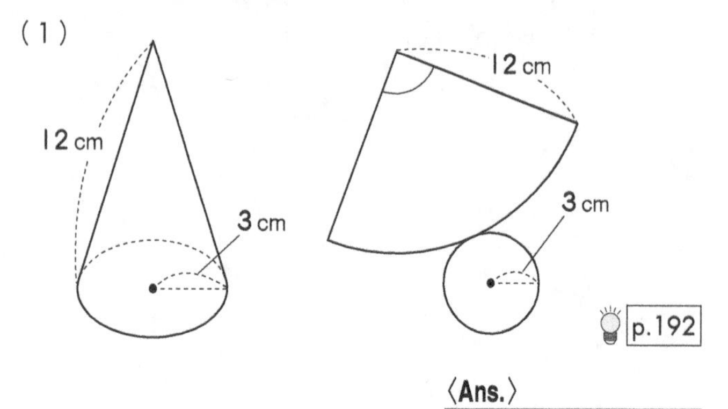

p.192

〈Ans.〉

(2)

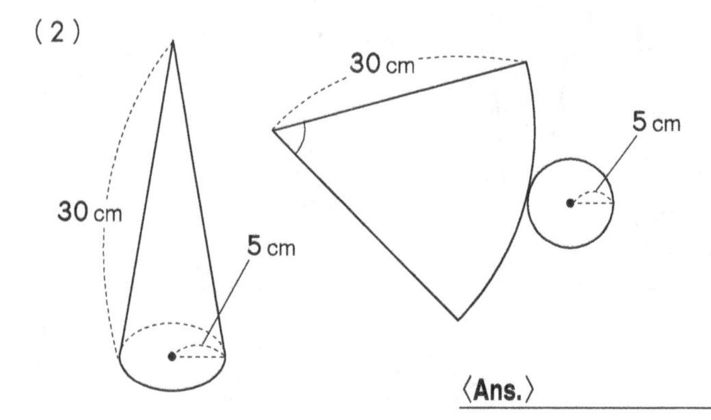

〈Ans.〉

3 **The figure on the right is the net of a cone. Using the information given, answer the questions below in terms of π.** 6 points per question

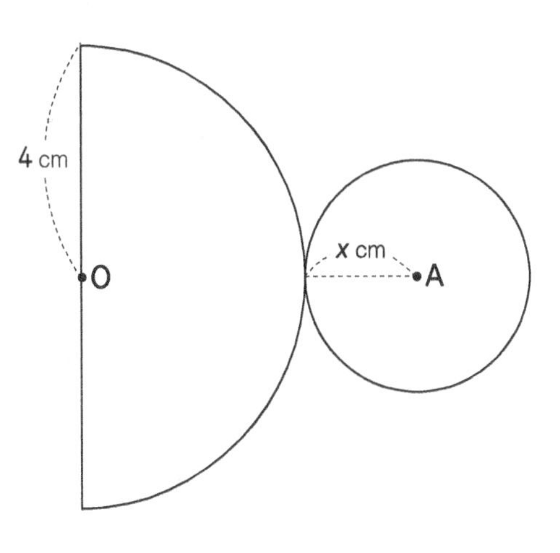

(1) Find the radius of the base.

⟨Ans.⟩ ____________

(2) Find the area of the base.

⟨Ans.⟩ ____________

(3) Find the lateral area of the cone.

⟨Ans.⟩ ____________

4 **The figure on the right is the net of a cone. Using the information given, answer the questions below in terms of π.** 6 points per question

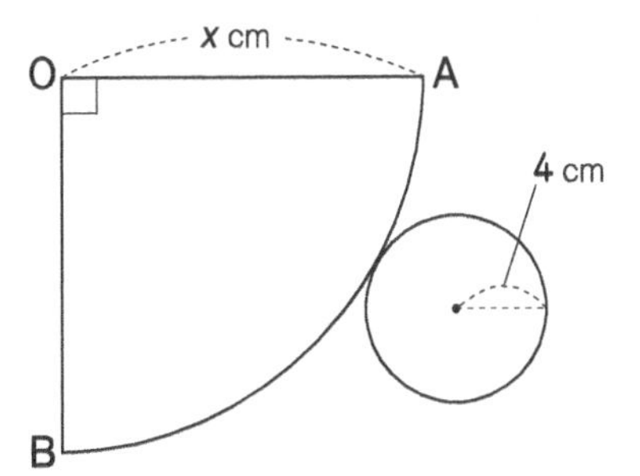

(1) Find the radius OA of the sector that is the side of the cone.

⟨Ans.⟩ ____________

(2) Find the lateral area of the cone.

⟨Ans.⟩ ____________

5 **Find the lateral area of the cones below in terms of π.** 6 points per question

(1)

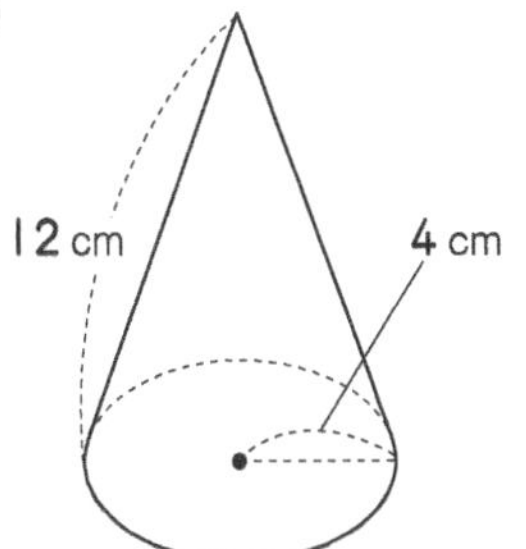

p.192

⟨Ans.⟩ ____________

(2)

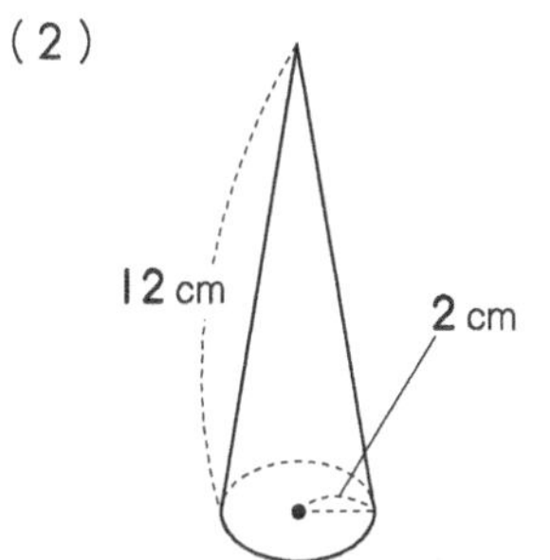

⟨Ans.⟩ ____________

(3)

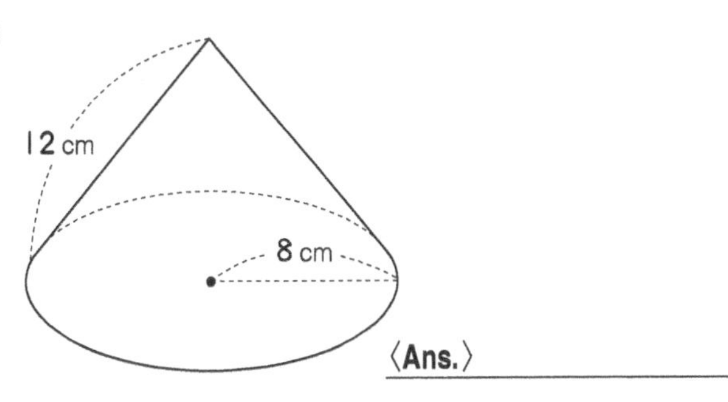

⟨Ans.⟩ ____________

(4)

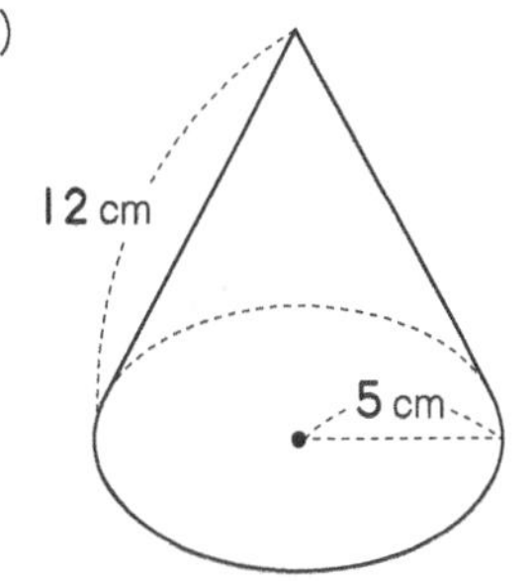

⟨Ans.⟩ ____________

19 Surface Area 2 (Cone & Pyramid)

Date / / Name

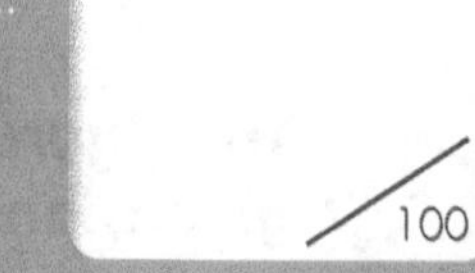

■ The Answer Key is on page 187.

1 **Answer the questions below about the triangular prism on the right.**

(1) 2 points (2)(3) 3 points per question

(1) Find the area of the base.

〈Ans.〉

(2) Find the lateral area of the triangular prism.

〈Ans.〉

(3) Find the surface area of the triangular prism.

〈Ans.〉

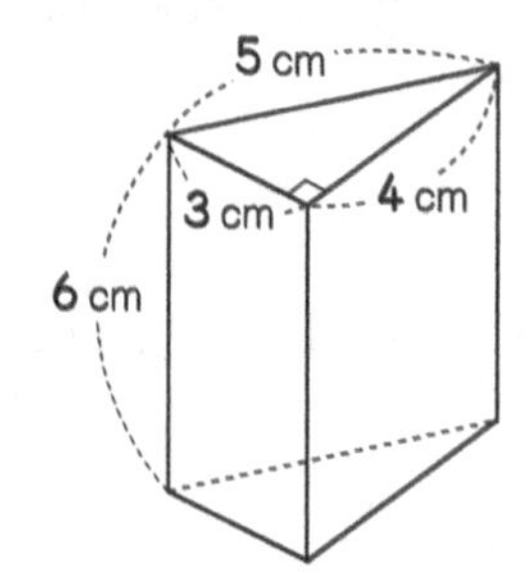

2 **Answer the questions below about the square pyramid on the right.**

(1) 2 points (2)(3) 3 points per question

(1) Find the area of the base.

〈Ans.〉

(2) Find the lateral area of the square pyramid.

〈Ans.〉

(3) Find the surface area of the square pyramid.

〈Ans.〉

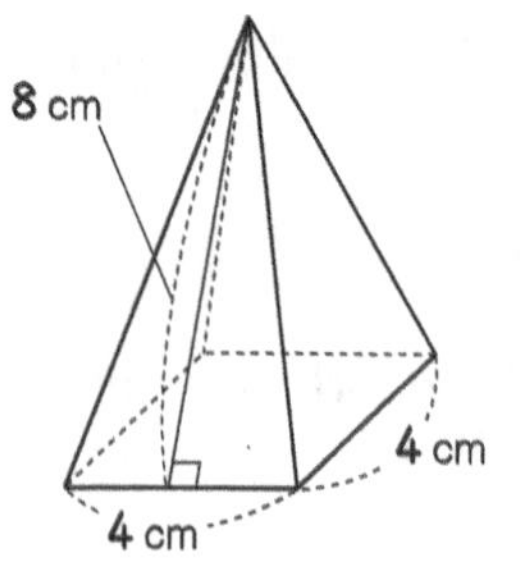

3 **Answer the questions below about the cylinder on the right in terms of π.**

6 points per question

(1) Find the area of the base.

〈Ans.〉

(2) Find the lateral area of the cylinder.

〈Ans.〉

(3) Find the surface area of the cylinder.

〈Ans.〉

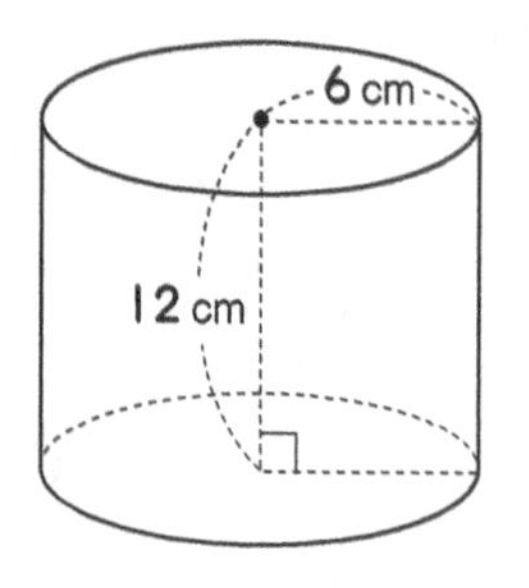

4 **Answer the questions below about the cone on the right in terms of π.**

6 points per question

(1) Find the area of the base.

〈Ans.〉

(2) Find the lateral area of the cone.

〈Ans.〉

(3) Find the surface area of the cone.

〈Ans.〉

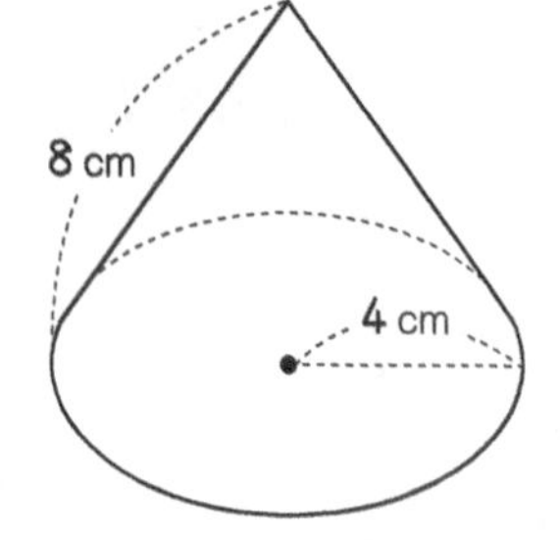

5 **Find the surface area of each solid below.** 8 points per question

(1)

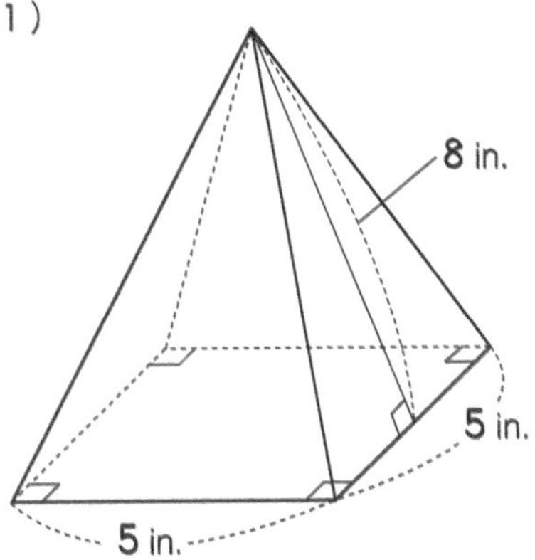

〈Ans.〉

(2)

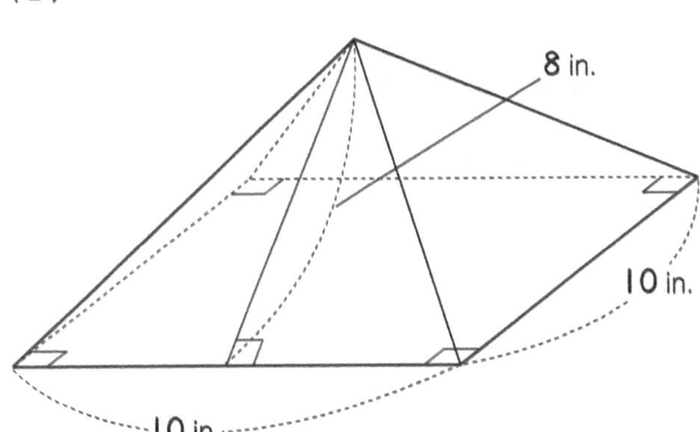

〈Ans.〉

(3)

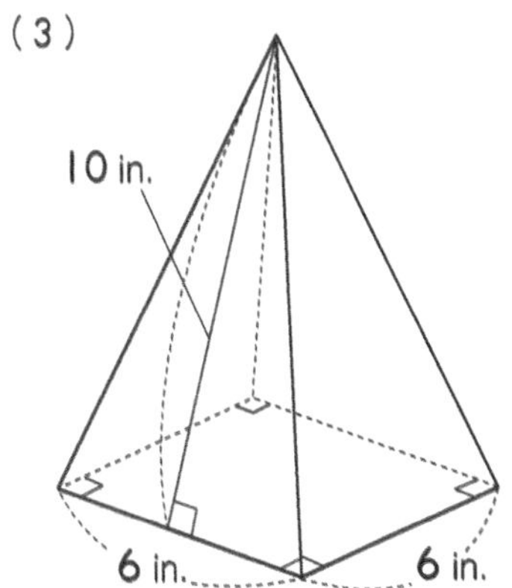

〈Ans.〉

(4)

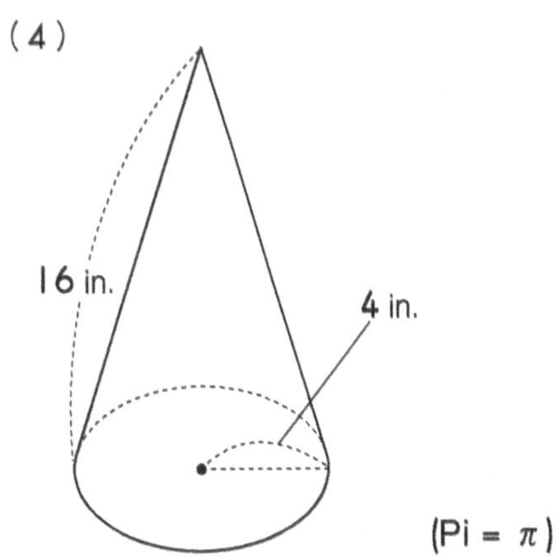

(Pi = π)

〈Ans.〉

(5)

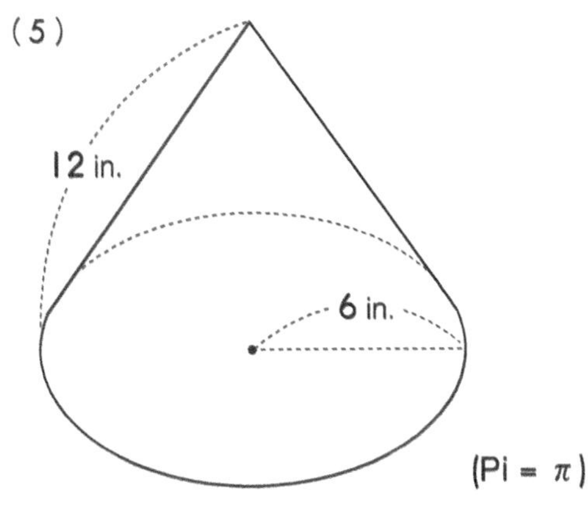

(Pi = π)

〈Ans.〉

(6)

15 in.

5 in.

(Pi = π)

〈Ans.〉

20 Volume & Surface Area (Sphere)

Level ★★

Date / / Name

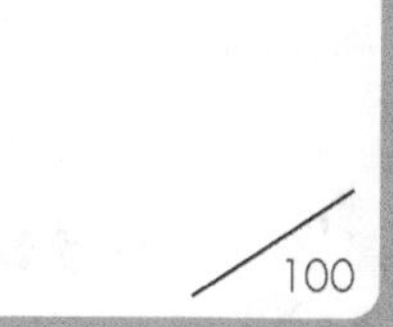

■ The Answer Key is on page 187.

Don't forget!

● The volume (V) of a sphere of radius (r) is calculated using the following formula.

Volume of sphere (V) = $\frac{4}{3}$ × (pi = π) × radius (r)^3 **$V = \frac{4}{3}\pi r^3$**

1 Find the volume of the spheres on the right.

7 points per question

(1) A sphere with a radius of 2 cm.

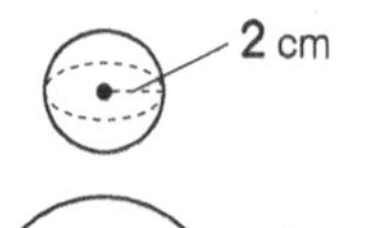

$V = \frac{4}{3} \times \pi \times 2 \times 2 \times 2$

〈Ans.〉 ______________

(2) A sphere with a radius of 4 cm.

4 cm

〈Ans.〉 ______________

(3) A sphere with a radius of 6 cm.

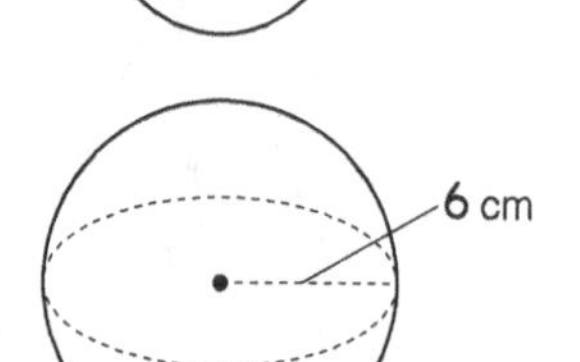

〈Ans.〉 ______________

Don't forget!

● Surface area (S) of a sphere of radius (r) is shown by the following formula.

Surface area of sphere (S) = 4 × (pi = π) × radius (r)^2 **$S = 4\pi r^2$**

2 Find the surface area of the spheres on the right.

7 points per question

(1) A sphere with a radius of 3 cm.

3 cm

$S = 4 \times \pi \times 3 \times 3$

〈Ans.〉 ______________

(2) A sphere with a radius of 6 cm.

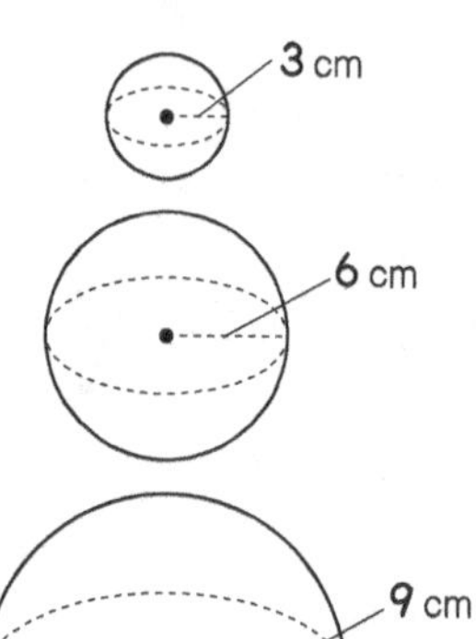

〈Ans.〉 ______________

(3) A sphere with a radius of 9 cm.

9 cm

〈Ans.〉 ______________

3 **The figure on the right is a sphere with a radius of 4 cm cut by a plane passing through the center. Answer the questions below.** 7 points per question

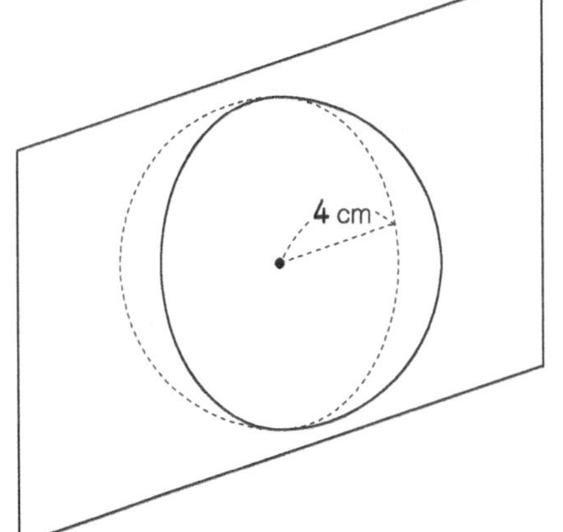

(1) What is the shape of the cross section?

⟨Ans.⟩

(2) Find the volume of the solid.

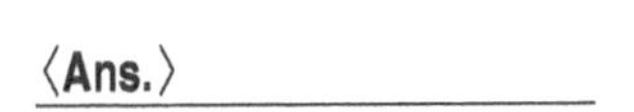

⟨Ans.⟩

(3) Find the surface area of the solid.
(Don't forget the area of the cross section.)

⟨Ans.⟩

4 **There is a semicircle with a diameter of 10 cm on the right. Answer the questions below about a solid which can be formed by one rotation around the diameter.** 7 points per question

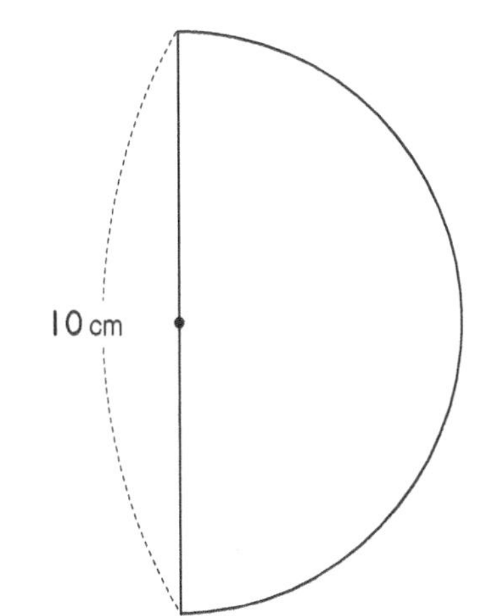

(1) What figure can be made by one rotation?

⟨Ans.⟩

(2) Find the volume of the solid.

⟨Ans.⟩

(3) Find the surface area of the solid.

⟨Ans.⟩

5 **There is a totally contained sphere in a cube with all edges of 2 cm. Answer questions below.** 8 points per question

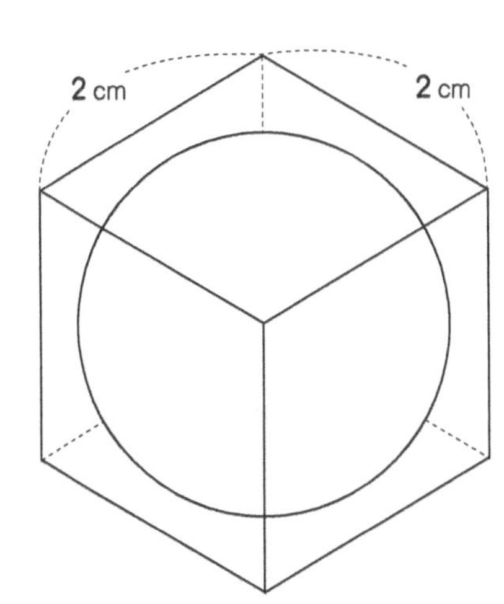

(1) Find the volume of the sphere.

⟨Ans.⟩

(2) How many times larger is the volume of the sphere than the volume of the cube?

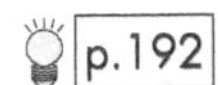
p.192

⟨Ans.⟩

21 Parallel Lines & Angles 1

Level ★★

Date / /

Name

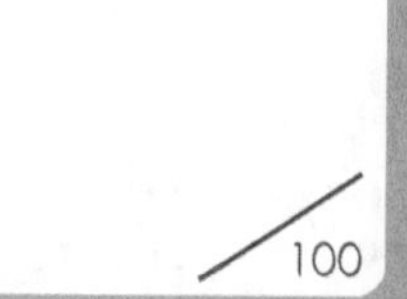

■ The Answer Key is on page 187.

Don't forget!

● The angles created when dividing a line into two are the **adjacent angles**.

$\angle a + \angle b = 180°$

● At the point where two lines intersect, the opposite angles are the vertical angles. **Vertical angles** are equal.

$\angle a = \angle b$

1 Find the measure of ∠*a* and ∠*b* below.

5 points per question

(1)

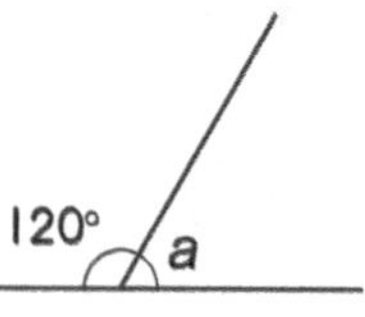

⟨Ans.⟩

(2)

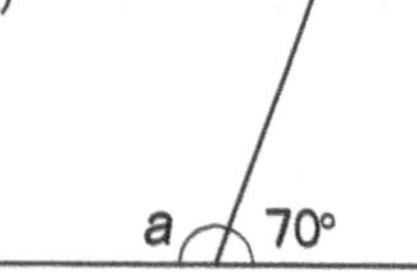

⟨Ans.⟩

(3) 140°, *a*, *b*

⟨Ans.⟩ *a*:

⟨Ans.⟩ *b*:

(4) 60°, *a*, *b*

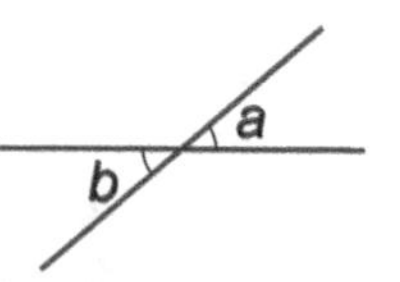

⟨Ans.⟩ *a*:

⟨Ans.⟩ *b*:

2 Answer the questions below using the angles in the figure on the right.

5 points per question

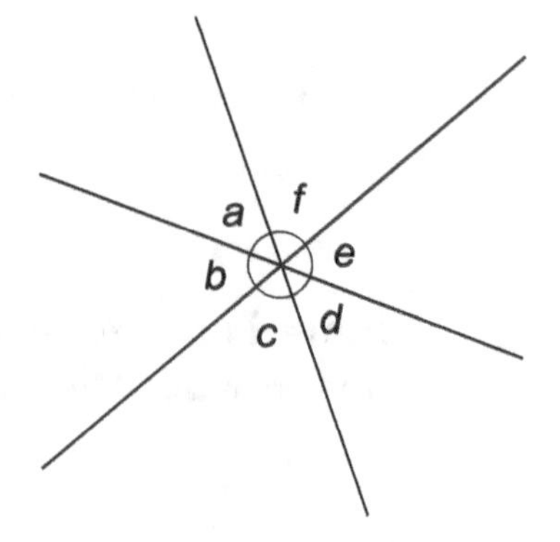

(1) What is the vertical angle of $\angle a$?

⟨Ans.⟩

(2) What is the vertical angle of $\angle b$?

⟨Ans.⟩

(3) $\angle c$ is shown by the following formula. $\angle c = 180° - (\angle a + \angle b)$
In the same way, show $\angle c$ with $\angle b$ and $\angle d$.

⟨Ans.⟩

3 Find the measures of the angles in the figure on the right.

5 points per question

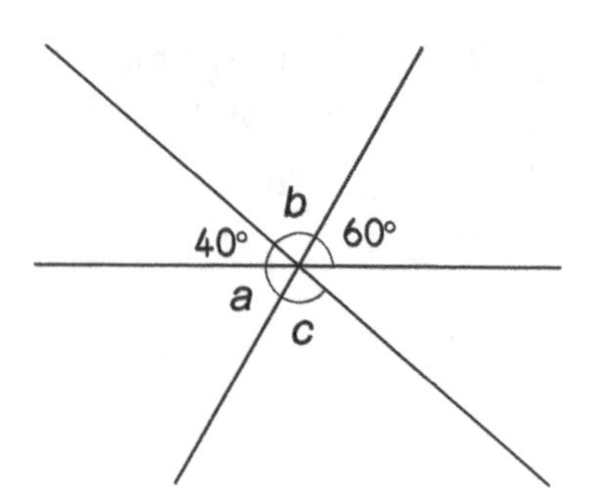

(1) What is the measure of $\angle a$?

⟨Ans.⟩

(2) What is the measure of $\angle b$?

⟨Ans.⟩

(3) What is the measure of $\angle c$?

⟨Ans.⟩

Don't forget!

●A line passing through two lines is the transversal. When the transversal intersects two lines, angles at the same position as shown on the right are the **corresponding angles**.

(Example) ∠Ⓐ and ∠Ⓑ, ∠Ⓒ and ∠Ⓓ

●In the same way, when the transversal intersects two lines, angles at the same position as shown on the right are the **alternate angles**. As shown in the figure, the inner angles are the **alternate interior angles**.

(Example) $\angle c$ and $\angle e$, $\angle b$ and $\angle h$

The outer angles are the **alternate exterior angles**.

(Example) $\angle a$ and $\angle g$, $\angle d$ and $\angle f$

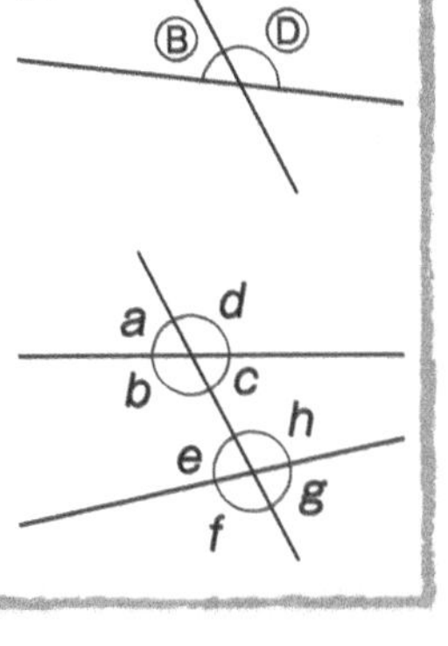

4 **Answer the questions below about the angles in the figure on the right.** 5 points per question

(1) What is the corresponding angle of $\angle a$?

〈Ans.〉 ______

(2) What is the corresponding angle of $\angle b$?

〈Ans.〉 ______

(3) What is the alternate exterior angle of $\angle a$?

〈Ans.〉 ______

(4) What is the alternate interior angle of $\angle b$?

〈Ans.〉 ______

5 **Answer the questions below using the angles in the figure on the right and find the measure of each angle.** 5 points per question

(1) What is the corresponding angle of $\angle c$?

〈Ans.〉 symbol ______, angle ______

(2) What is the alternate interior angle of $\angle c$?

〈Ans.〉 symbol ______, angle ______

(3) What is the corresponding angle of $\angle a$?

〈Ans.〉 symbol ______, angle ______

(4) What is the alternate exterior angle of $\angle a$?

〈Ans.〉 symbol ______, angle ______

(5) What is the corresponding angle of $\angle b$?

〈Ans.〉 symbol ______, angle ______

(6) What is the vertical angle of $\angle e$?

〈Ans.〉 symbol ______, angle ______

22 Parallel Lines & Angles 2

Level ★★

Date / / Name

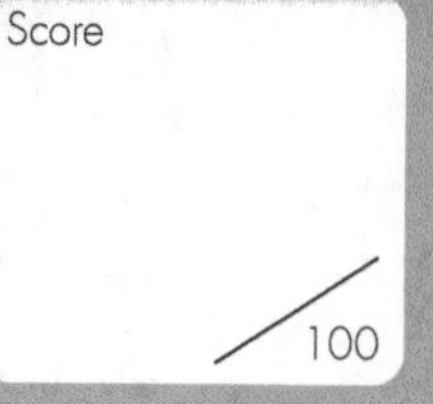

■ The Answer Key is on page 187.

1 **The transversal *n* intersects with the parallel lines ℓ and *m*. Answer the questions below using the angles in the figure on the right.** 6 points per question

(1) What is the relationship between ∠*b* and ∠*f*?

〈Ans.〉

(2) What is the relationship between ∠*b* and ∠*h*?

〈Ans.〉

(3) Lines ℓ and *m* are parallel (ℓ ∥ *m*). Choose a letter from (*e*, *f*, *g*, and *h*) and fill in the blanks.

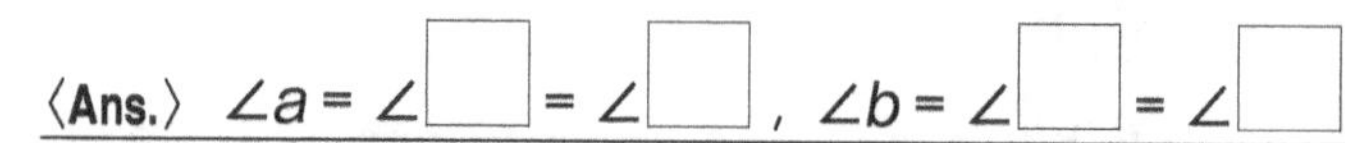

〈Ans.〉 ∠*a* = ∠☐ = ∠☐ , ∠*b* = ∠☐ = ∠☐

(4) When ∠*d* = 50°, find the measure of ∠*h*.

〈Ans.〉

Don't forget!

- When the transversal intersects with two parallel lines, the corresponding angles are equal. (If ℓ ∥ *m*, ∠*a* = ∠*b*)
- In the same way, when the corresponding angles are equal, the two lines are parallel. (If ∠*a* = ∠*b*, ℓ ∥ *m*)

2 **Lines ℓ and *m* are parallel. When ∠*a* is 150°, find the measure of ∠*b*, ∠*c*, and ∠*d*.** 6 points per question

(1) What is the measure of ∠*b*?

〈Ans.〉

(2) What is the measure of ∠*c*?

〈Ans.〉

(3) What is the measure of ∠*d*?

〈Ans.〉

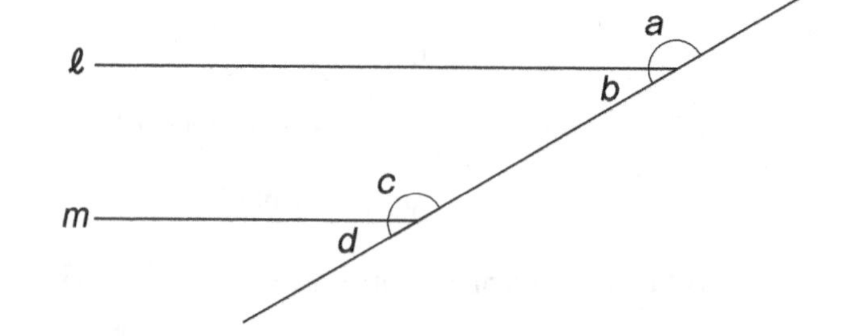

3 **Answer the questions below using the figure on the right.** 6 points per question

(1) Choose which two sets of lines are parallel. Show your answer with the correct symbol.

〈Ans.〉 ,

(2) From ∠*a*, ∠*b*, ∠*c*, and ∠*d*, choose two pairs of angles that are equal angles.

〈Ans.〉 = , =

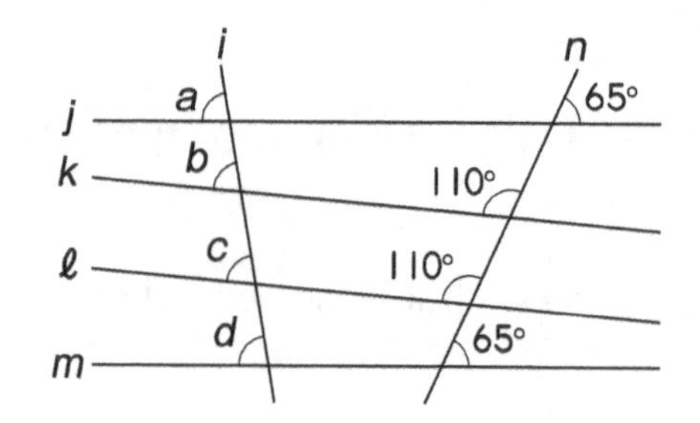

4 The transversal *n* intersects with the parallel lines ℓ and *m*. Answer the questions below about the angles in the figure on the right.

6 points per question

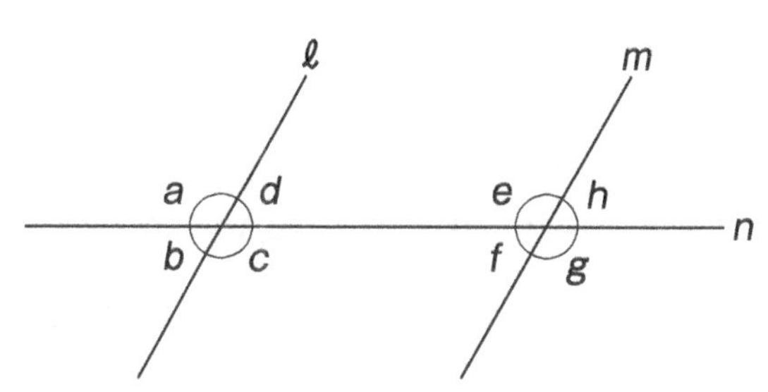

(1) What type of angles are $\angle c$ and $\angle g$?

〈Ans.〉

(2) What type of angles are $\angle b$ and $\angle h$?

〈Ans.〉

(3) Lines ℓ and m are parallel ($\ell \parallel m$). Choose a letter from (e, f, g, and h) and fill in the blanks.

〈Ans.〉 $\angle a = \angle \square = \angle \square$, $\angle b = \angle \square = \angle \square$

(4) When $\angle f = 60°$, find the measure of $\angle d$.

〈Ans.〉

Don't forget!

- When the transversal intersects with two parallel lines, the alternate angles are equal. (If $\ell \parallel m$, $\angle a = \angle b$ $\angle c = \angle d$)
- In the same way, when the alternate angles are equal, the two lines are parallel. (If $\angle a = \angle b$ $\angle c = \angle d$, $\ell \parallel m$)

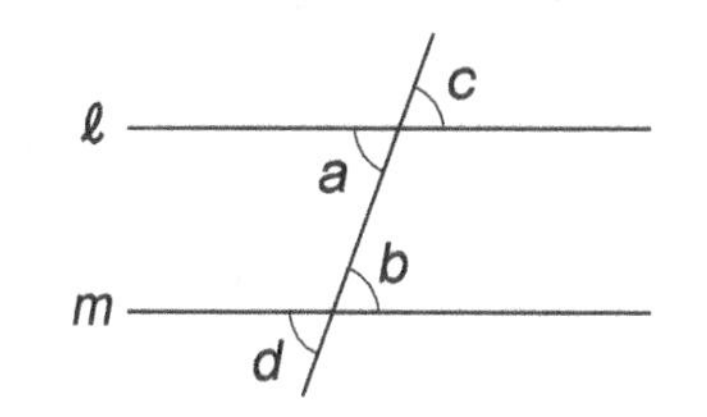

5 Lines *j* and *k* are parallel. When ∠*f* is 55°, find the measure of ∠*e* and ∠*g*.

5 points per question

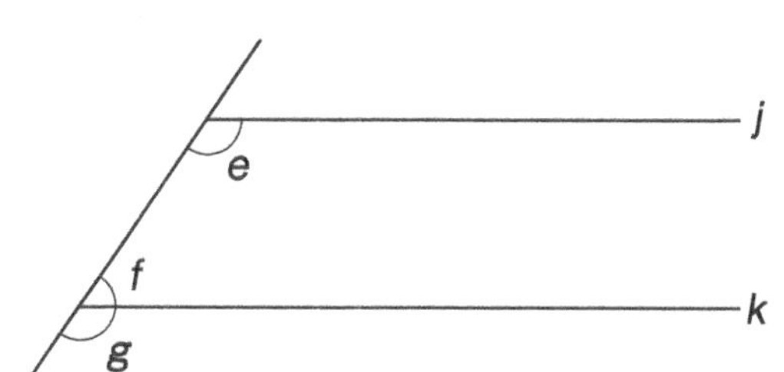

(1) What is the measure of $\angle g$?

〈Ans.〉

(2) What is the measure of $\angle e$?

〈Ans.〉

6 Answer the questions below using the figure on the right.

6 points per question

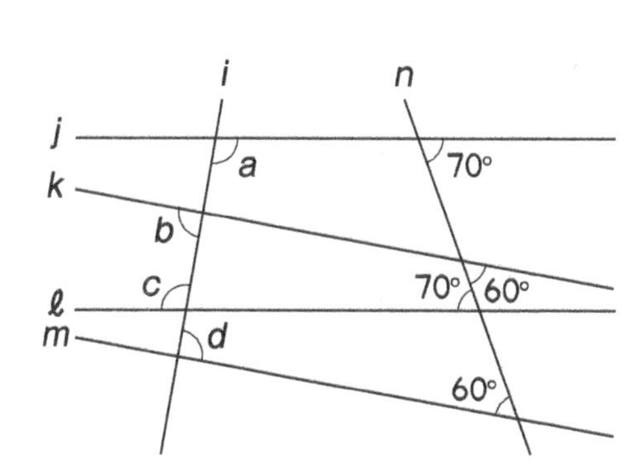

(1) Choose which two sets of lines are parallel. Show your answer with the correct symbol.

〈Ans.〉 ,

(2) From $\angle a$, $\angle b$, $\angle c$, and $\angle d$, choose two pairs of angles are equal angles.

〈Ans.〉 = , =

Parallel Lines & Angles 3

Level ★ 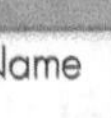 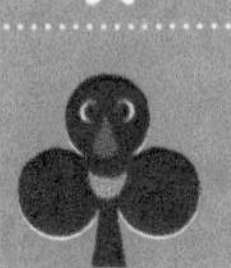

Date / / Name

■ The Answer Key is on page 187.

1 **In the figure on the right, two pairs of parallel lines intersect. Answer the questions below using the angles in the figure.** 4 points per question

(1) Which angles are the corresponding angles of $\angle a$?

〈Ans.〉 ,

(2) Which angles are the corresponding angles of $\angle c$?

〈Ans.〉 ,

(3) Which angles are the alternate angles of $\angle h$?

〈Ans.〉 ,

(4) Which angles are the alternate angles of $\angle t$?

〈Ans.〉 ,

(5) When $\angle a = 105°$, find the measure of the angles below.

〈Ans.〉 $\angle d =$, $\angle g =$, $\angle q =$, $\angle v =$

2 **In the figures below $\ell \parallel m$, find the measure of $\angle x$ and $\angle y$.** 5 points per question

(1)

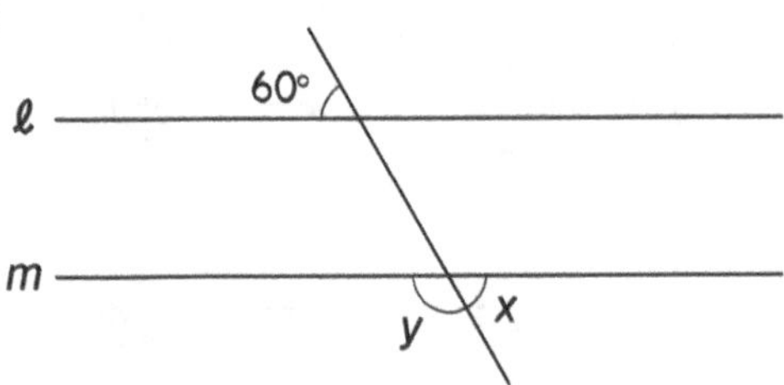

〈Ans.〉 $\angle x =$, $\angle y =$

(2)

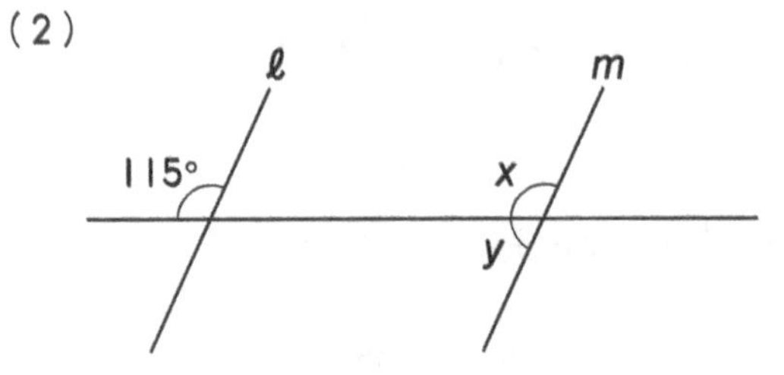

〈Ans.〉 $\angle x =$, $\angle y =$

(3)

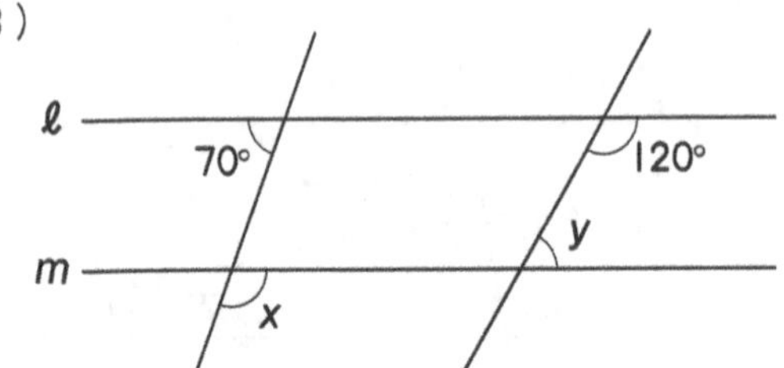

〈Ans.〉 $\angle x =$, $\angle y =$

(4)

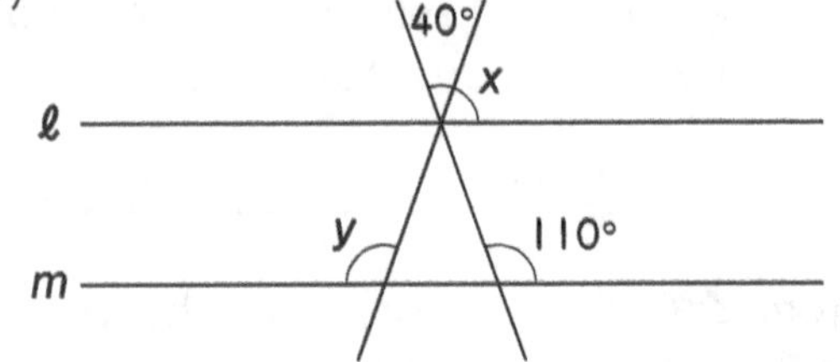

〈Ans.〉 $\angle x =$, $\angle y =$

3 In the figures below $\ell \parallel m$, find the measure of $\angle x$.

5 points per question

(1)

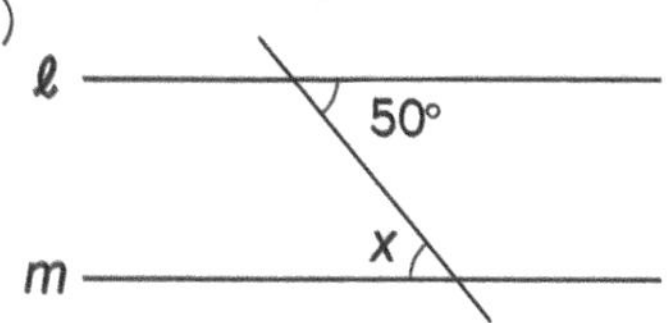

⟨Ans.⟩ ______

(2)

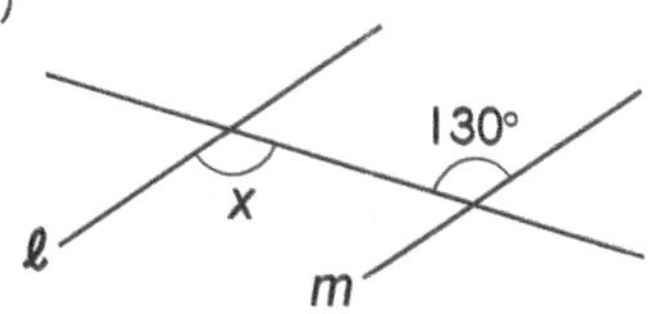
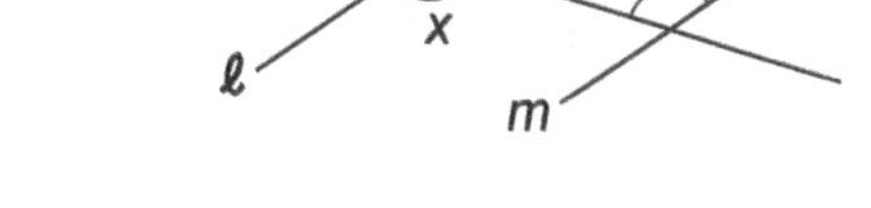

⟨Ans.⟩ ______

(3)

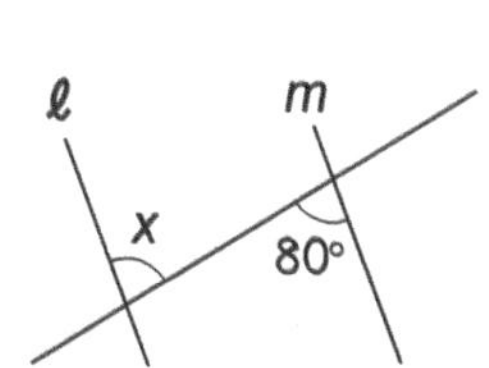

⟨Ans.⟩ ______

(4)

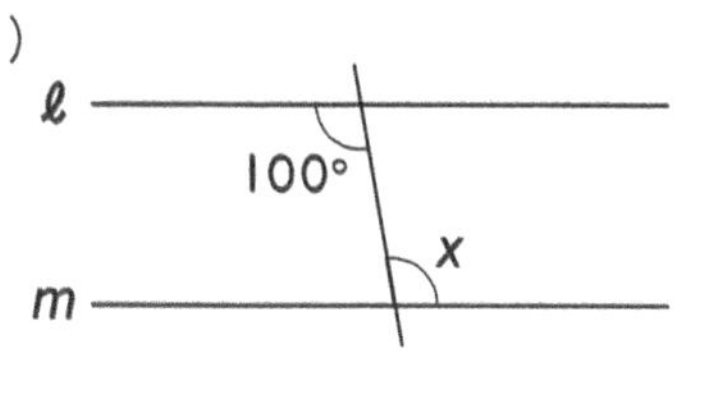

⟨Ans.⟩ ______

(5)

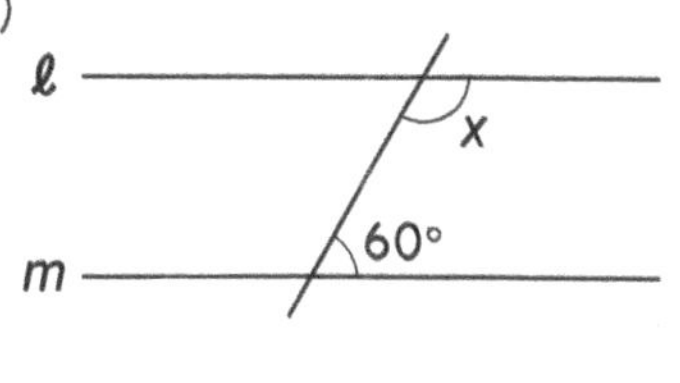

⟨Ans.⟩ ______

(6)

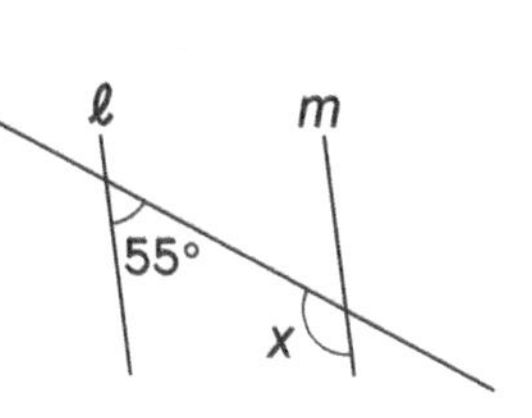

⟨Ans.⟩ ______

4 In the figures below $\ell \parallel m \parallel n$, find the measure of $\angle x$ and $\angle y$.

5 points per question

(1)

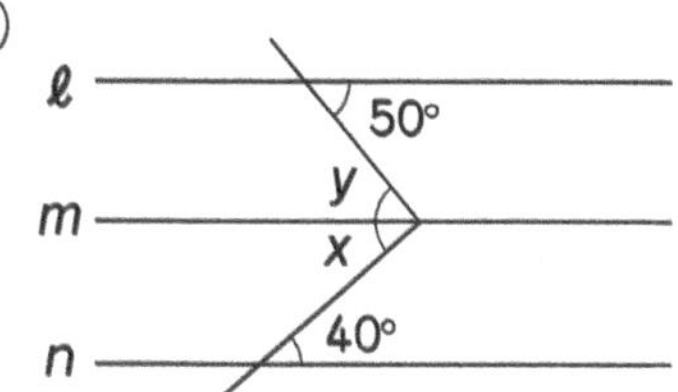

⟨Ans.⟩ $\angle x =$ ______ , $\angle y =$ ______

(2)

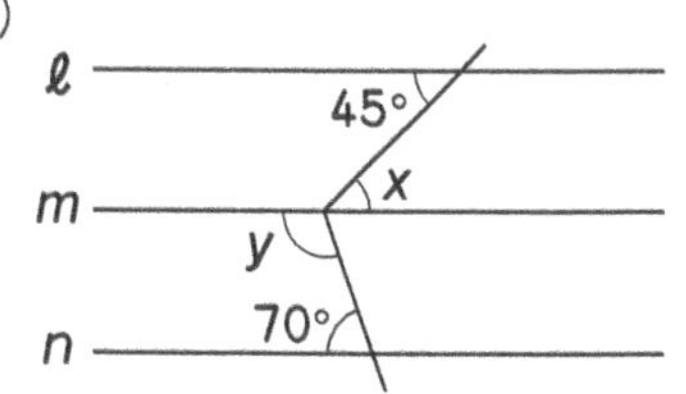

⟨Ans.⟩ $\angle x =$ ______ , $\angle y =$ ______

(3)

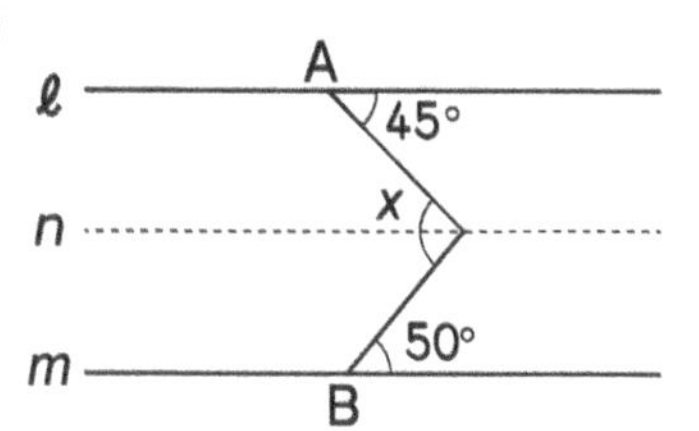

⟨Ans.⟩ $\angle x =$ ______

(4)

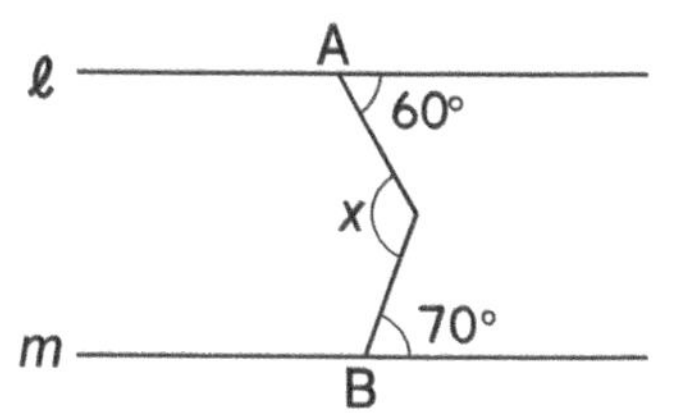

⟨Ans.⟩ $\angle x =$ ______

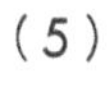

(5)

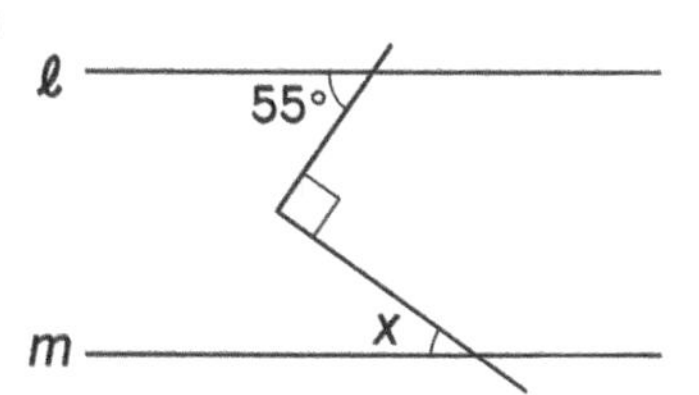

⟨Ans.⟩ $\angle x =$ ______

(6)

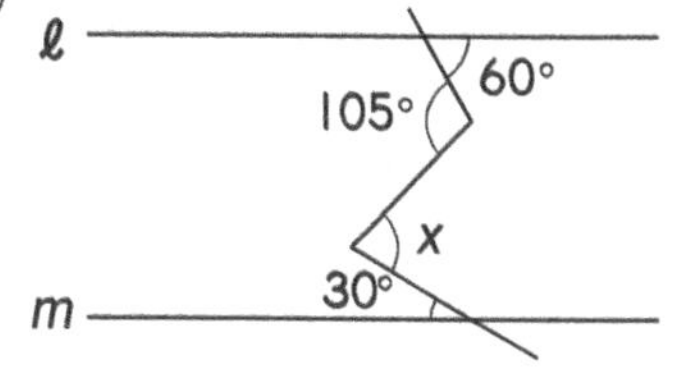

⟨Ans.⟩ $\angle x =$ ______

24 Parallel Lines & Angles 4

Level 

Date / / Name

Score /100

■ The Answer Key is on page 188.

1 **In the figures below, mark ○ if line ℓ and line *m* are parallel and mark × if they are not parallel.** 4 points per question

(1)

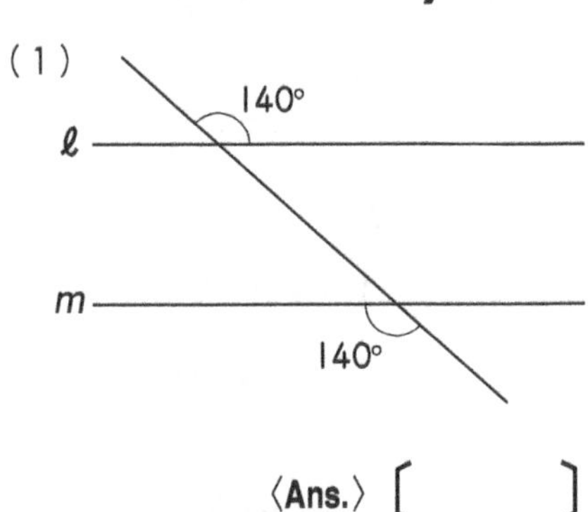

〈Ans.〉 [ ]

(2)

ℓ
110°
70°
m

〈Ans.〉 [ ]

(3)

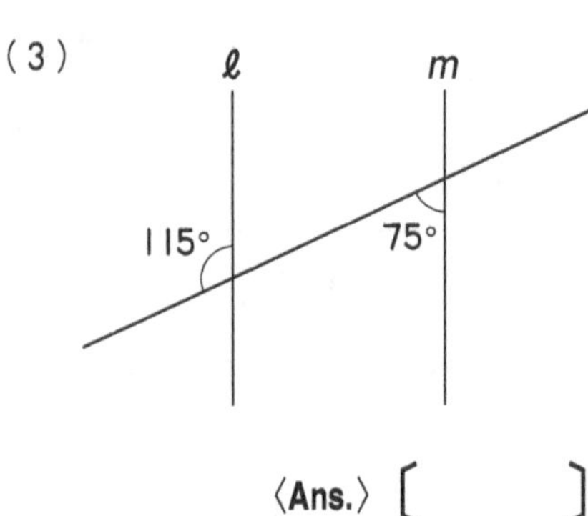

〈Ans.〉 [ ]

2 **In the figures below *a* ∥ *b* and ℓ ∥ *m* , what is the measure of ∠*x* and ∠*y*?** 8 points per question

(1)

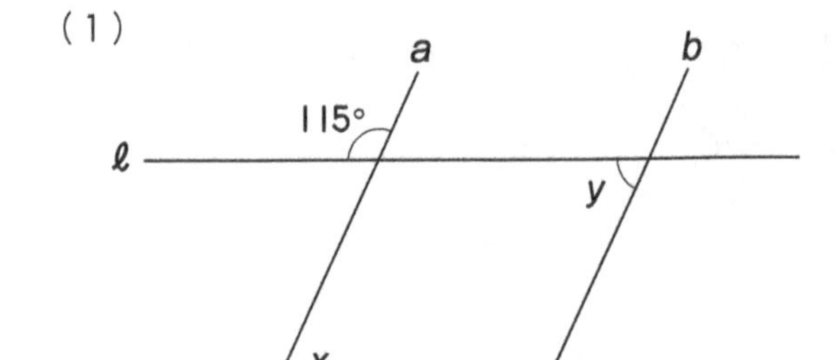

〈Ans.〉 ∠*x* = , ∠*y* =

(2)

a
b
ℓ
y
x
135°
m

〈Ans.〉 ∠*x* = , ∠*y* =

(3)

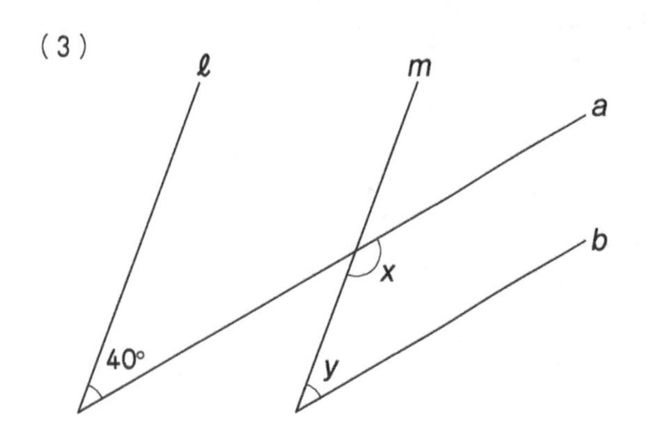

〈Ans.〉 ∠*x* = , ∠*y* =

(4)

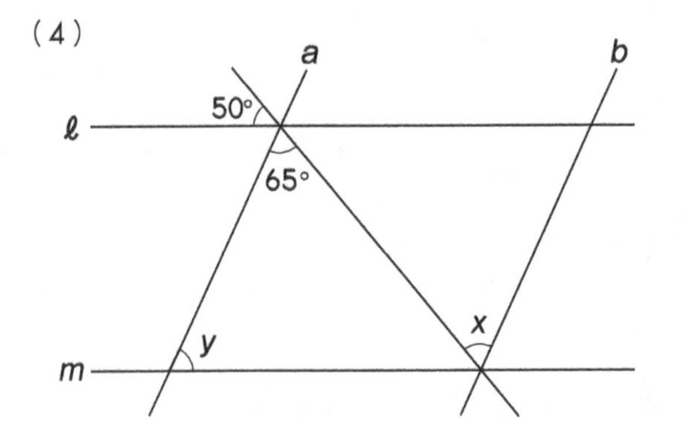

〈Ans.〉 ∠*x* = , ∠*y* =

3 In the figure on the right, line ℓ and line *m* are parallel. Answer the questions below.

8 points per question

(1) Show the relationship between $\angle a$ and $\angle b$ with an equation.

〈Ans.〉

(2) Show the relationship between $\angle a$ and $\angle c$ with an equation.

〈Ans.〉

(3) When $\angle a = 55°$, find the measure of $\angle c$.

〈Ans.〉

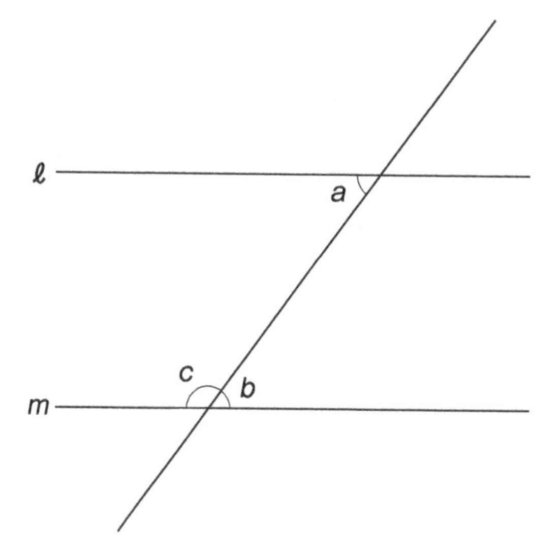

Don't forget!

●In the figure on the right, if line ℓ and line *m* are parallel (ℓ ∥ *m*),

$$\angle a + \angle b = 180°$$

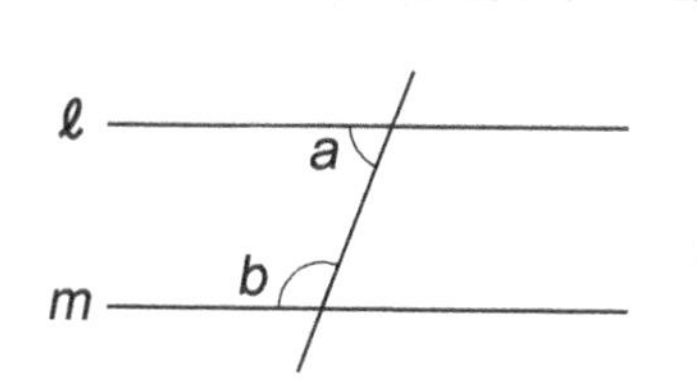

4 In the figures below ℓ ∥ *m*, what is the measure of $\angle x$ and $\angle y$?

8 points per question

(1)

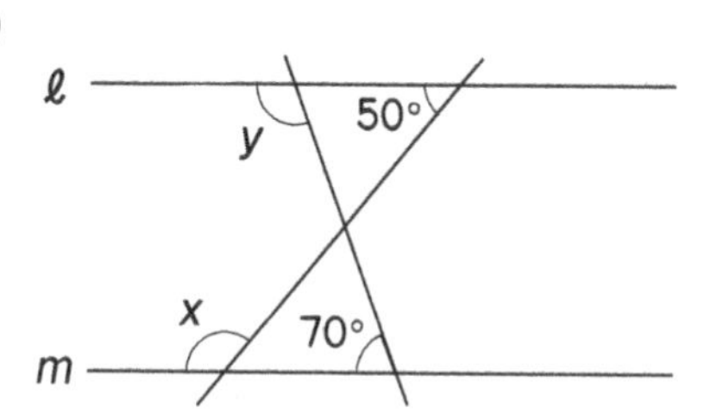

〈Ans.〉 $\angle x =$, $\angle y =$

(2)

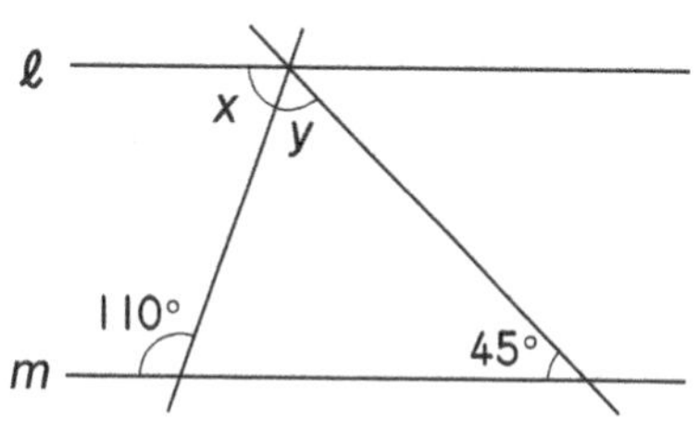

〈Ans.〉 $\angle x =$, $\angle y =$

(3)

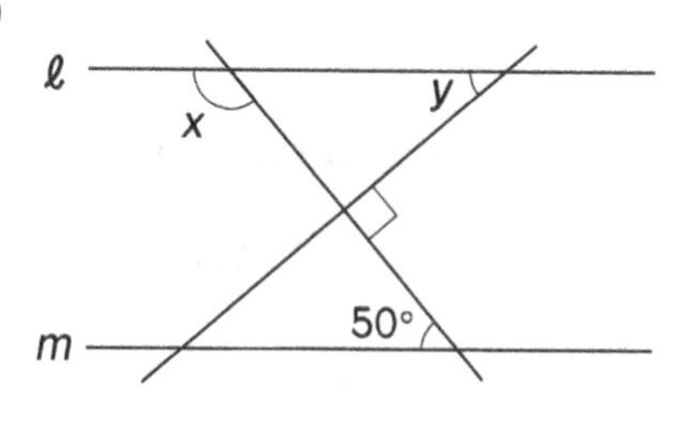

〈Ans.〉 $\angle x =$, $\angle y =$

(4)

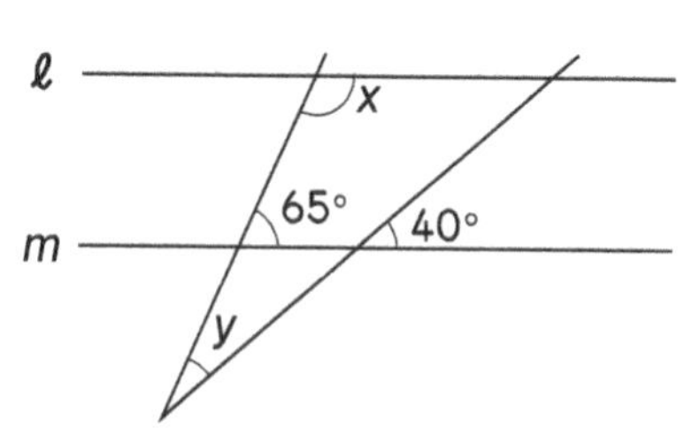

〈Ans.〉 $\angle x =$, $\angle y =$

25 Opposite Angle & Opposite Side

Level ★

Date / / Name

Score /100

■ The Answer Key is on page 188.

Don't forget!

● A triangle has three sides and three angles. There is an angle between each of the two sides. The side opposite of the angle is called the **opposite side**. Alternatively, the angle opposite of the side is called the **opposite angle**.

opposite side

opposite angle

1 **Answer the questions below about the opposite sides and the opposite angles of △ABC on the right.** 6 points per question

(1) What are the opposite sides of ∠A, ∠B, and ∠C?

〈Ans.〉 ∠A ⇨ , ∠B ⇨ , ∠C ⇨

(2) What are the opposite angles of sides AB, AC, and BC?

〈Ans.〉 AB ⇨ , AC ⇨ , BC ⇨

A

B C

2 **Answer the questions below about the opposite sides and the opposite angles of △PQR on the right.** 5 points per question

(1) What are the opposite sides of ∠P, ∠Q, and ∠R?

〈Ans.〉 ∠P ⇨ , ∠Q ⇨ , ∠R ⇨

(2) What are the opposite angles of sides PQ, QR, and PR?

〈Ans.〉 PQ ⇨ , QR ⇨ , PR ⇨

P

R

Q

3 **Answer the questions below about the opposite sides and the opposite angles of △DEF.** 6 points per question

(1) What is the opposite side of ∠D?

〈Ans.〉

(2) What is the opposite side of ∠E?

〈Ans.〉

(3) What is the opposite side of ∠F?

〈Ans.〉

(4) What is the opposite angle of side DF?

〈Ans.〉

(5) What is the opposite angle of side DE?

〈Ans.〉

(6) What is the opposite angle of side EF?

〈Ans.〉

Don't forget!

① A right triangle has one angle of 90°.
② An acute triangle has three angles less than 90°.
③ An obtuse triangle has one angle greater than 90°.

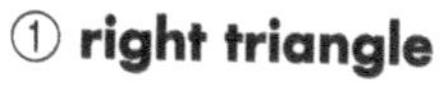

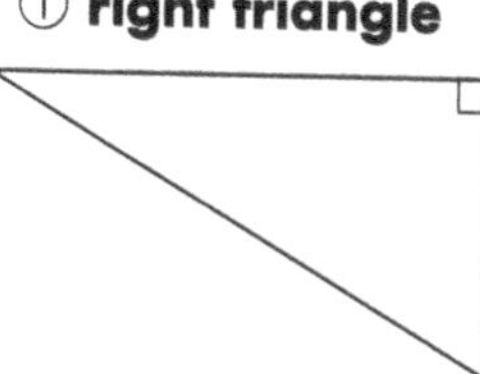

② acute triangle

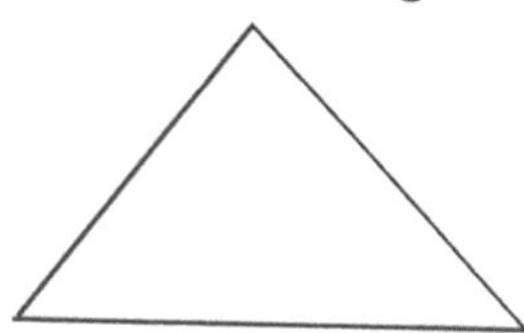

③ obtuse triangle

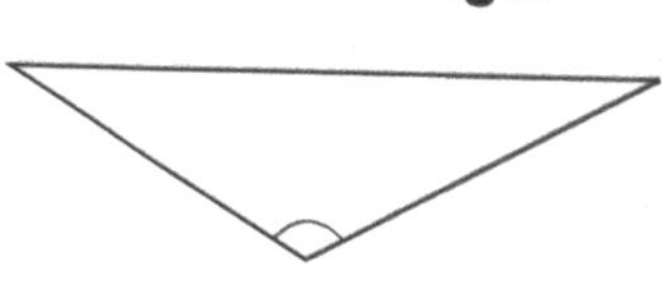

4 Classify the triangles below as ① right triangle, ② acute triangle, or ③ obtuse triangle.

14 points for completion

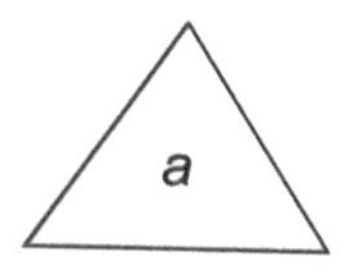

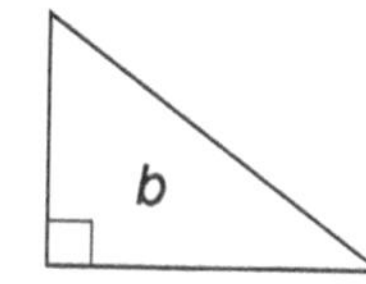

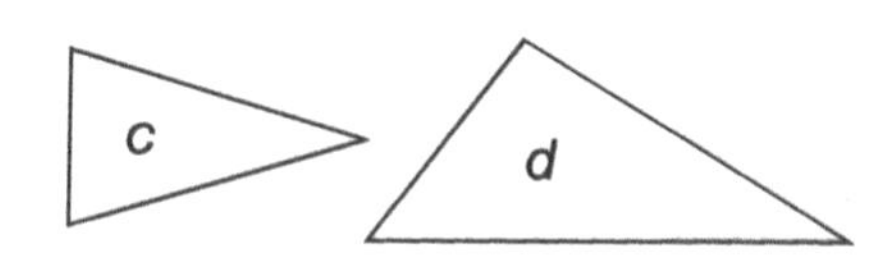

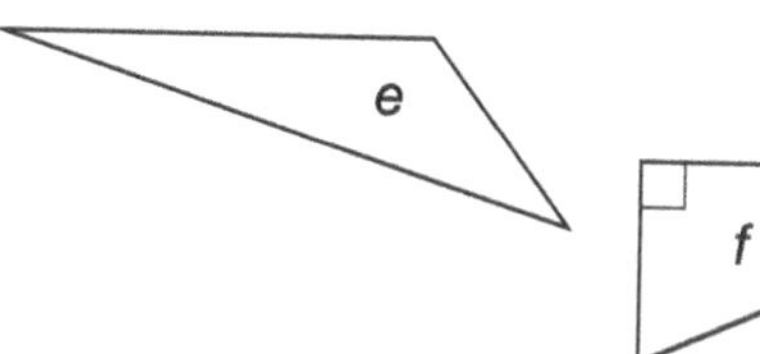

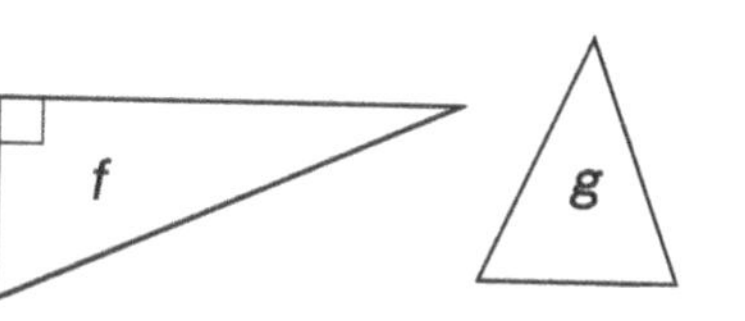

h

〈**Ans.**〉 right triangle: , ,

〈**Ans.**〉 acute triangle: , ,

〈**Ans.**〉 obtuse triangle: ,

5 Number the sides and the angles of the following triangles in descending size order.

14 points per question

(1)

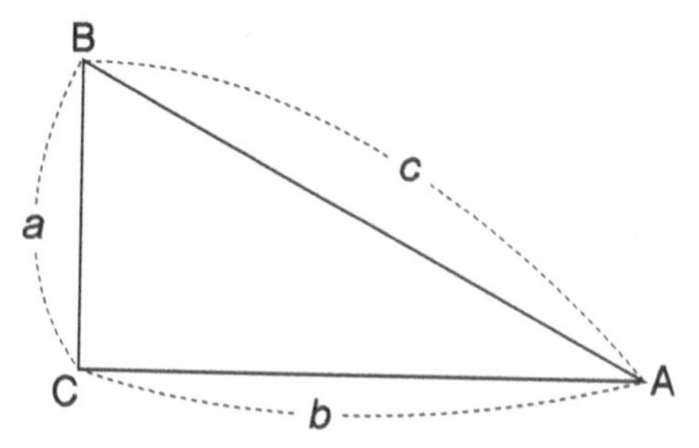

〈**Ans.**〉 side: *a* [ ], *b* [ ], *c* [ ]

angle: A [ ], B [ ], C [ ]

(2)

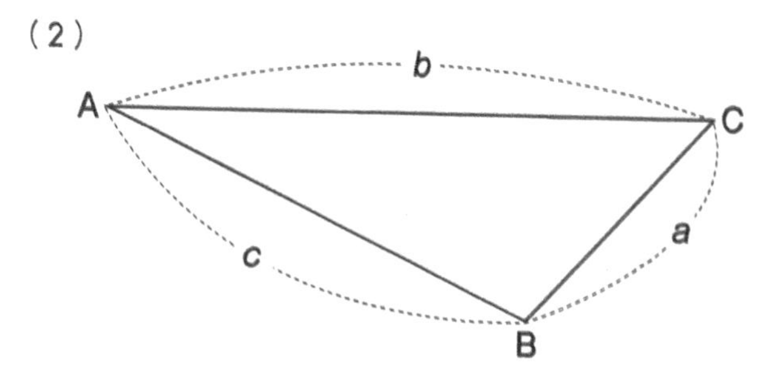

〈**Ans.**〉 side: *a* [ ], *b* [ ], *c* [ ]

angle: A [ ], B [ ], C [ ]

Don't forget!

● In a triangle, the longest side is opposite to the largest angle and the smallest angle is opposite to the shortest sides.

26 Interior & Exterior Angles of a Triangle 1

Level ★★

Date / /   Name

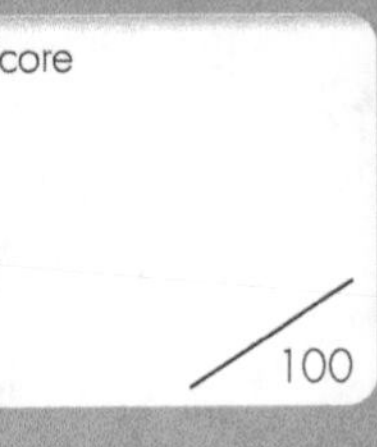

■ The Answer Key is on page 188.

1 **Fill in the blanks below to complete the explanation of why, "the sum of three interior angles is 180°".** 24 points for completion

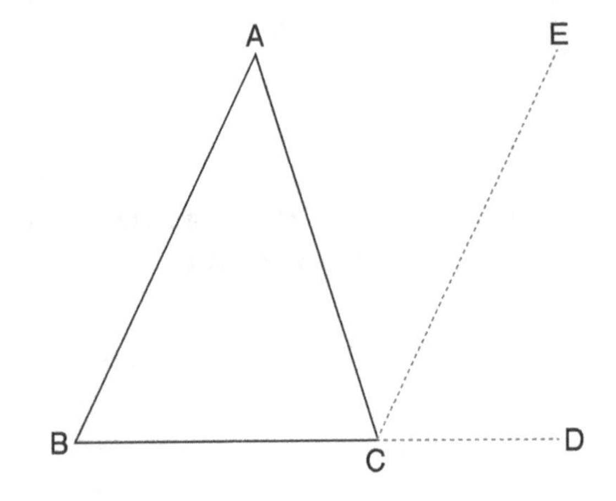

Draw an extension line of side BC of $\triangle ABC$ to make line CD. Then, draw the line CE from point C to be parallel to side AB. Since BA ∥ CE, the alternate (interior) angles are equal.

$\angle A = \angle$ ☐ .

Since BA ∥ CE, the corresponding angles are equal.

$\angle B = \angle$ ☐ .

Therefore,

$\angle A + \angle B + \angle ACB = \angle ACE + \angle$ ☐ $+ \angle ACB = \angle BCD$

Since three points B, C, and D are on the same straight line, $\angle$ ☐ $= 180°$.
Therefore, the sum of interior angles of a triangle is 180°.

Don't forget!

● When one side at one vertex of a triangle is extended, the angle formed by the adjacent side is called an **exterior angle**.

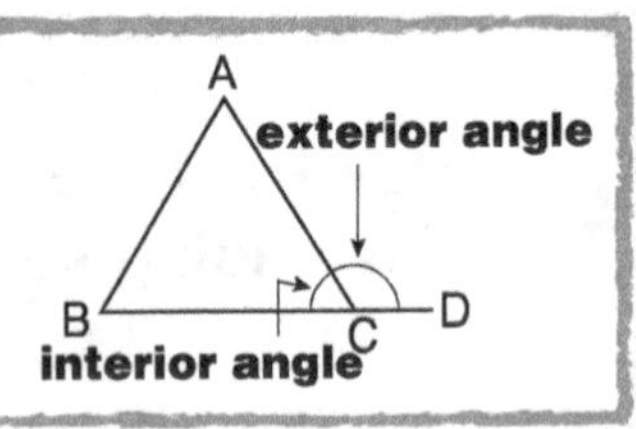

2 **Find the measure of the following angles in $\triangle ABC$ in the right figure.** 7 points per question

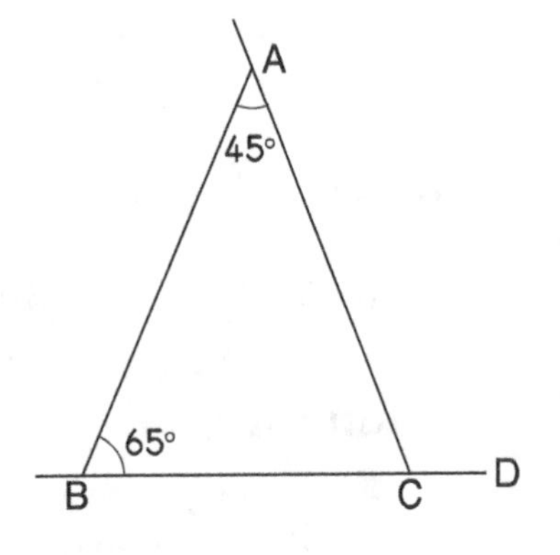

(1) What is the measure of $\angle ACB$?

〈Ans.〉

(2) What is the measure of $\angle ACD$?

〈Ans.〉

(3) What is the measure of the exterior angle of vertex A in $\triangle ABC$?

〈Ans.〉

(4) What is the measure of the exterior angle of vertex B in $\triangle ABC$?

〈Ans.〉

3 **In the figure of $\triangle ABC$ on the right, $\angle ACD = \angle A + \angle B$ is explained as follows. Fill in the blanks.** 12 points for completion

Since the sum of the interior angle is 180°,

$\angle A + \angle B + \angle$ [ ] $= 180°$ .... (1)

Since three points B, C, and D are on the same straight line,

$\angle$ [ ] $+ \angle ACB = 180°$ .... (2)

From (1) and (2),

$\angle$ [ ] $= \angle A + \angle B$

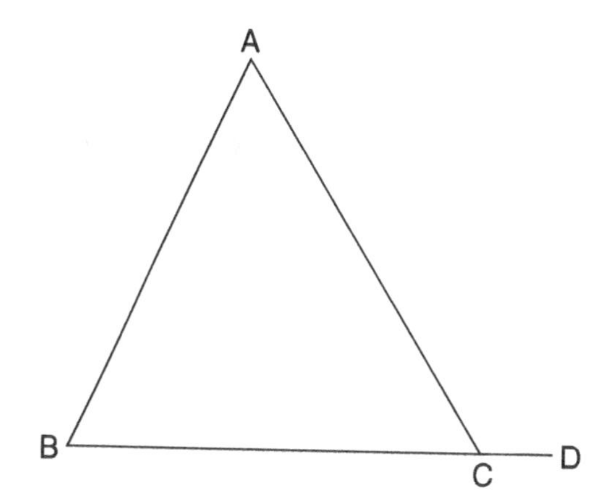

Don't forget!

● The measure of an exterior angle in a triangle is equal to the sum of the two interior angles, not next to the exterior angle.

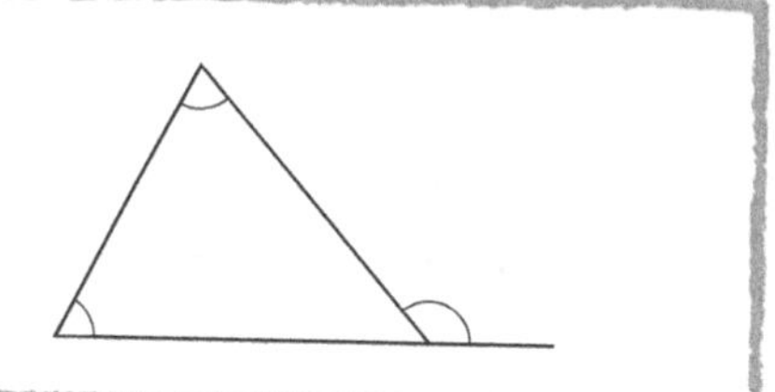

4 **Find the measure of $\angle x$ in the following figures.** 6 points per question

(1)

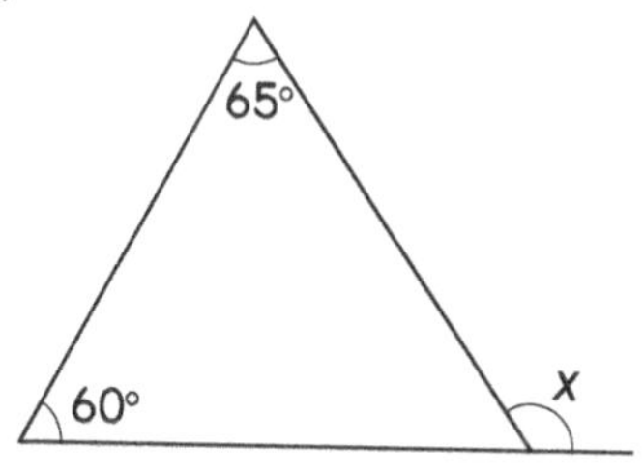

⟨Ans.⟩ $\angle x =$

(2)

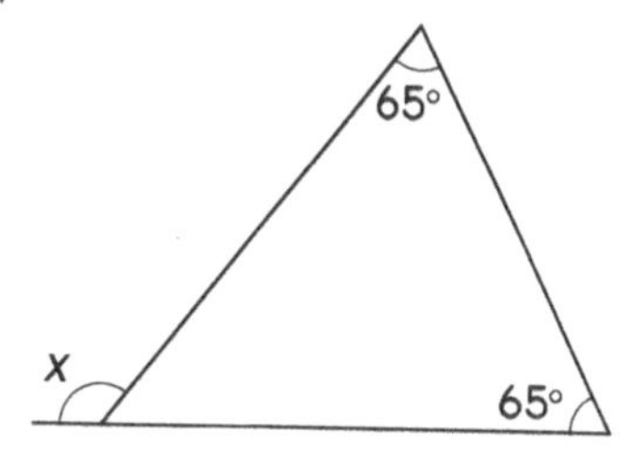

⟨Ans.⟩ $\angle x =$

(3)

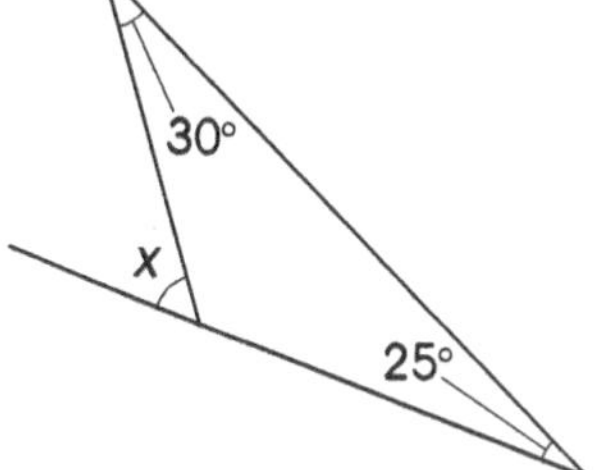

⟨Ans.⟩ $\angle x =$

(4)

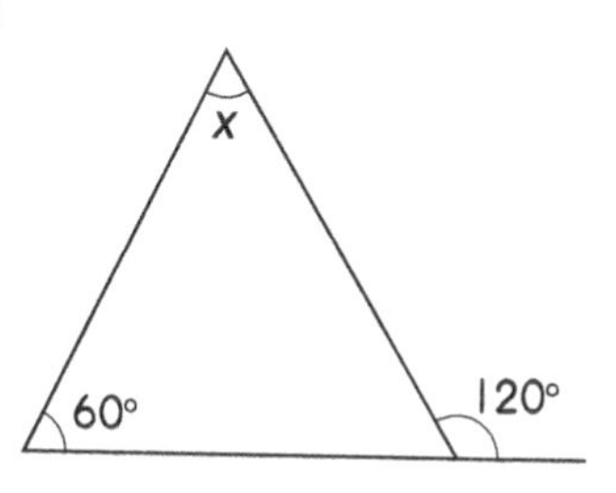

⟨Ans.⟩ $\angle x =$

(5)

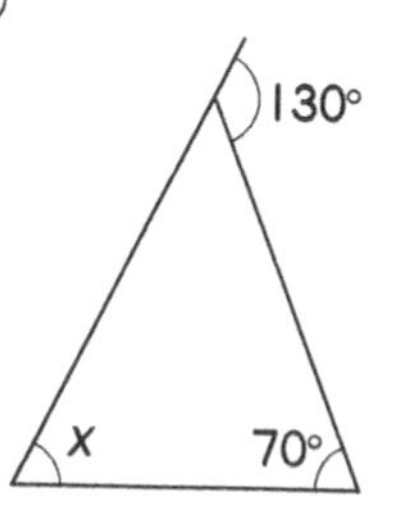

⟨Ans.⟩ $\angle x =$

(6)

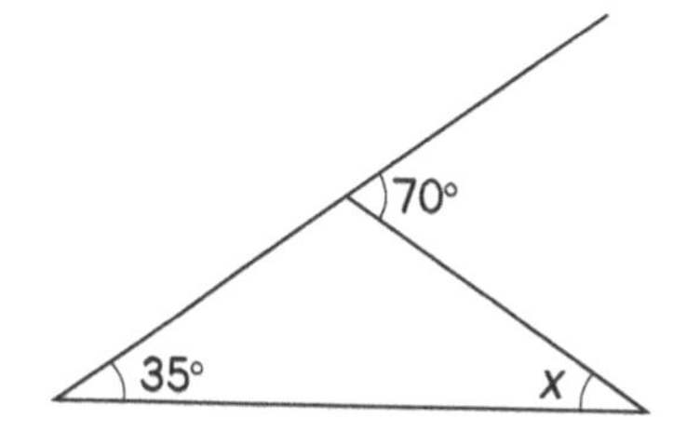

⟨Ans.⟩ $\angle x =$

27 Interior & Exterior Angles of a Triangle 2

Date / / Name

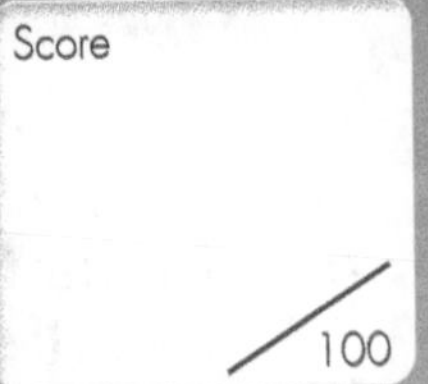

■ The Answer Key is on page 188.

1 **The following figures are isosceles triangles with AB = AC. Find the measure of ∠x.** 6 points per question

(1)

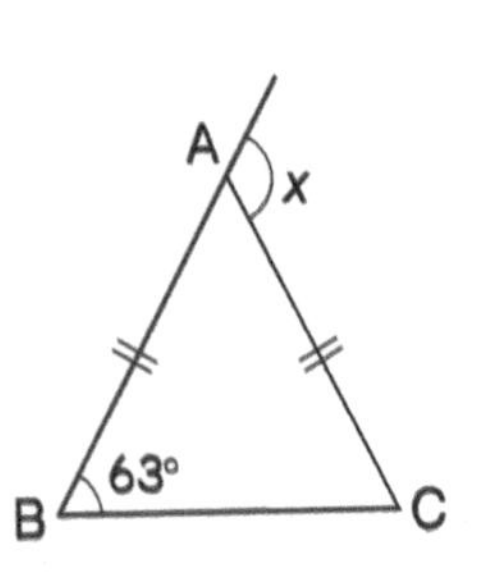

〈Ans.〉 $\angle x =$

(2)

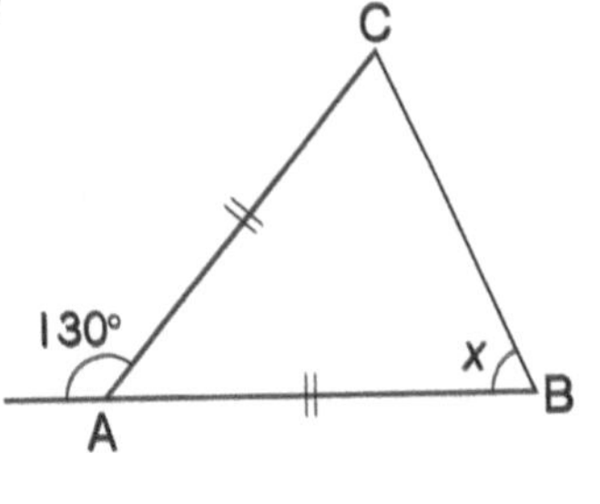

〈Ans.〉 $\angle x =$

(3)

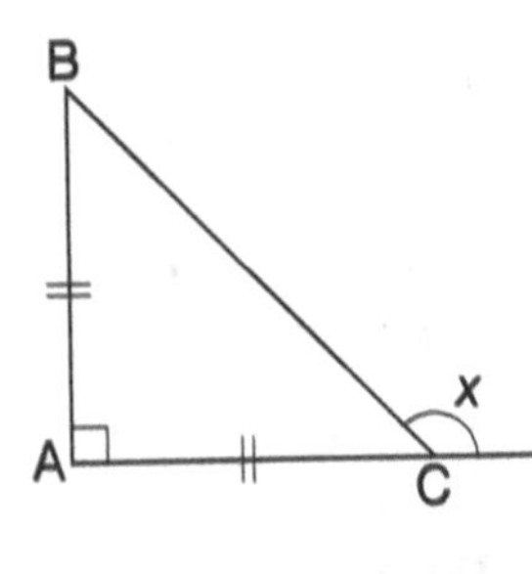

〈Ans.〉 $\angle x =$

2 **∠A + ∠B = ∠C + ∠D in the figure on the right, is explained as follows. Fill in the blanks.** 24 points for completion

Since the measure of an exterior angle in a triangle is equal to the sum of the two interior angles, not next to the exterior angle, in △AOB,

∠AOC = ∠A + ∠☐ .... (1)

Likewise, in △COD,

∠AOC = ∠C + ∠☐ .... (2)

From (1) and (2),

∠A + ∠B = ∠☐ + ∠☐

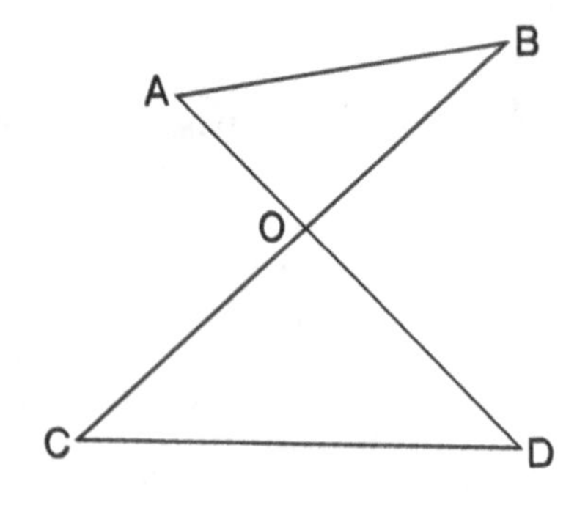

3 **Find the measure of ∠x and ∠y in the following figures.** 6 points per question

(1)

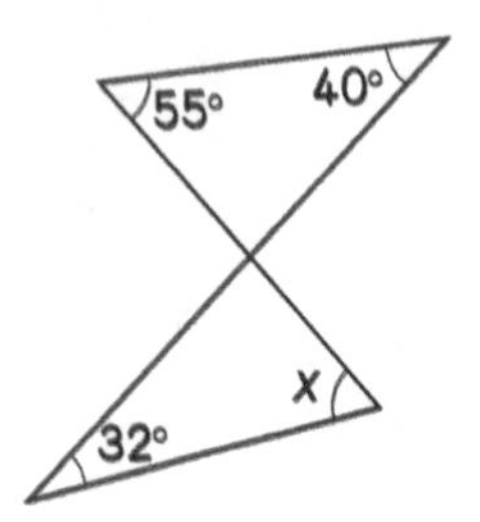

〈Ans.〉 $\angle x =$

(2)

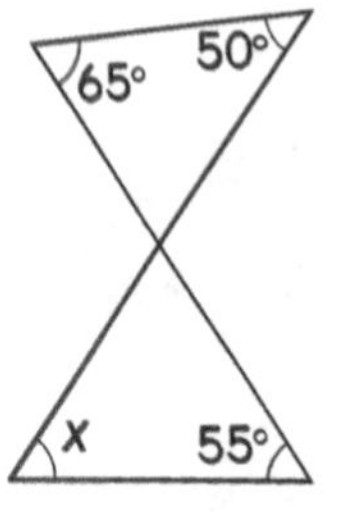

〈Ans.〉 $\angle x =$

(3)

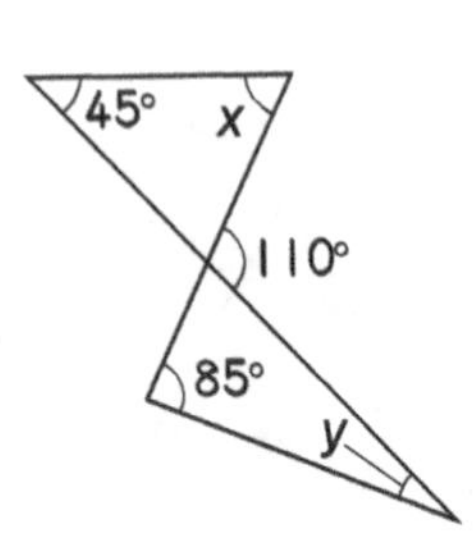

〈Ans.〉 $\angle x =$

〈Ans.〉 $\angle y =$

4 Find the measure of ∠x and ∠y in the following figures.

6 points per question

(1)

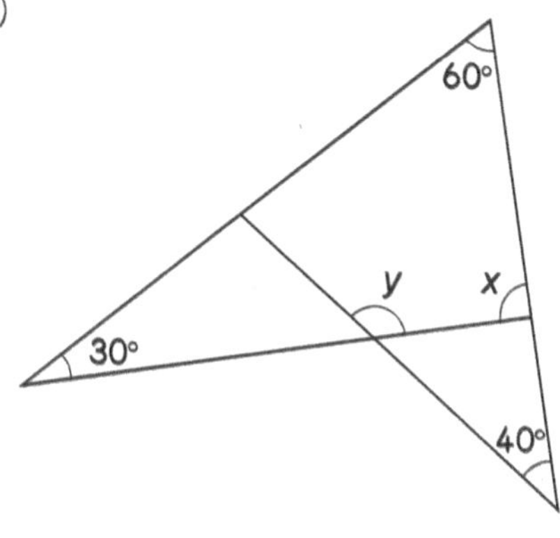

〈Ans.〉 ∠x = , ∠y =

(2)

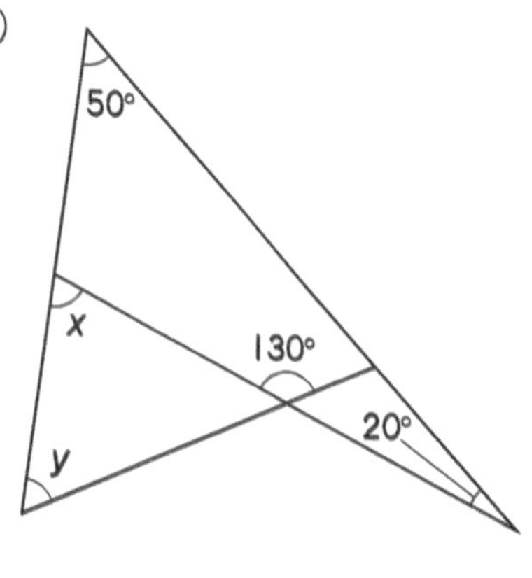

〈Ans.〉 ∠x = , ∠y =

5 Use the figure on the right to answer the following questions to find ∠x.

7 points per question

(1) Find the measure of ∠DEC. In the figure on the right, let line BD be extended to make intersection point E with line AC. ∠BDC is an exterior angle of ∠CDE.

〈Ans.〉 ∠DEC =

(2) By using the result of (1), find the measure of ∠x. ∠DEC is the exterior angle of △ABE.

〈Ans.〉 ∠x =

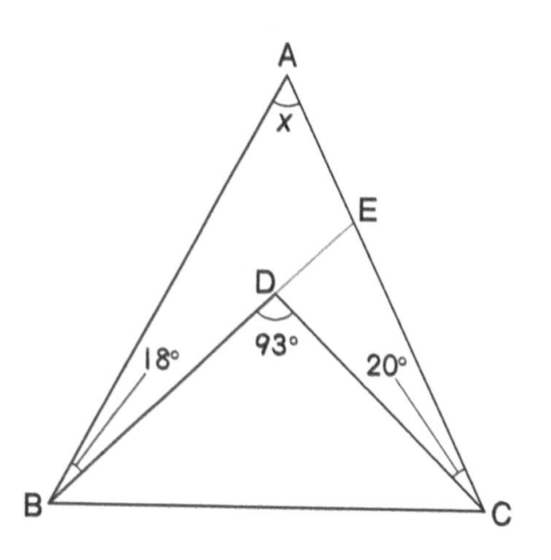

6 Find the measure of ∠x in the following figures. Draw an appropriate extension line like in question 5.

7 points per question

(1)

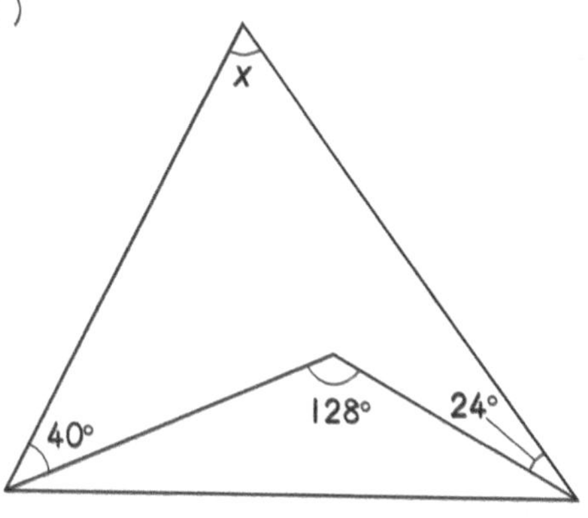

〈Ans.〉 ∠x =

(2)

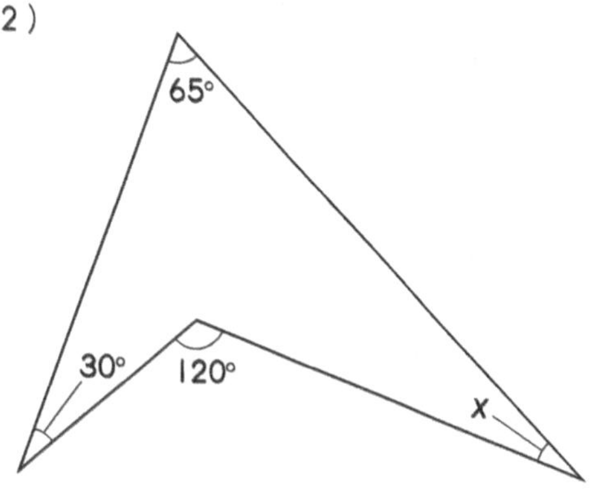

〈Ans.〉 ∠x =

28 The Sum of the Interior Angles of a Polygon

Level ★

Date / /

Name

Score /100

■ The Answer Key is on page 188.

1 **Referring to the following figures, fill in the blanks in (1) ~ (4) below and answer the (5) ~ (7).**

6 points per question

〔 quadrilateral 〕 〔 pentagon 〕 〔 hexagon 〕

(1) A quadrilateral is divided into ☐ triangles by a diagonal line from one vertex.

The sum of the interior angles of a triangle is 180°, so the sum of the interior angles of a qudrilateral is 180° × ☐ .

(2) A pentagon is divided into ☐ triangles by diagonal lines drawn from one vertex.

Therefore, the sum of the interior angles of a pentagon is 180° × ☐ .

(3) A hexagon is divided into ☐ triangles by diagonal lines from one vertex.

Therefore, the sum of the interior angles of a hexagon is 180° × ☐ .

(4) n - angle polygon is divided into (n − ☐) triangles by diagonal lines drawn from one vertex.

Therefore, the sum of the interior angles of n - angle polygon is 180° × (☐).

Don't forget!

● The sum of interior angles of an n - angle polygon is 180° × (n −2).

(5) Find the sum of the interior angles of an octagon.

〈Ans.〉

(6) Find the sum of the interior angles of a nonagon.

〈Ans.〉

(7) Find the sum of the interior angles of a twelve - angle polygon (dodecagon).

〈Ans.〉

2 **You can draw a regular polygon using a circle. Using this information, the sum of interior angles of regular polygon can be found. Referring to the figure on the right, fill the correct words and numbers in the blanks below.**

10 points per question

(1) To make a regular pentagon, divide the center angle of the circle into [ ] equal parts and connect the intersection with the circumference.
They are straight lines connecting the center of the circle and each vertex, and it becomes five [ ] triangles.
The angle at which the vertex of each triangle overlaps the center of the circle is $360 \div 5 =$ [ ]°.
The sum of base angles of each triangle is $180° - 72° =$ [ ]°.
The sum of the interior angles of each regular pentagons is [ ]° because the two base angles of a triangle make one interior angle.

(2) In the case of a regular decagon, the angle of vertex is [ ]°.
The interior angle is [ ]°. The sum of the interior angles is [ ]°.

Don't forget!

- All vertices of a regular polygon are on one circumference.

3 **Answer the following questions below.**

6 points per question

(1) Find the sum of the interior angles of an octagon.

〈Ans.〉

(2) Find an interior angle of a regular octagon.

〈Ans.〉

(3) Find an interior angle of a regular dodecagon.

〈Ans.〉

(4) Find n of the n - angle polygon whose sum of interior angles is 900°.

〈Ans.〉

(5) Find n of the n - angle polygon whose sum of interior angles is 720°.

〈Ans.〉

4 **Find the sum of the interior angles of hexagon in the figure below and find the measure of $\angle x$.**

8 points for completion

150°
128°
140°
x

〈Ans.〉 the sum =

〈Ans.〉 $\angle x$ =

The Sum of the Exterior Angles of a Polygon

Level ★★

Date / /

Name

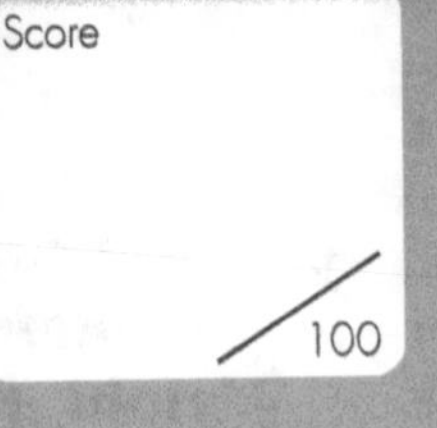

Score /100

■ The Answer Key is on page 188.

1 **Fill in the following blanks to find the sum of the exterior angles of a pentagon.** 35 points for completion

In the figure on the right, the sum of the exterior angles in a pentagon is $\angle a' + \angle b' + \angle c' + \angle d' + \angle e'$.
Since the sum of the interior angles in a pentagon is

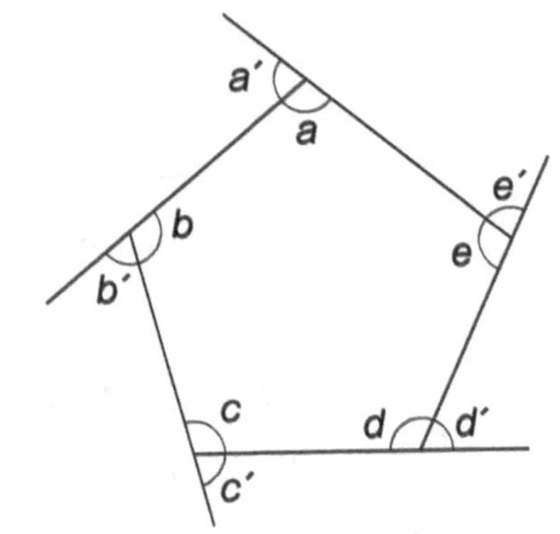

$180° \times (\square - \square)$,

$\angle a + \angle b + \angle c + \angle d + \angle e = 180° \times \square$

Also, since the sum of the interior angles and exterior angles at a vertex is 180°,

$(\angle a + \angle a') + (\angle b + \angle b') + (\angle c + \angle c') + (\angle d + \angle d') + (\angle e + \angle e') = 180° \times \square$

Therefore,

$\angle a' + \angle b' + \angle c' + \angle d' + \angle e' = 180° \times (\square - \square) = \square°$.

The same explaination can be applied to other polygons.

2 **By referring to question 1, the sum of the exterior angles in the *n* - angle polygon can be found as follows. Fill in the blanks.** 20 points for completion

The sum of the interior angles of n - angle polygon is expressed using n as follows.

(The sum of the interior angles) $= 180° \times (\square)$

Since the sum of the interior angles and exterior angles at a vertex is 180°,

(The sum of the interior angles) + (the sum of exterior angles) $= 180° \times \square$

Therefore, the sum of the exterior angles in an n - angle polygon is,

(the sum of the exterior angles) $= 180° \times n - 180° \times (n-2) = 180° \times \square = \square°$

Don't forget!

● The sum of the exterior angles is 360°.

3 Answer the questions below.

7 points per question

(1) Find the sum of the exterior angles of an octagon.

〈Ans.〉

(2) Find an exterior angle of a regular octagon.

〈Ans.〉

(3) Find n of an n - angle polygon when the exterior angle is 18°

〈Ans.〉

4 Find $\angle x$ in the following figures.

6 points per question

(1)

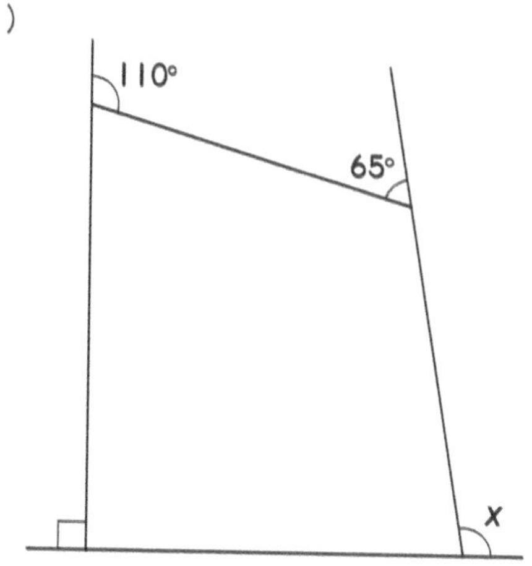

〈Ans.〉 $\angle x =$

(2)

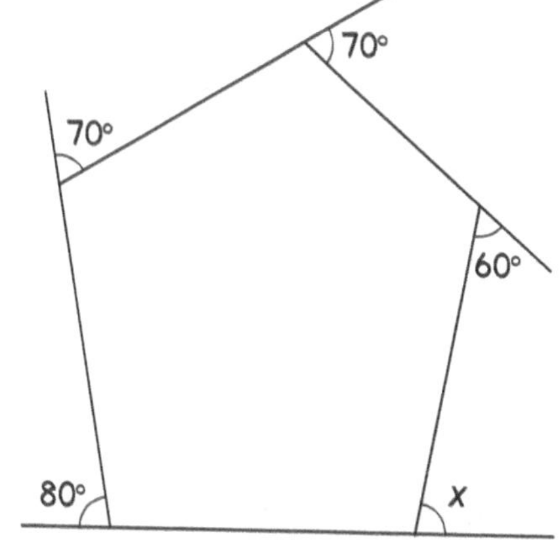

〈Ans.〉 $\angle x =$

(3)

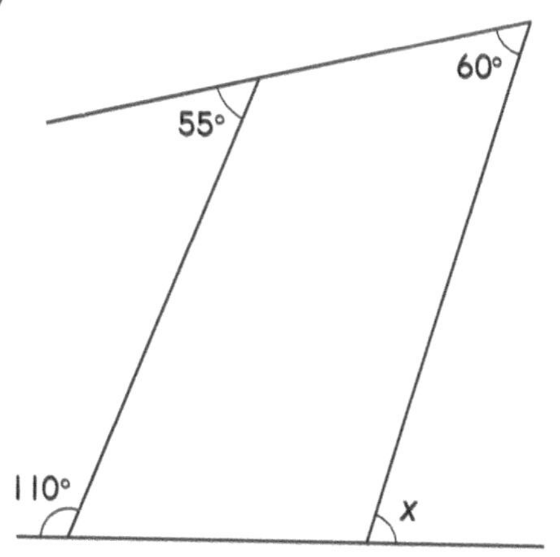

〈Ans.〉 $\angle x =$

(4)

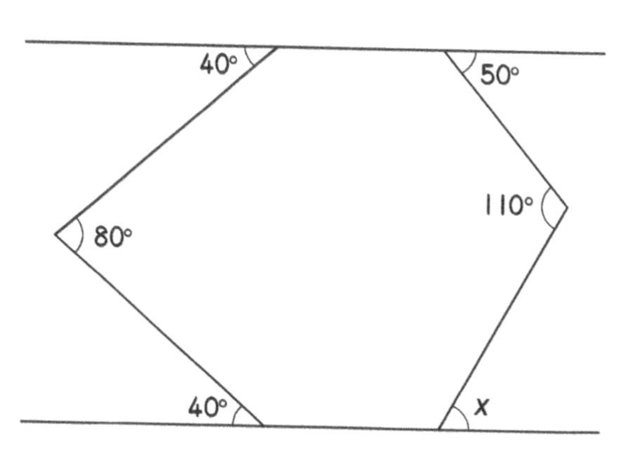

〈Ans.〉 $\angle x =$

30 Summary of Parallel Lines & Angles 1

Date / / Name

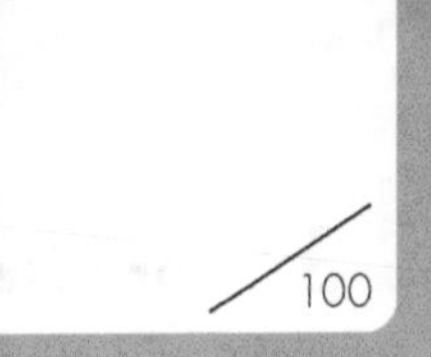

■ The Answer Key is on page 188.

1 Find $\angle x$ in the following figures.

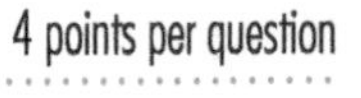

(1)

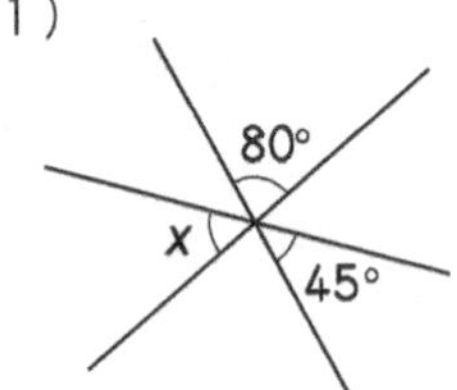

⟨Ans.⟩ $\angle x =$

(2)

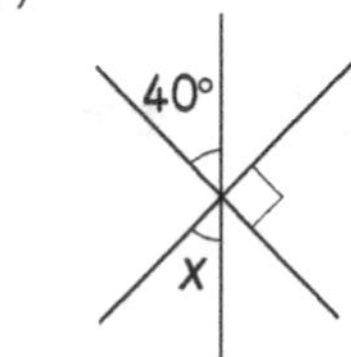

⟨Ans.⟩ $\angle x =$

(3) $\ell \parallel m$

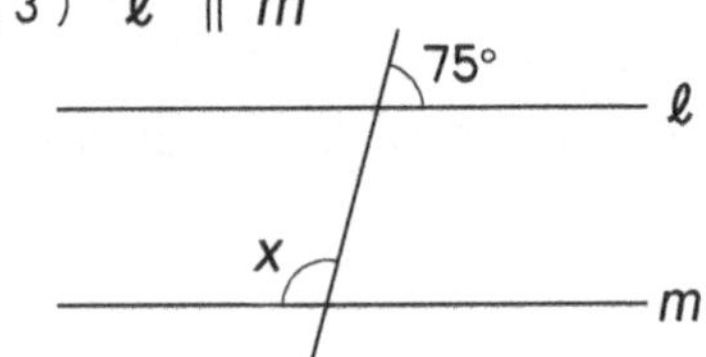

⟨Ans.⟩ $\angle x =$

(4) $\ell \parallel m$

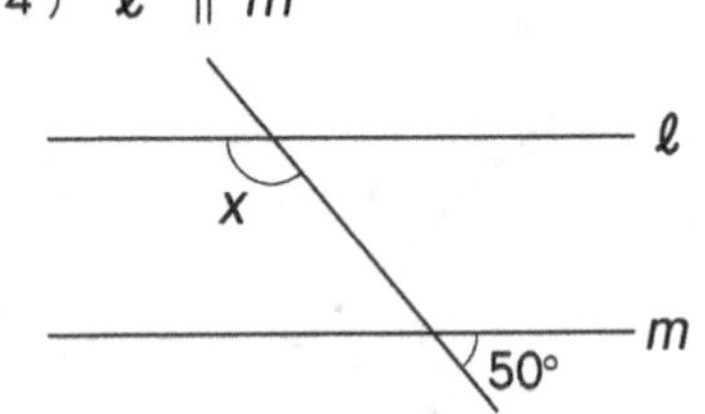

⟨Ans.⟩ $\angle x =$

(5) $\ell \parallel m$

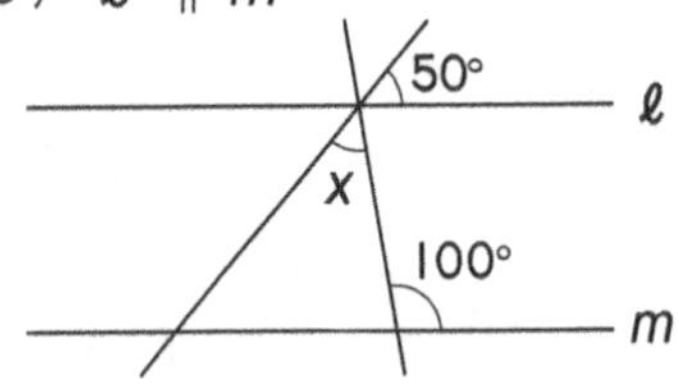

⟨Ans.⟩ $\angle x =$

2 Find $\angle x$ and $\angle y$ in the following figures.

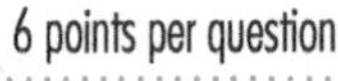

(1)

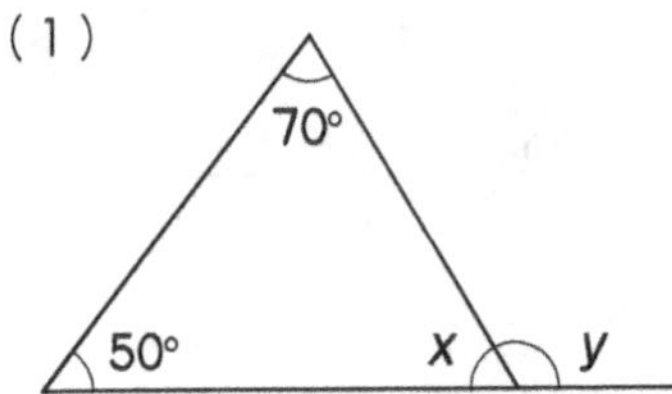

⟨Ans.⟩ $\angle x =$

$\angle y =$

(2)

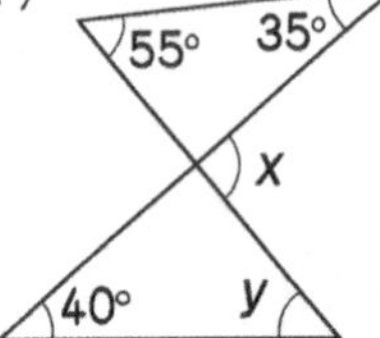

⟨Ans.⟩ $\angle x =$

$\angle y =$

(3)

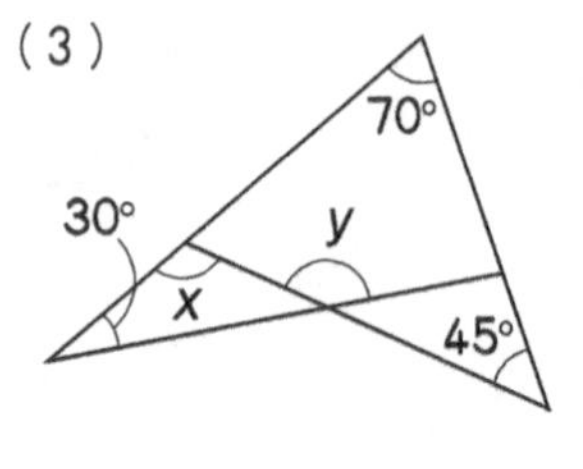

⟨Ans.⟩ $\angle x =$

$\angle y =$

(4)

⟨Ans.⟩ $\angle x =$

$\angle y =$

(5)

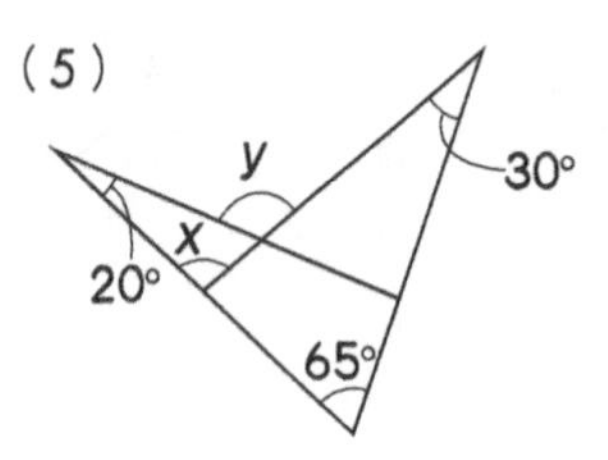

⟨Ans.⟩ $\angle x =$

$\angle y =$

3 Find $\angle x$ and $\angle y$ in the following figures when $\ell \parallel m$.

6 points per question

(1)

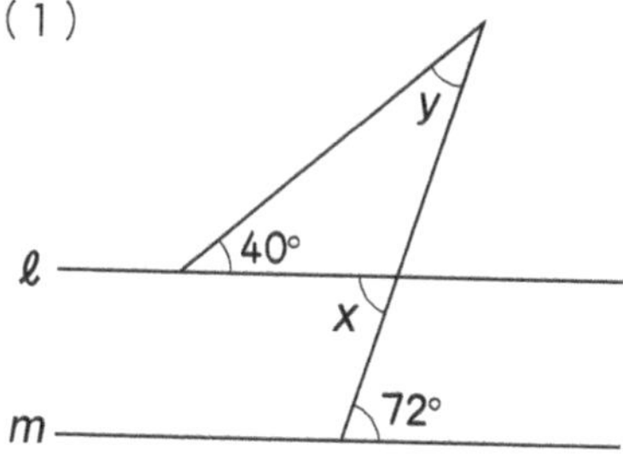

〈Ans.〉 $\angle x =$ ______

$\angle y =$ ______

(2)

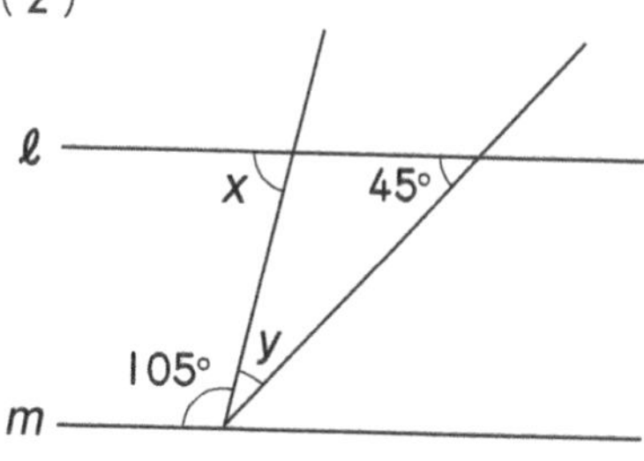

〈Ans.〉 $\angle x =$ ______

$\angle y =$ ______

(3)

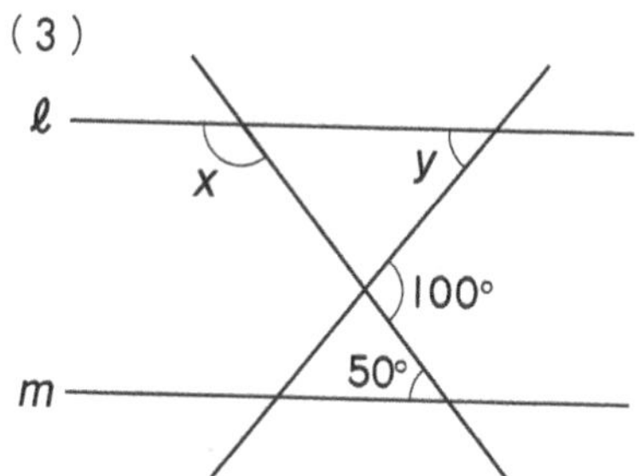

〈Ans.〉 $\angle x =$ ______

$\angle y =$ ______

(4)

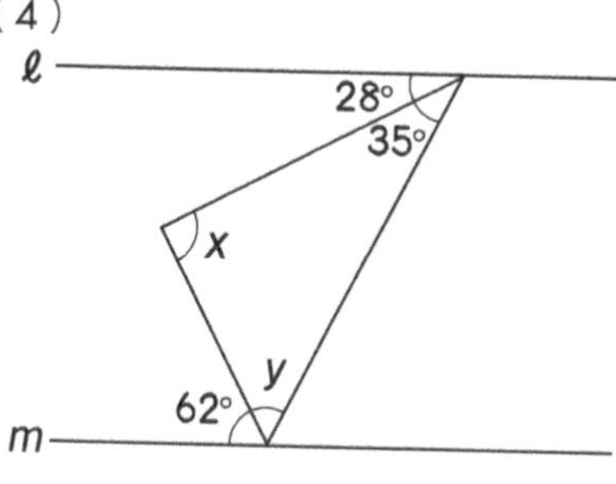

〈Ans.〉 $\angle x =$ ______

$\angle y =$ ______

(5)

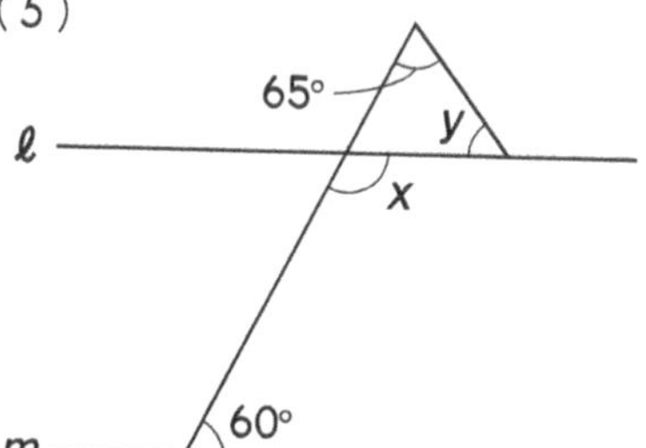

〈Ans.〉 $\angle x =$ ______

$\angle y =$ ______

4 Find $\angle x$ in the following figures.

5 points per question

(1)

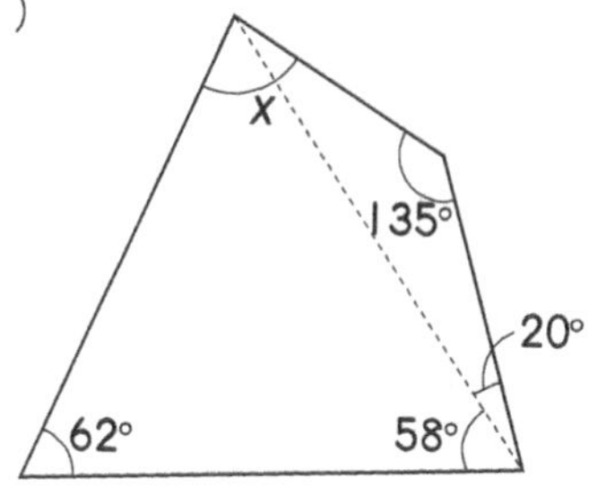

〈Ans.〉 $\angle x =$ ______

(2)

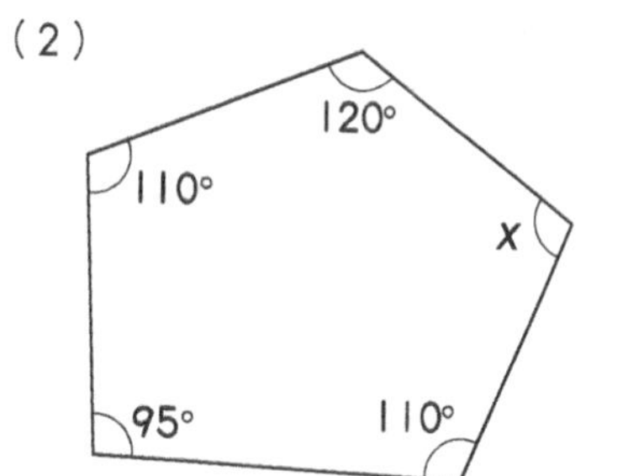

〈Ans.〉 $\angle x =$ ______

(3)

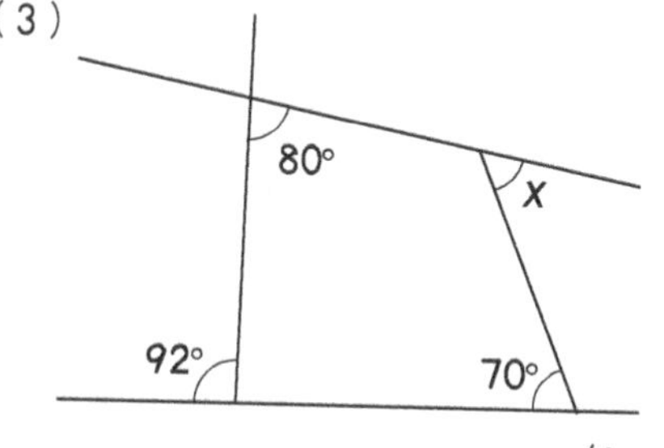

〈Ans.〉 $\angle x =$ ______

(4)

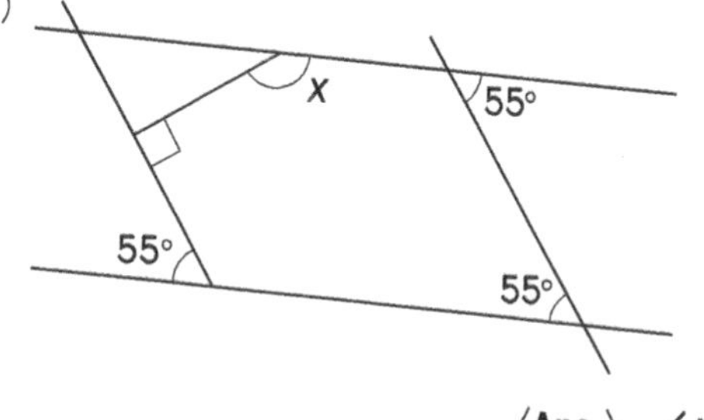

〈Ans.〉 $\angle x =$ ______

31 Summary of Parallel Lines & Angles 2

Level ★★

Date / /

Name

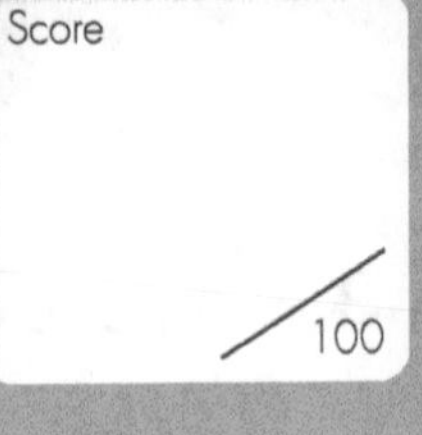

■ The Answer Key is on page 189.

1 Find ∠x in the following figures.

6 points per question

(1) $\ell \parallel m$

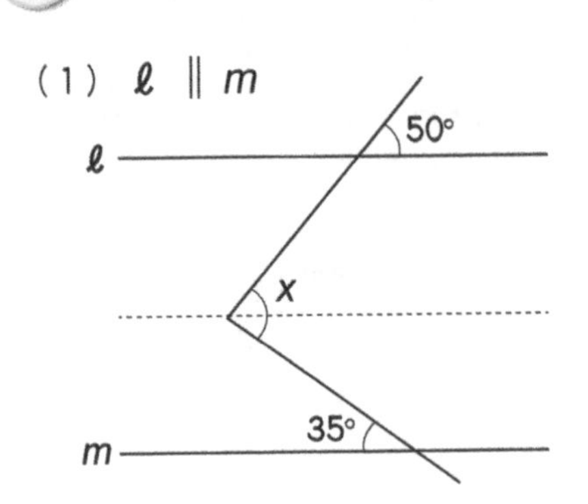

(2) $\ell \parallel m$

ℓ m x 40° 60°

(3) $\ell \parallel m$

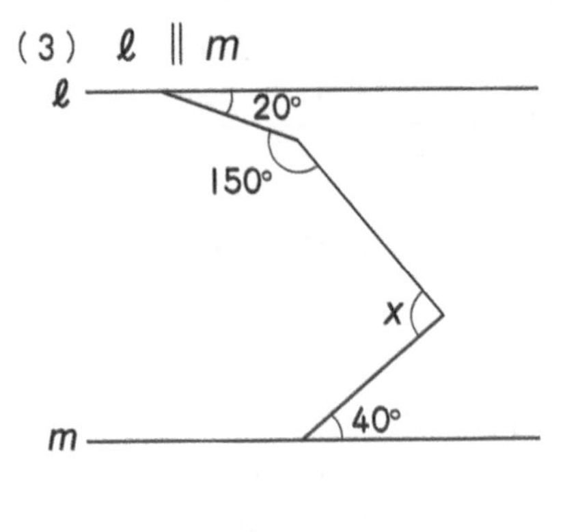

〈Ans.〉 ∠x =

〈Ans.〉 ∠x =

〈Ans.〉 ∠x =

2 Answer the following questions to find ∠x in the figure on the right.

5 points per question

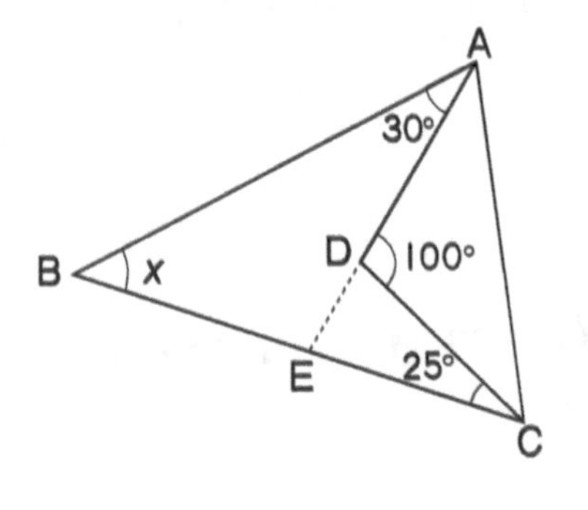

(1) Given that line segment AD was extended, and intersects side BC at point E. Find the measure of ∠DEC.

〈Ans.〉 ∠DEC =

(2) By using the result of (1), find ∠x.

〈Ans.〉 ∠x =

3 Find ∠x in the following figures.

6 points per question

(1)

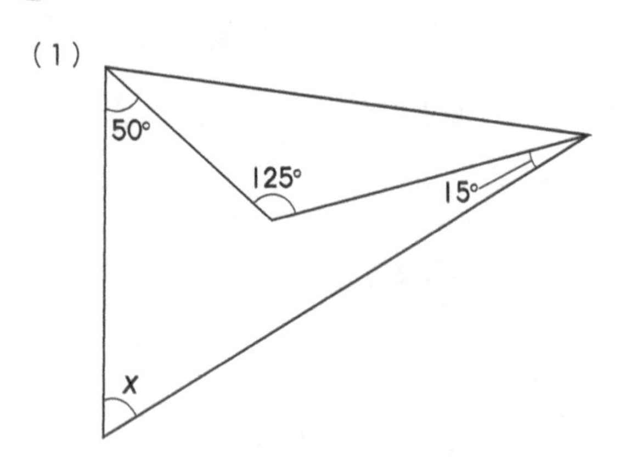

(2)

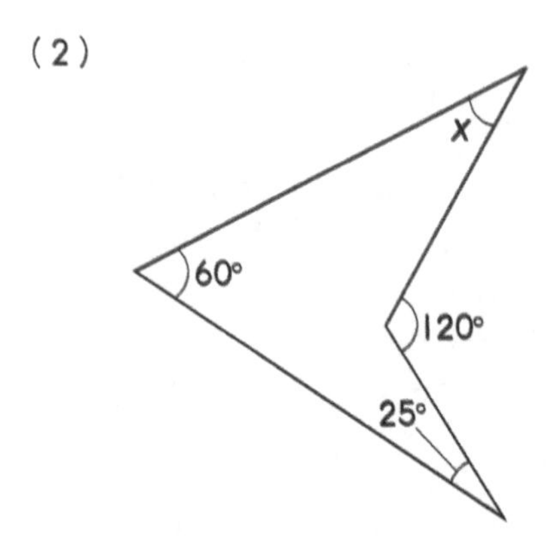

〈Ans.〉 ∠x =

〈Ans.〉 ∠x =

4 Answer the following questions about polygons.

6 points per question

(1) Find the sum of the exterior angles of a rectangle.

⟨Ans.⟩

(2) Find the sum of the exterior angles of a decagon.

⟨Ans.⟩

(3) Find the sum of the interior angles of a decagon.

⟨Ans.⟩

(4) What is one of the interior angles of a regular decagon?

⟨Ans.⟩

(5) What is a regular polygon with an interior angle of 160°?

⟨Ans.⟩

5 The figure on the right is a regular octagon circumscribed in circle O. Answer the following questions based on the figure.

5 points per question

(1) What is the measure of ∠AOB?

⟨Ans.⟩

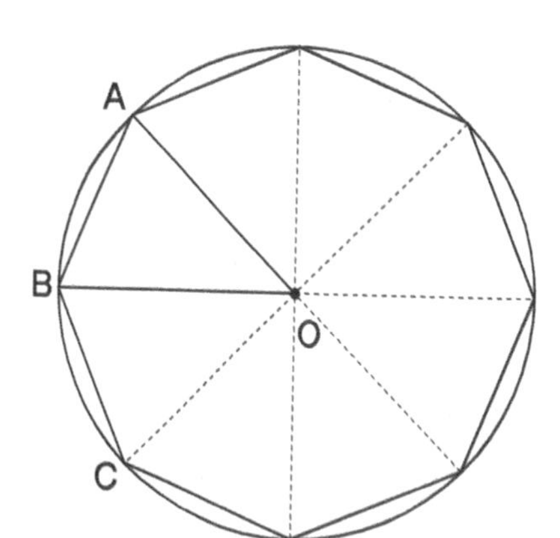

(2) What kind of triangle is △OAB?

⟨Ans.⟩

(3) What is the measure of ∠ABO?

⟨Ans.⟩

(4) What is the measure of ∠ABC?

⟨Ans.⟩

6 Find ∠x and ∠y in the following figures.

5 points per question

(1)

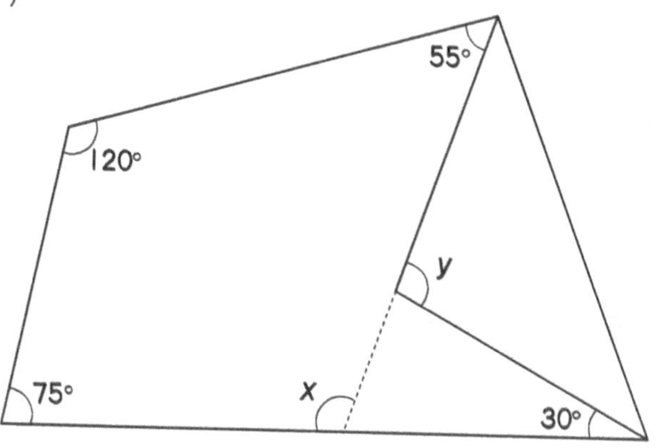

(2)

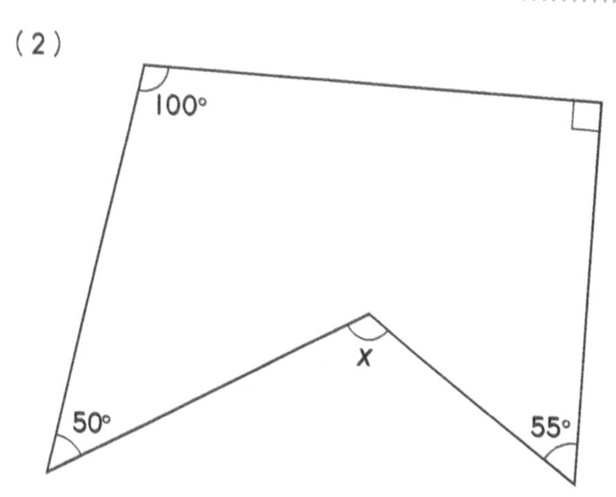

⟨Ans.⟩ ∠x = , ∠y =

p.192 ⟨Ans.⟩ ∠x =

32 Circles & Triangles 1 (Tangent)

Level ★★

Date / /　Name

■ The Answer Key is on page 189.

1 **Below is point O with the distance d to straight line ℓ. By drawing a circle with a radius *r* around point O, answer the following questions.**

(1) 20 points for completion (2)(3) 10 points per question

(1) When the numeric values of distance d and radius r are set as below, what is the number of intersection points between circle O and line ℓ?

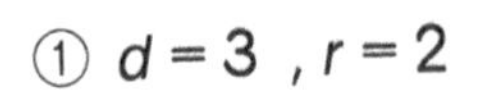

① $d = 3$, $r = 2$ 〈**Ans.**〉

② $d = 3$, $r = 5$ 〈**Ans.**〉

③ $d = 3$, $r = 3$ 〈**Ans.**〉

④ $d = 4$, $r = \frac{9}{2}$ 〈**Ans.**〉

(2) In ① to ④ in question (1), which line ℓ is tangent to circle O?

〈**Ans.**〉

Don't forget!

●The following figures show the relative position between a circle and a line.
There is circle O with the radius r. When the distance from center O to line ℓ is d, the following is true.

① **two intersection points**

② **one intersection point**

③ **no intersection**

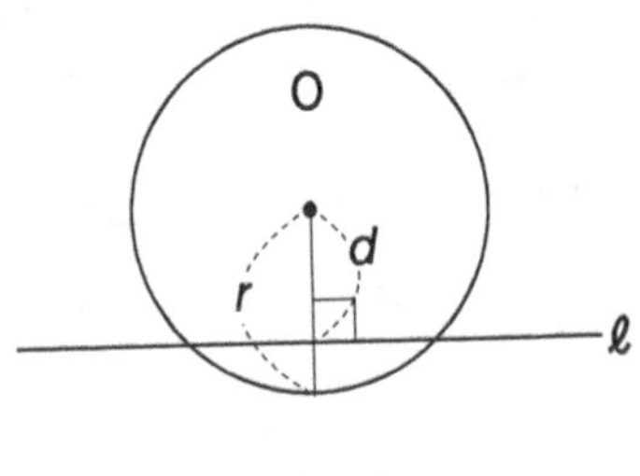

$r > d$

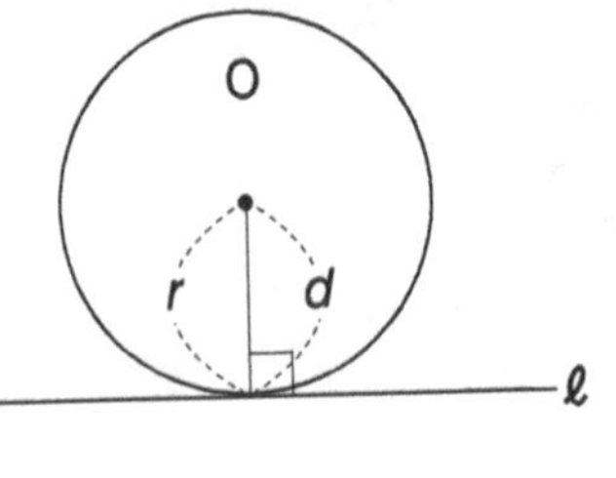

$r = d$

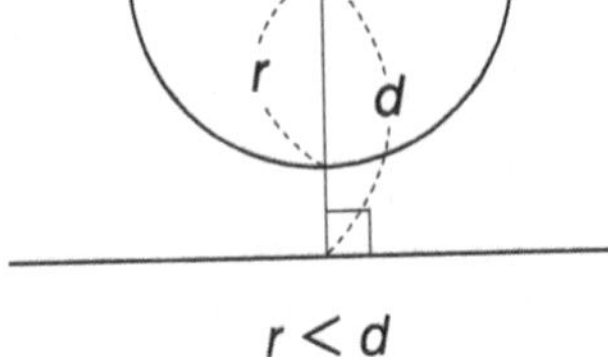

$r < d$

As in ②, when there is only one intersection, the point touching the circle is the **point of tangency** and the line tangent to the circle is called the **tangent line**. At that time, the distance between the tangent and the center of the circle is called the same as the radius of the circle.

(3) When $d = 5$ and $r = 5$, what is the distance between the center of the circle and the tangent line?

〈**Ans.**〉

2 Using the figure on the right, draw the tangent passing through point P on the circumference of circle O in the following order.

10 points per question

(Drawing order)

(1) Extend line segment OP in the direction of P and then set point P′ on it so that OP = PP′.

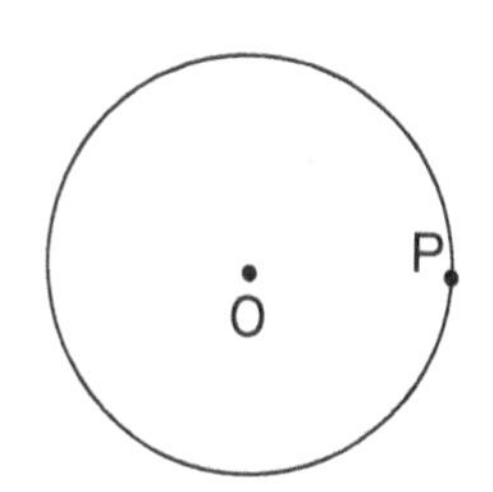

(2) Draw the perpendicular bisector of line segment OP′.

Don't forget!

●The tangent line of a circle is perpendicular to the radius through the point of tangency.

$\ell \perp OP$

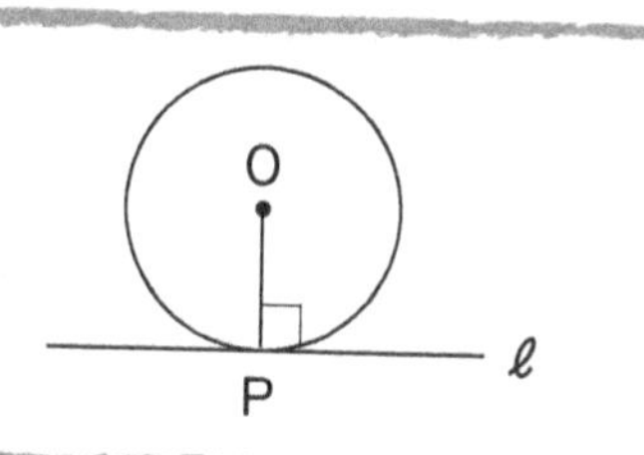

3 Answer the question below.

20 points

In the figure on the right, draw center P of a circle on side OA tangent to the point of tangency X on side OB.
(At the same time, center P is on a line perpendicular to side OB passing through the point of tangency X.)

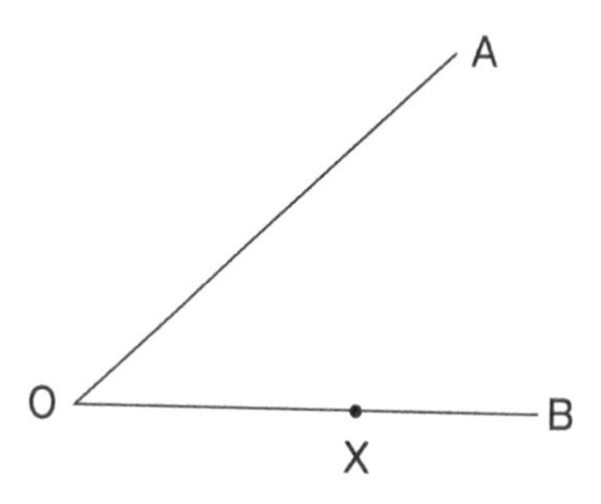

4 Answer the question below.

20 points

In the figure on the right, draw the center P of a circle tangent to the point of tangency X on side OB.
(At the same time, center P of the circle is on the bisector of ∠AOB.)

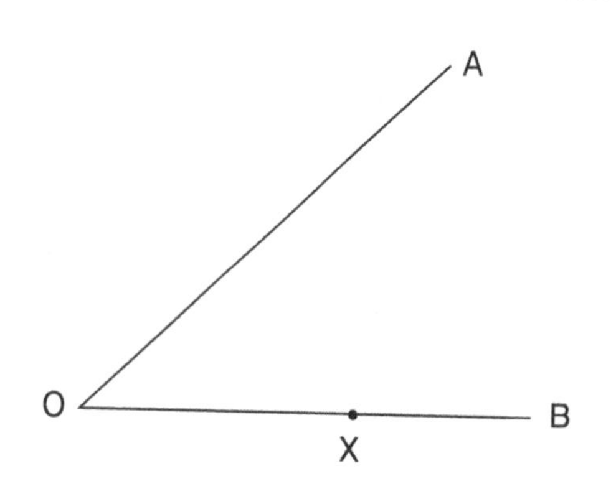

33 Circles & Triangles 2 (Inscribed in a Circle)

Date / / Name

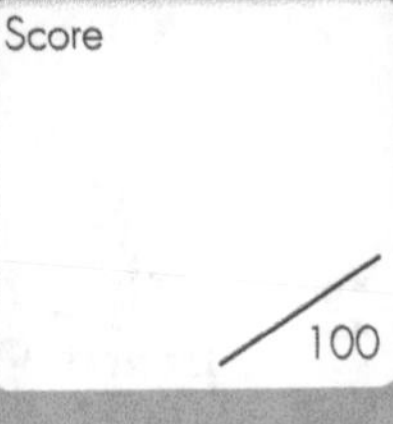

■ The Answer Key is on page 189.

1 **Divide the circumference of circle O into six equal parts in the following order and fill the correct words and numbers in the blanks below.**

5 points per question

(1) Draw a circle P around point P on circle O with the same radius as the circle. Then let the ☐ intersection points with circle O be Q and R.

(2) Draw a circle Q with the same radius around point Q and let an intersection point with circle O be S. Another intersection point overlaps with point ☐ .

(3) Draw a circle S with the same radius around point S and let an intersection point with circle O be T. Another intersection point overlaps with point ☐ .

O

P

(4) A line connecting point T and point P passes through point ☐ .

(5) Similarly for point R, draw a circle R with the same radius around point R and let an intersection point with circle O be U. Another intersection point overlaps with point ☐ .

(6) Through the above drawing, ☐ intersection points are formed on the circumference of circle O.

(7) When connecting the intersection points on the circumference of circle O next to it, a regular ☐ is formed.

(8) When connecting the intersection points on the circumference of circle O one after another, a ☐ is formed.

(9) When connecting the intersection points on the circumference of circle O to center point O, ☐ equilateral triangles are formed.

Don't forget!

- A part of the circumference between two points on the circumference is called an **arc**.
- A line segment connecting two points on the circumference is called a **chord**.

The diameter of a circle is the longest chord of the circle.

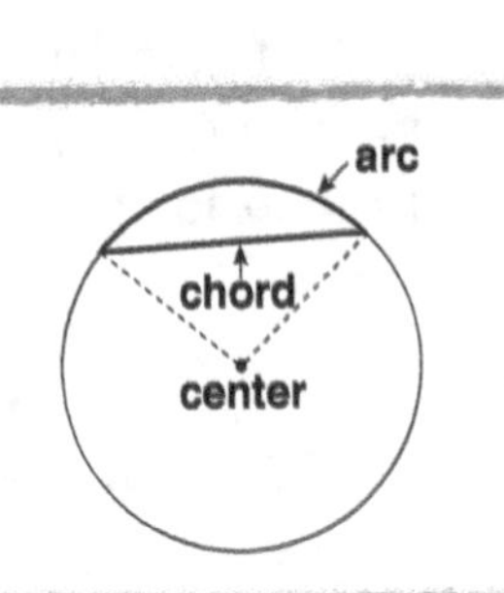

2 **Draw an equilateral triangle in the following order. Then fill the correct words and numbers in the blanks below.**

4 points per question

(1) Draw circle O with any radius.

(2) Draw circle P with the same radius as circle O. At that time, let the [ ] intersection points with circle O be A and B.

(3) Draw chord AB.

(4) Draw the circle [ ] around the intersection point A with the length of chord AB as the radius.

(5) Let C be the point that is not B among the two intersection points of circle O and circle A.

(6) When chord BC and chord AC are connected, an [ ] ABC is formed.

O

※Confirm that it intersects the original circle O at two points A and C, when drawing a circle B of radius AB around point B.

Don't forget!

●A circle passing through the three vertices of △ABC is called a **circumscribed circle**.
At this time, △ABC is called an **inscribed triangle** in a circle.

A
B
C

3 **On the right is equilateral triangle ABC inscribed in circle O. Answer the questions below and fill in the blanks.**

(1)-(5) 5 points per question (6) 6 points for completion

(1) Connect vertices A and B to center O of circle. What is the measure of central angle ∠AOB for arc AB?

〈Ans.〉

(2) What is the measure of ∠ACB?

〈Ans.〉

(3) Extend $\overline{CO}$. Next, let D be the intersection point with circle O. At that time, $\overline{CD}$ divides equilateral triangle CAB symmetrically. Show the relationship between $\overline{CD}$ and $\overline{AB}$ with the correct symbol.

C
O
A
B
D

〈Ans.〉

(4) What is the measure of ∠AOD and ∠ACD? 〈Ans.〉 ∠AOD = ∠ACD =

(5) What is the measure of ∠BOD and ∠BCD? 〈Ans.〉 ∠BOD = ∠BCD =

(6) ∠AOB = ∠AOD + ∠[ ], arc AB = arc AD + arc [ ], ∠ACB = [ ] ∠AOB

34 Circles & Triangles 3 (Inscribed Angles 1)

Level ★★

Date / /　　Name

■ The Answer Key is on page 189.

Don't forget!

● In one circle, the central angles of arcs of the same length are equal.
And the arcs of central angles of the same angle are equal.
Furthermore, the central angles are proportional to the length of the arc.

1 **Find the measure of $\angle x$ or the measure of arc x of the figure below.** 5 points per question

(1)

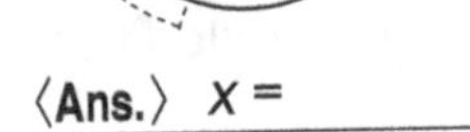

(2)

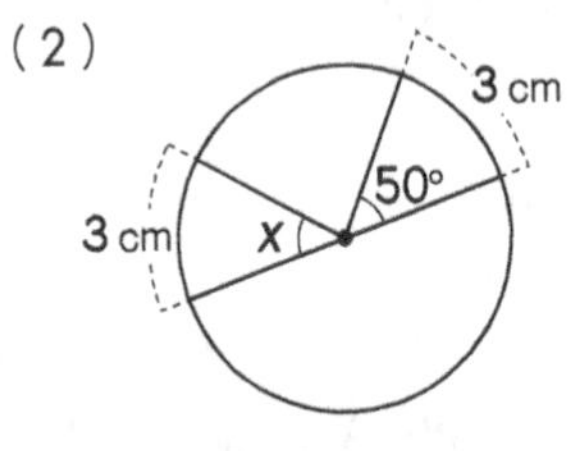

(3)

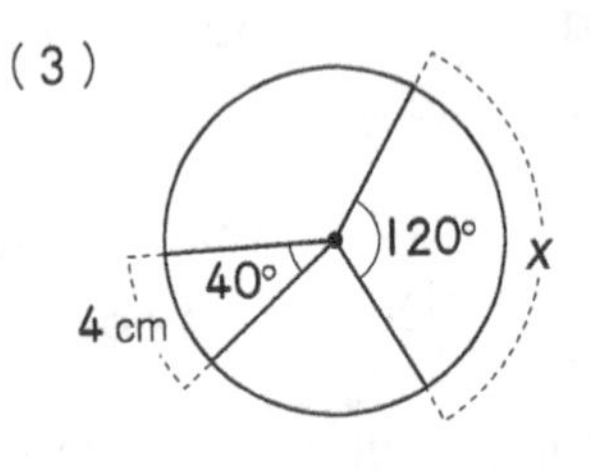

(4)

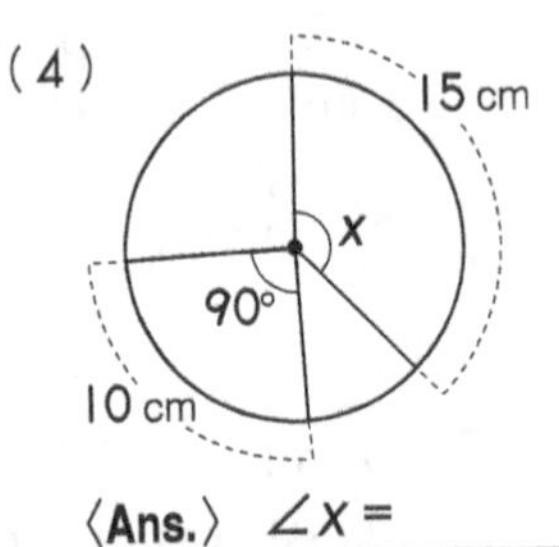

⟨Ans.⟩ $x=$ ______　⟨Ans.⟩ $\angle x=$ ______　⟨Ans.⟩ $x=$ ______　⟨Ans.⟩ $\angle x=$ ______

2 **The figure below are isosceles triangles connecting the diameter of semicircle with one point on the circumference and the center point of the circle. Find the measure of $\angle x$.** 5 points per question

(1)

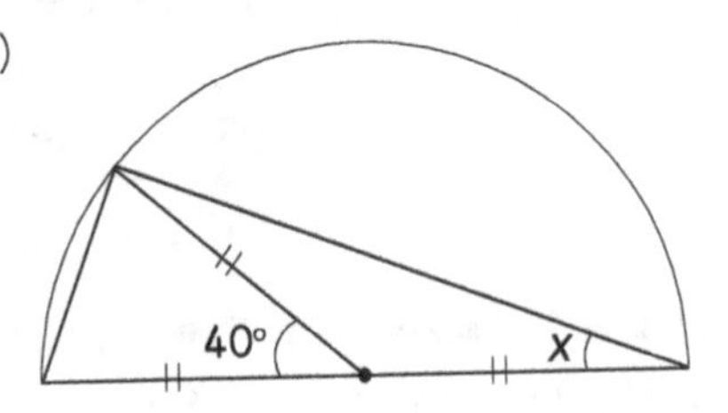

⟨Ans.⟩ $\angle x=$ ______

(2)

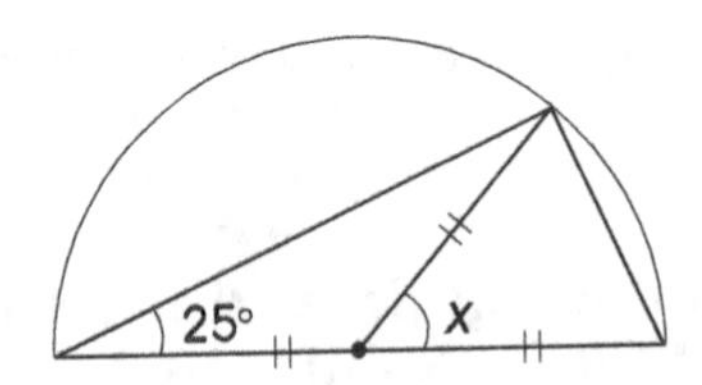

⟨Ans.⟩ $\angle x=$ ______

3 **Put point C anywhere on the circumference of the circle O, other than on diameter AB. In $\triangle ABC$ where three points A, B, and C are connected by a line, let the measure of $\angle CAO$ be x and the measure of $\angle COB$ be y. Answer the questions below and fill in the blanks.** (1) (2) 7 points per question　(3) 6 points

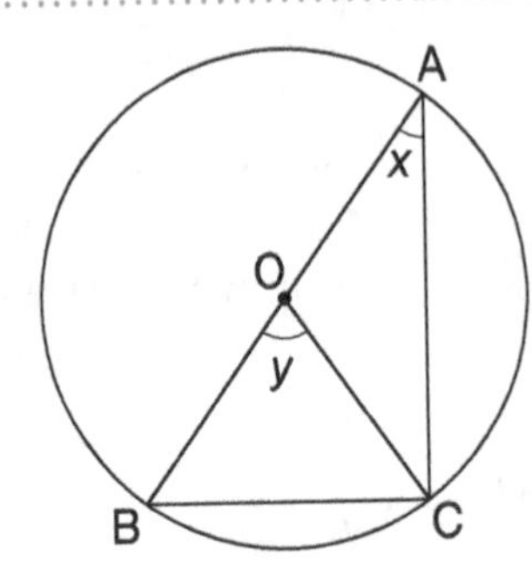

(1) Show the measure of y in terms of x.

By CO = AO,　　$\angle ACO = \angle$ ☐ $= x$,

Therefore　$y =$ ☐ x

(2) Show the measure of $\angle OCB$ in terms of x.

$\angle OCB = \angle OBC = \frac{1}{2}(180 - y)$

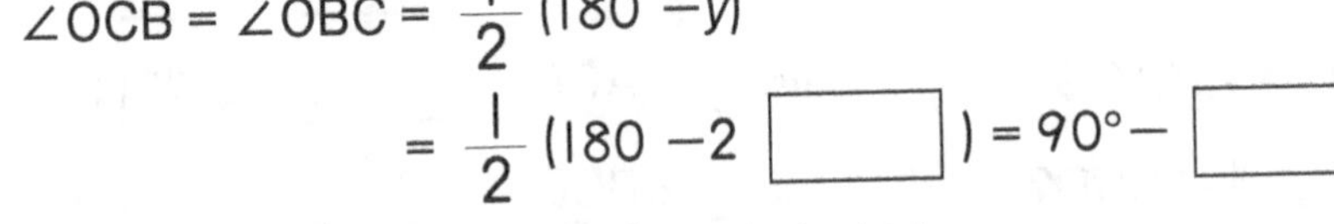

$= \frac{1}{2}(180 - 2$ ☐ $) = 90° -$ ☐

(3) Find the measure of $\angle ACB$ from (1), (2).

$\angle ACB = \angle ACO + \angle OCB =$ ☐ °

Don't forget!

- Both ends of the diameter of a circle have two vertices, and the triangle where the other vertex is on the circumference of the circle is the right triangle. At that time, the diameter is the longest side of the right triangle.

4 **In the figure on the right, put three points A, B, and P on the circumference of circle O, and let the diameter pass through PO be PR. Answer the questions below.**

5 points per question

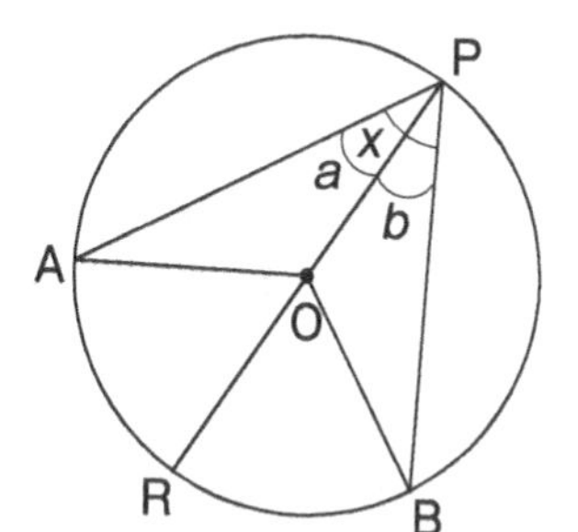

(1) Find the central angle of arc ARB.

〈Ans.〉 ∠

(2) Given ∠APO = a, show the measure of ∠AOR in terms of a.

〈Ans.〉 ∠AOR =

(3) Given ∠BPO = b, show the measure of ∠BOR in terms of b.

〈Ans.〉 ∠BOR =

(4) ∠APB is called the inscribed angle in terms of respect to arc ARB. Show the measure of ∠APB with a and b.

〈Ans.〉 ∠APB =

(5) Show the measure of ∠AOB in terms of a and b. 〈Ans.〉 ∠AOB =

(6) What is the relationship between ∠APB and ∠AOB? 〈Ans.〉 ∠APB =

Don't forget!

- ∠ACB formed by arc AB of circle O and point C on the circumference not on arc AB is called the **inscribed angle** with respect to arc AB.
- The measure of the inscribed angle with respect to arc AB is one-half of the central angle with respect to arc AB.

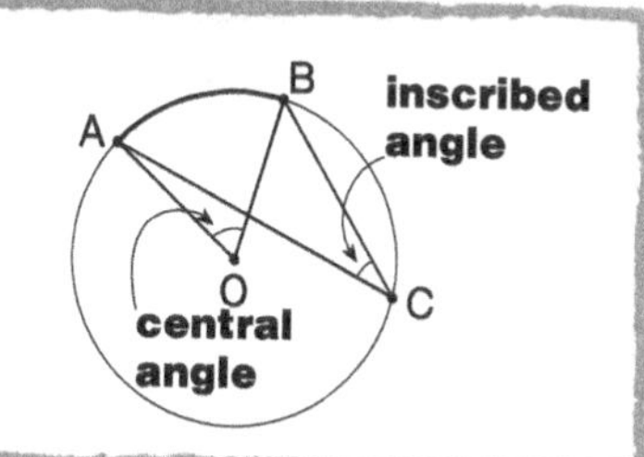

5 **Find the measure of the inscribed angle *x* or the central angle *y* in the circles below.**

5 points per question

(1)

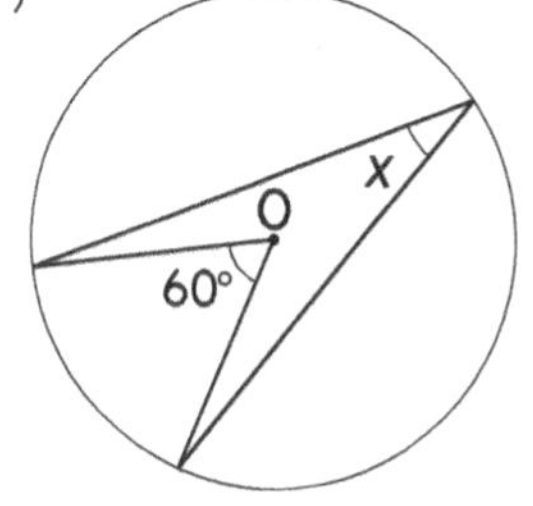

〈Ans.〉 ∠x =

(2)

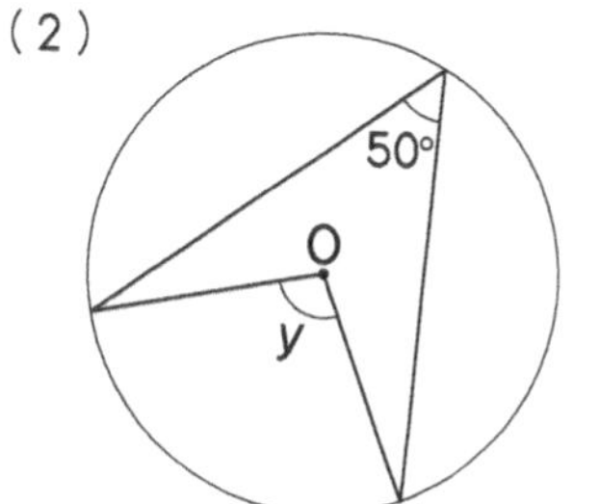

〈Ans.〉 ∠y =

(3)

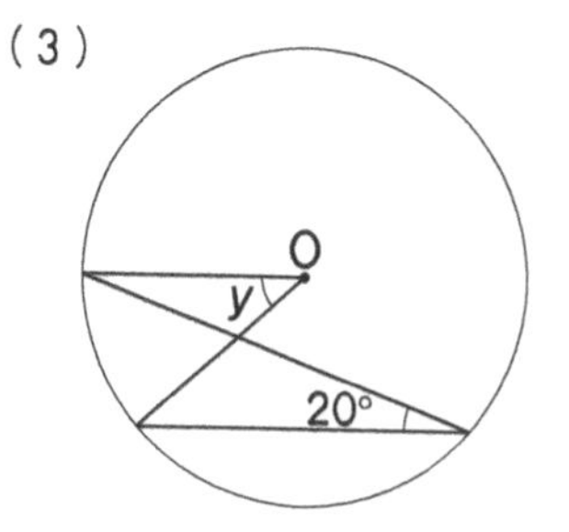

〈Ans.〉 ∠y =

(4)

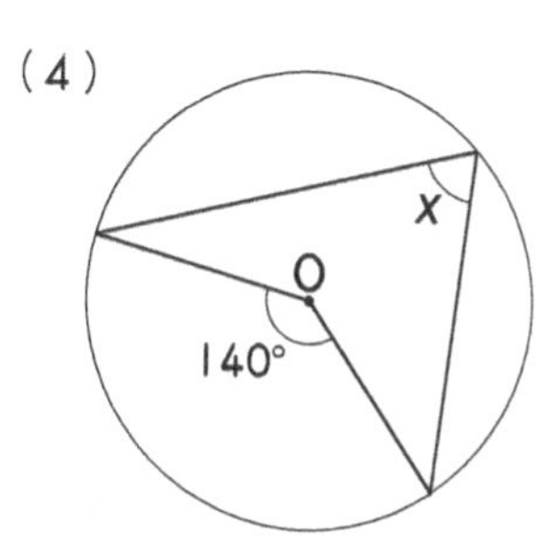

〈Ans.〉 ∠x =

35 Circles & Triangles 4 (Inscribed Angles 2)

Level ★★

Date / /

Name

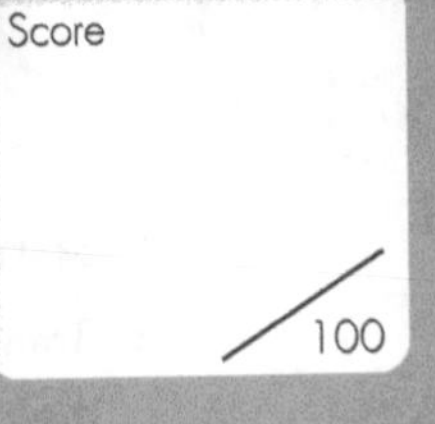

■ The Answer Key is on page 190.

Don't forget!

- Since the central angles of arcs of the same length are equal in a circle, the inscribed angles of arcs of the same length are also equal.
- The inscribed angle of arc AB is one-half of the central angle of arc AB.

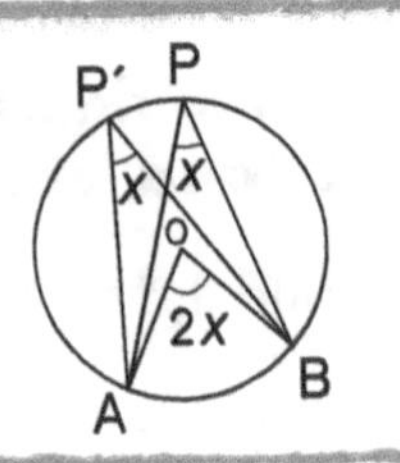

1 **Find the measure of inscribed angle x and central angle y in the circles below.** 5 points per question

(1)

Q P 60° x O B A

〈Ans.〉 $\angle x =$

(2)

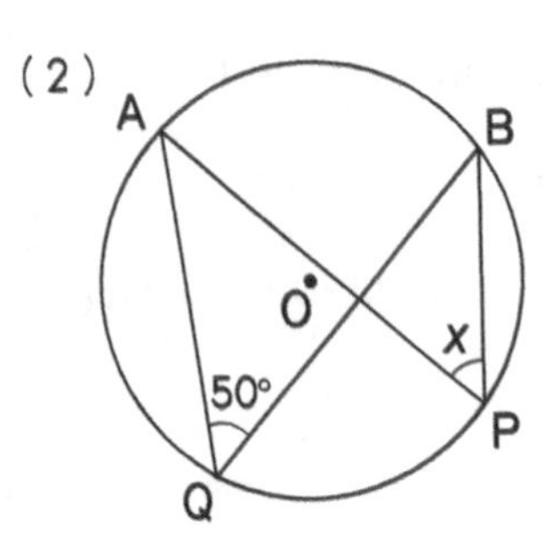

〈Ans.〉 $\angle x =$

(3)

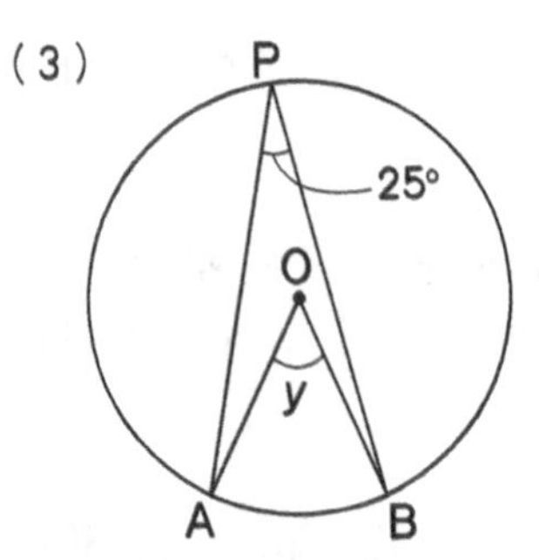

〈Ans.〉 $\angle y =$

(4)

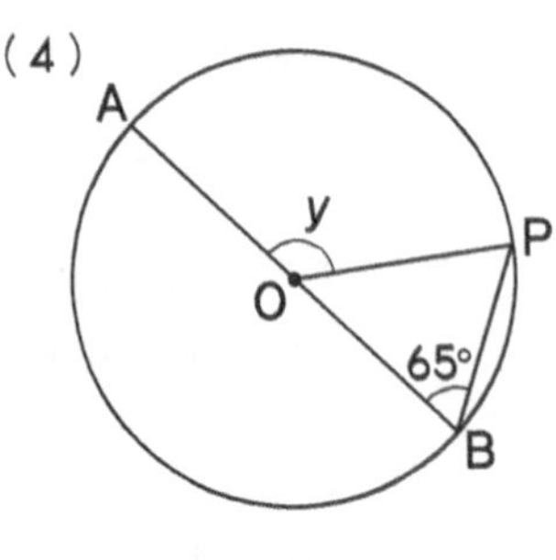

〈Ans.〉 $\angle y =$

Don't forget!

- The length of the arc is proportional to the measure of the inscribed angle in a circle. When the arc length doubles, the inscribed angle also doubles.

2 **Find the measure of $\angle x$ or the measure of arc x in the figures below.** 4 points per question

(1)

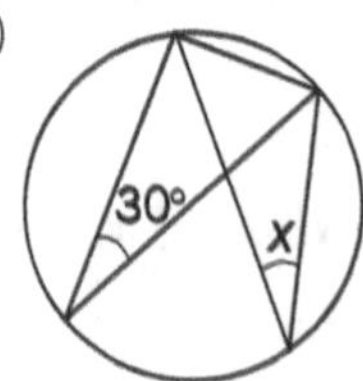

〈Ans.〉 $\angle x =$

(2)

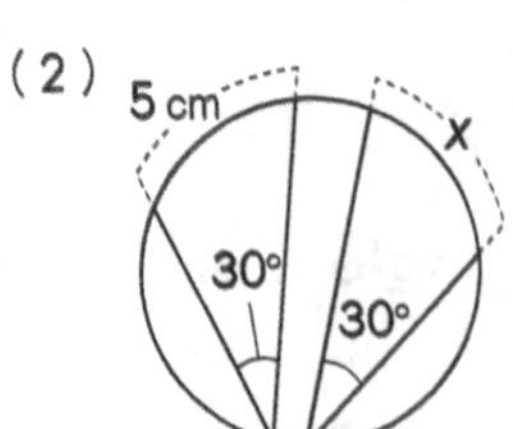

〈Ans.〉 $x =$

(3)

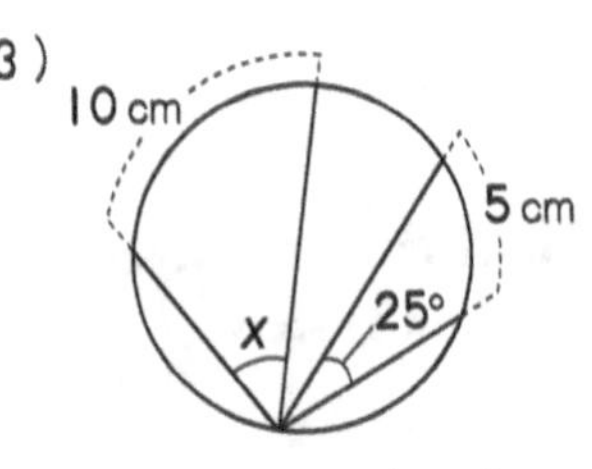

〈Ans.〉 $\angle x =$

(4)

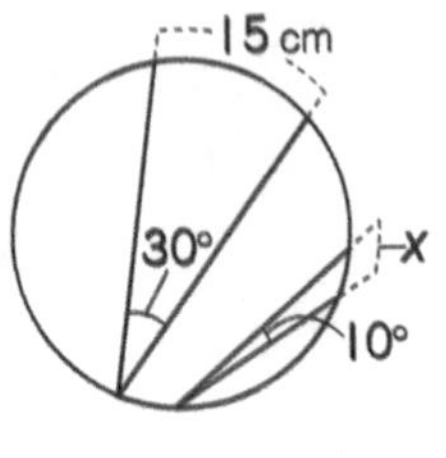

〈Ans.〉 $x =$

(5)

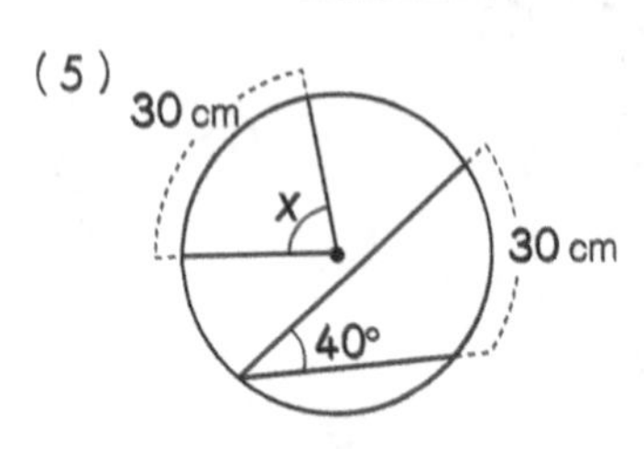

〈Ans.〉 $\angle x =$

(6)

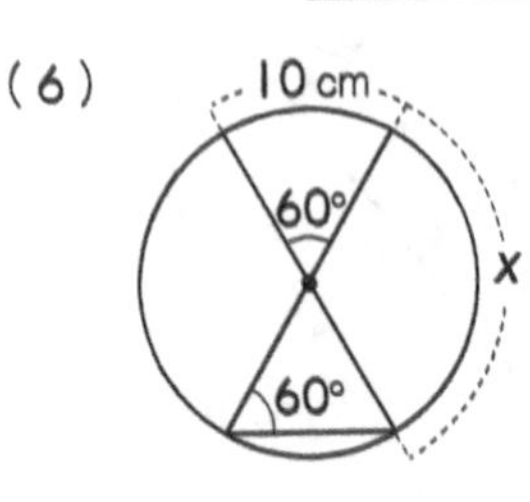

〈Ans.〉 $x =$

3 Put point C on the circumference of circle O with $\overline{AB}$ as its diameter. And, connect points A, B, and C to make △ABC. Answer the questions below.

(1)(2) 2 points per question (3) 4 points

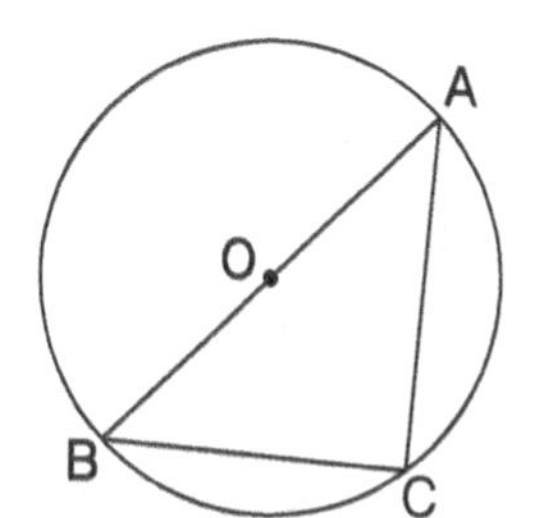

(1) Find the measure of the central angle of arc AB, which is half of the circumference.

〈Ans.〉 ____________

(2) Find the measure of ∠ACB.

〈Ans.〉 ∠ACB = ____________

(3) What kind of triangle is △ACB?

〈Ans.〉 ____________

Don't forget!

● The center of a circumscribed circle of a right triangle maps onto the midpoint of the longest side.

4 Find the measure of ∠x in the figures below.

4 points per question

(1)

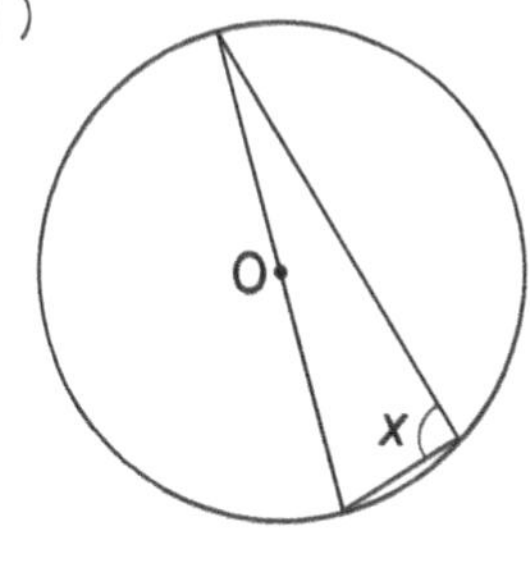

〈Ans.〉 ∠x = ____________

(2)

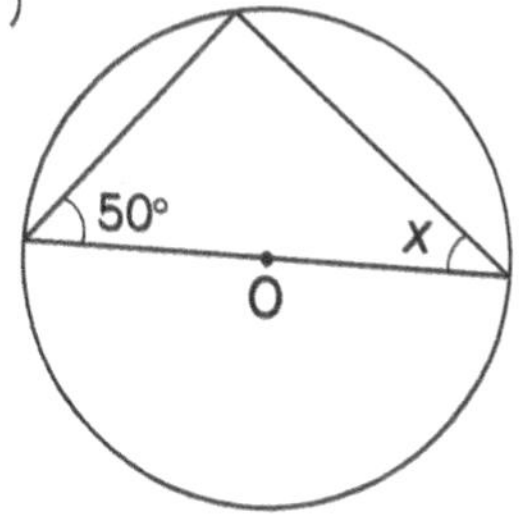

〈Ans.〉 ∠x = ____________

(3)

〈Ans.〉 ∠x = ____________

(4)

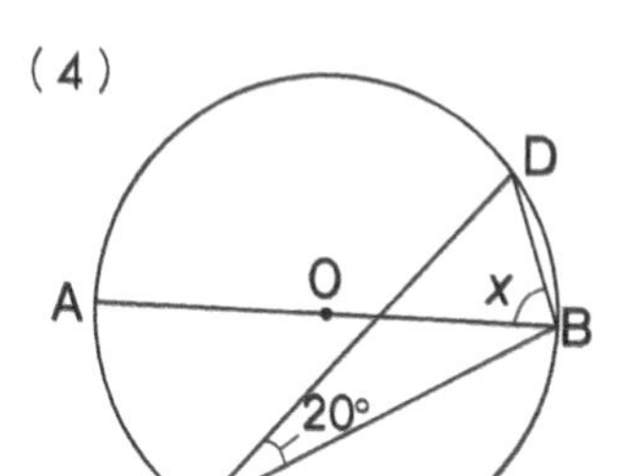

〈Ans.〉 ∠x = ____________

p.192

5 Find the measure of ∠x in the figures below.

4 points per question

(1)

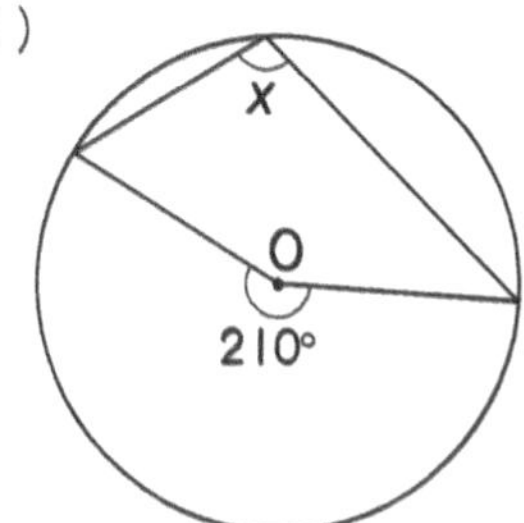

〈Ans.〉 ∠x = ____________

(2)

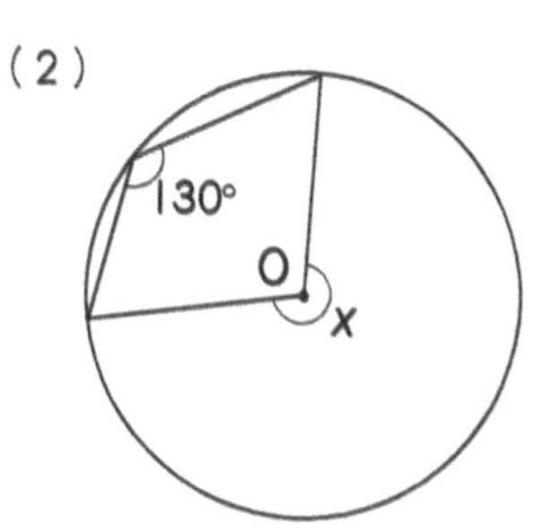

〈Ans.〉 ∠x = ____________

(3)

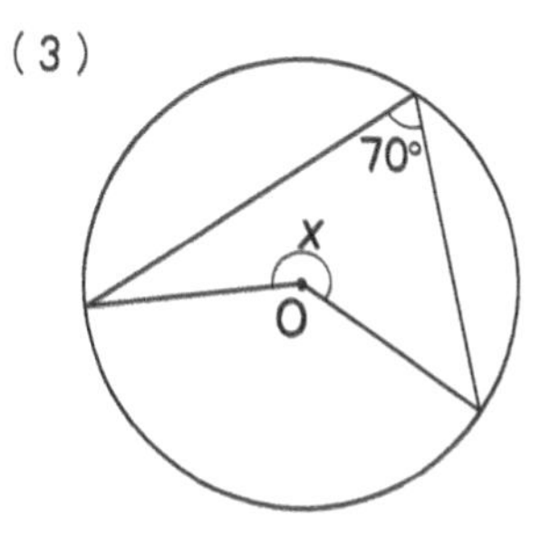

〈Ans.〉 ∠x = ____________

(4)

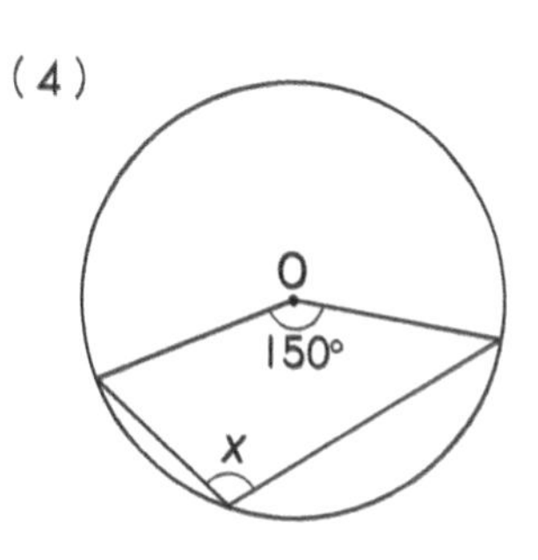

〈Ans.〉 ∠x = ____________

(5)

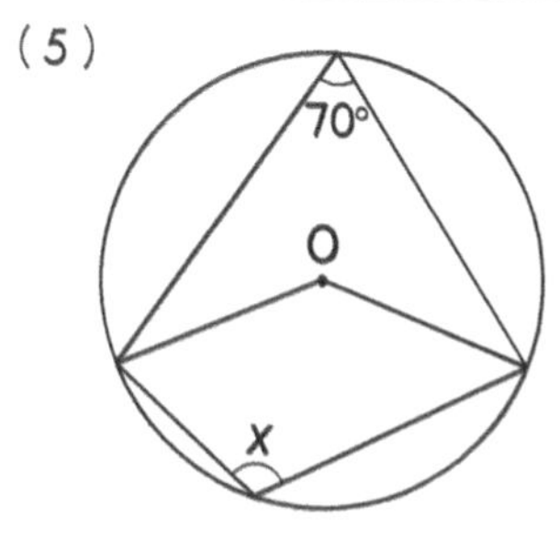

〈Ans.〉 ∠x = ____________

(6)

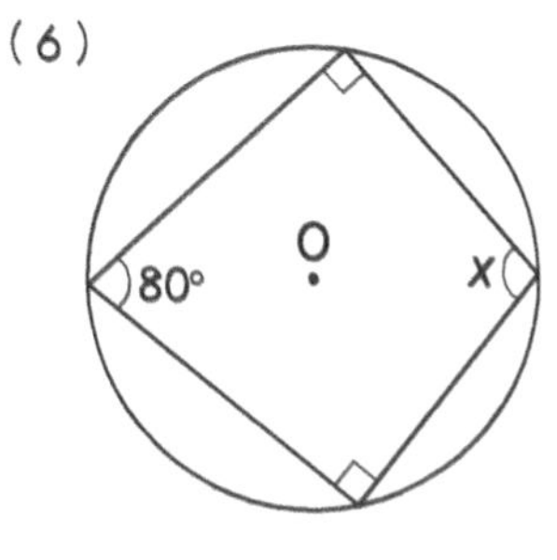

〈Ans.〉 ∠x = ____________

(7)

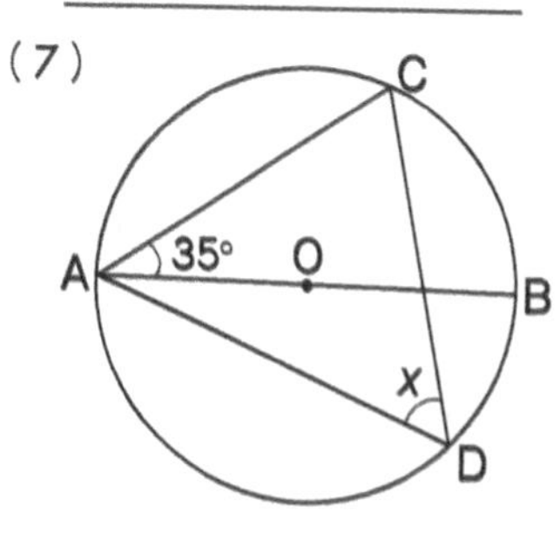

〈Ans.〉 ∠x = ____________

(8)

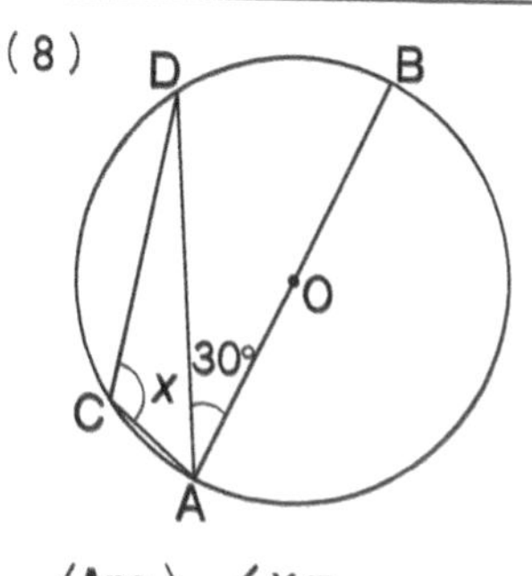

〈Ans.〉 ∠x = ____________

36 Isosceles Triangles 1

Level ★★ 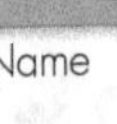

Date / / Name

■ The Answer Key is on page 190.

Don't forget!

● An isosceles triangle is a triangle with two equal sides. The angles and edges of an isosceles triangle can be described as below.

① vertex angle .... The angle between equal sides.
② base angle .... Two angles other than the vertex angle. The two base angles are equal.
③ base .... The opposite side of the vertex angle.

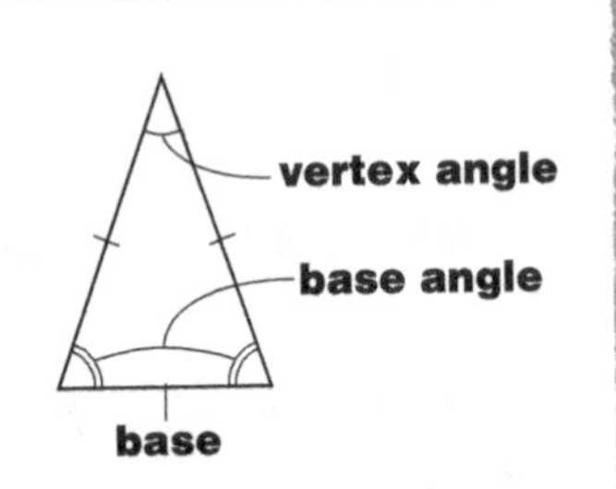

1 **The figure on the right is an isosceles triangle with AB = AC. Answer the questions below.** 6 points per question

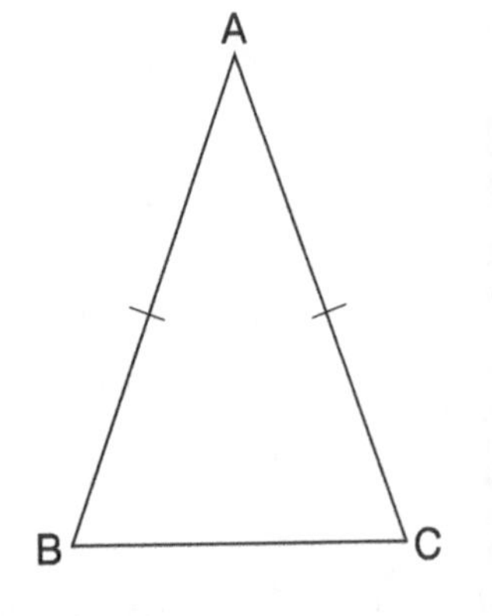

(1) Find the length of side AC, given AB = 6 cm.

〈Ans.〉

(2) Find the measure of base angle C, given base angle B = 75°.

〈Ans.〉

(3) Find the measure of vertex angle A, given base angle B = 75°.

〈Ans.〉

(4) Find the measure of vertex angle A, given base angle C = 50°.

〈Ans.〉

2 **Find the length of *x* in the triangles below.** 6 points per question

(1)

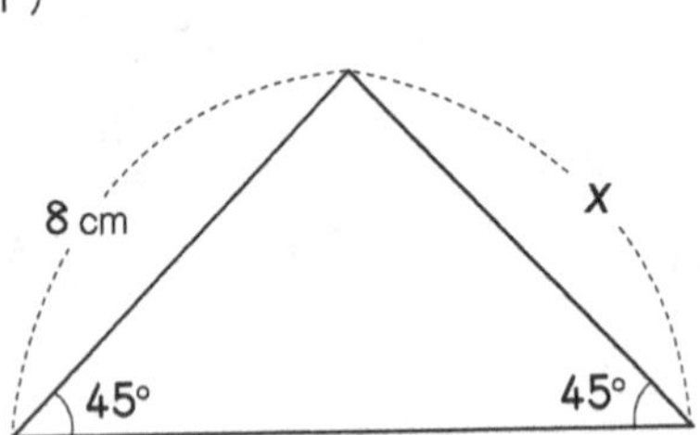

〈Ans.〉

(2)

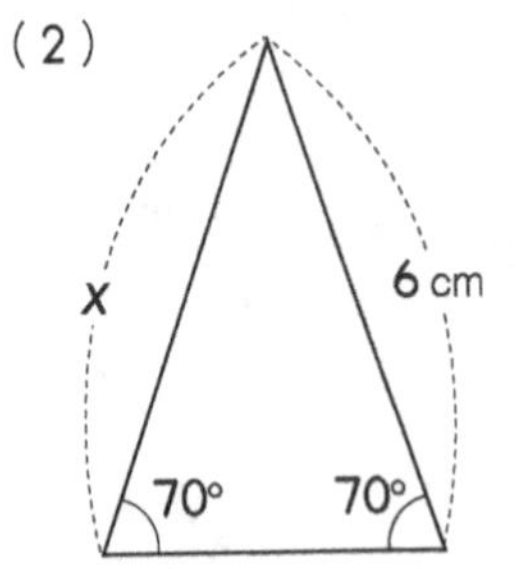

〈Ans.〉

(3)

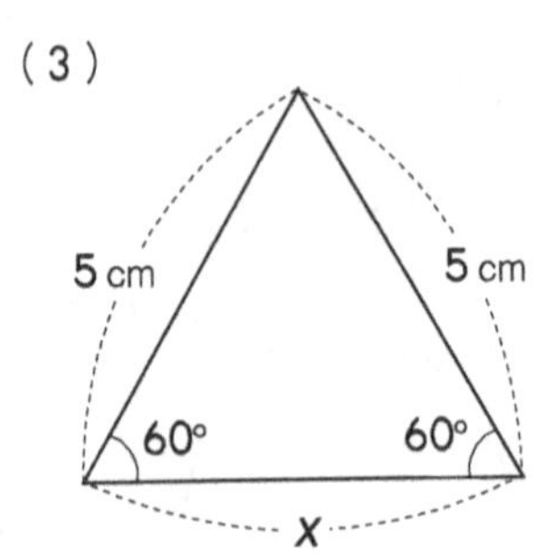

〈Ans.〉

3 The figure on the right is an isosceles triangle with AB = AC and ∠B = 55°. Answer the questions below to find the measure of ∠A and fill in the blanks.

6 points per question

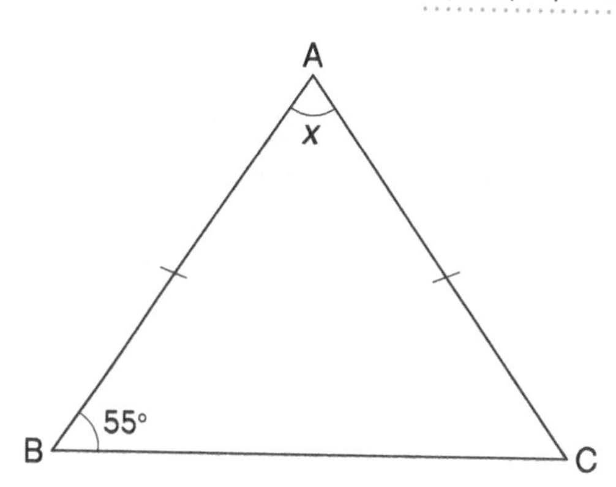

(1) In △ABC,

Since AB = ☐ ,

the base angle B = ∠ ☐ = ☐ °

Since the sum of the interior angles of a triangle is 180°,

Given the measure of ∠A as x,

x°= 180° − ☐ °× 2 = ☐ °

(2) Find the measure of ∠A when base angle ∠B = 70°.

〈Ans.〉 ______

(3) Show the measure of ∠A in terms of a, given ∠B is a°.

〈Ans.〉 ______

4 Below are isosceles triangles with AB = AC. Find the measure of ∠x. (3) is expressed in terms of a.

8 points per question

(1)

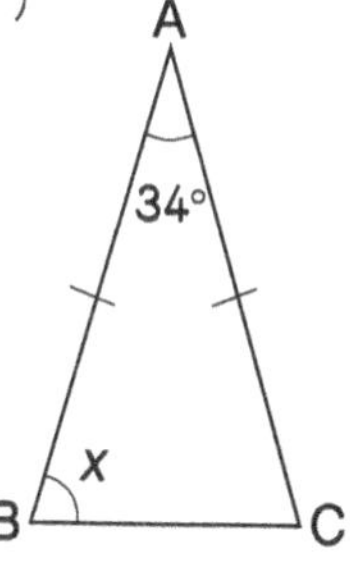

〈Ans.〉 ______

(2)

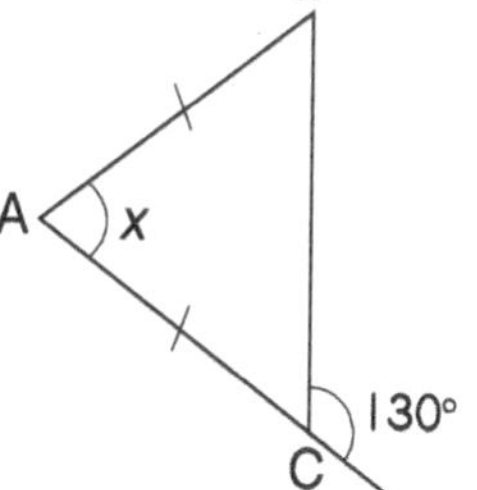

〈Ans.〉 ______

(3)

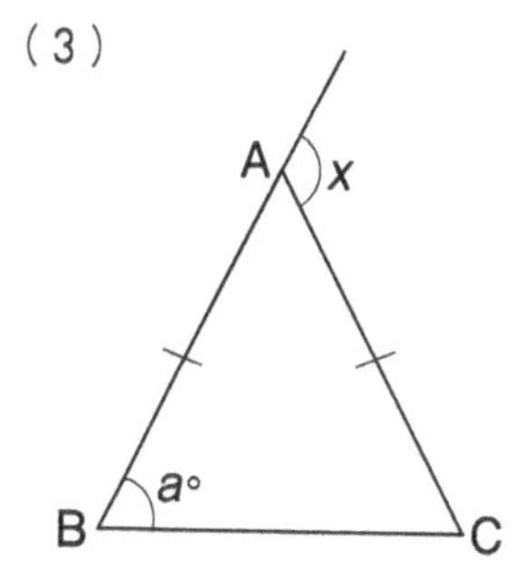

〈Ans.〉 ______

5 Answer the question about the angle measure of the isosceles triangles below.

8 points per question

(1) Find the measure of a base angle, given the vertex angle is 50°.

〈Ans.〉 ______

(2) Find the measure of vertex angle, given a base angle is 35°.

〈Ans.〉 ______

Don't forget!

●The bisector of a vertex angle bisects the base prependicularly in an isosceles triangle.

37 Isosceles Triangles 2

Level ★

Date / /

Name

Score /100

■ The Answer Key is on page 190.

1 **Draw an isosceles triangle with sides AB = AC = 3 cm and vertex angle A = 40° in the following order, then answer the question below.** 3 points per question

(1) Draw ray AP and ray AQ with an angle of 40 °.

(2) Draw a circle with a radius of 3 cm around point A. Then let B and C be the intersection points with AP and AQ.

(3) Connect point B and point C, to make △ABC.

(4) Find the measure of a base angle of isosceles triangle ABC.

〈**Ans.**〉

2 **Draw an isosceles triangle with base BC = 4 cm and base angle B and C = 50° in the following order, then answer the question below.** 3 points per question

(1) Draw ray BP from point B so that ∠CBP = 50°.

(2) Draw ray CQ from point C so as to intersect ray BP and be ∠BCQ = 50°.

(3) Let A be the intersection point of ray BP and ray CQ.

(4) Connect point A and point B, and point A and point C to make △ABC.

(5) What is the measure of vertex angle of isosceles triangle ABC?

〈**Ans.**〉

B C

3 **Draw an isosceles triangle with base BC = 2.5 cm and sides AB = AC = 5 cm in the following order.** 3 points per question

(1) Draw an arc with a radius of 5 cm around point B.

(2) Draw an arc with a radius of 5 cm around point C so as to intersect with the arc from (1).

(3) Let A be the intersection point of two arcs.

(4) Connect point A and point B, and connect point A and point C to make △ABC.

B

C

4 Draw the figures in the following order. Then, fill the appropriate words and numbers in the blanks below.

4 points per question

(1) Draw a semicircle centered on point O with $\overline{AB}$ of 5 cm in length as the diameter.

(2) Draw a line perpendicular to $\overline{AB}$ from point O, and let C be the intersection point with semicircle O.

(3) Connect point B and point C to make △OCB.

(4) Connect point A and point C to make △OCA.

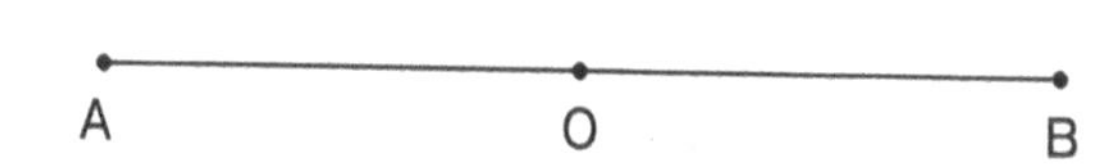

(5) Since $\overline{OC}$ = ☐ ,

△OCB is an ☐ triangle.

(6) ∠OBC = ∠☐ = ☐ °.

(7) In the same way, △OCA is an isosceles triangle. ∠OAC = ☐ °.

(8) In △ABC, ∠ACB = ☐ °.

5 Draw the two equilateral triangles in the following order. Then, fill the appropriate words and numbers in the blanks below.

(1)-(5) 4 points per question (6)-(8) 3 points per question

(1) Draw an arc with the radius of 4 cm in length, centered on point A and point B, $\overline{AB}$ = 4 cm.

(2) Take C and D to be two intersection points on both sides of $\overline{AB}$.

(3) When connecting each point, it becomes equilateral triangle ABC and equilateral triangle ADB.

(4) When connecting point C and point D, it becomes △ACD.

Since $\overline{AC}$ = ☐ in △ACD, △ACD is an isosceles triangle.

(5) ∠DCA = ∠CDA = ☐ °

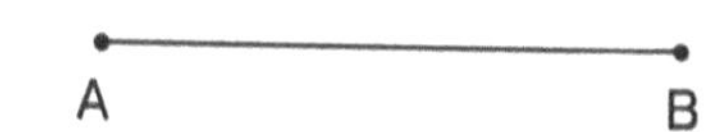

(6) In the same way, since $\overline{BC}$ = ☐ in △BCD, △BCD is an isosceles triangle.

(7) ∠BCD = ∠DCB = ☐ °

(8) ∠CAD is the vertex angle of △ACD. ∠CAD = ☐ °

38 Right Triangles

Level ★

Date / / Name

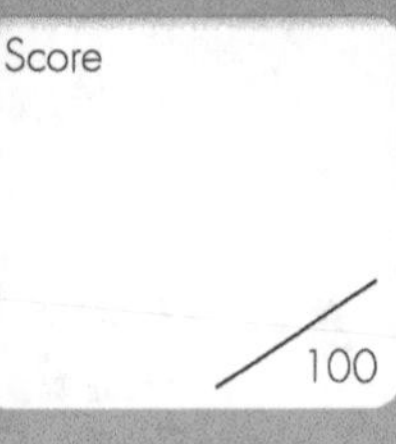

■ The Answer Key is on page 190.

Don't forget!

● In a right triangle, the opposite side of the right angle is called the **hypotenuse**. The hypotenuse is the longest side of a right triangle.

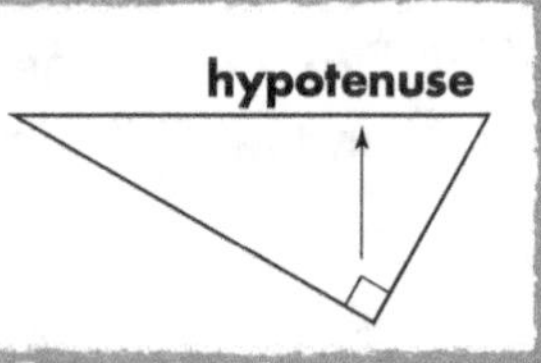

1 **Answer the questions below about the opposite sides and the opposite angles in right triangle ABC on the right.** 5 points per question

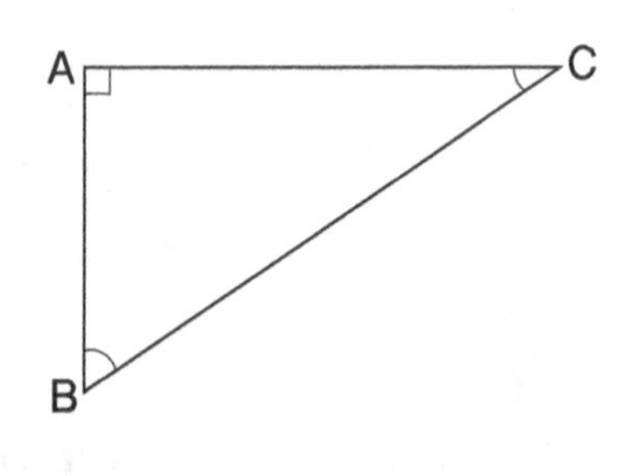

(1) Find the opposite sides of ∠A, ∠B, and ∠C.

〈Ans.〉 ∠A ⇨ , ∠B ⇨ , ∠C ⇨

(2) Find the opposite angles of sides AB, AC, and BC.

〈Ans.〉 AB ⇨ , AC ⇨ , BC ⇨

(3) Which side is the hypotenuse?

〈Ans.〉

2 **Answer the questions below about the hypotenuse, the opposite sides and the opposite angles in right triangle PQR on the right.** 5 points per question

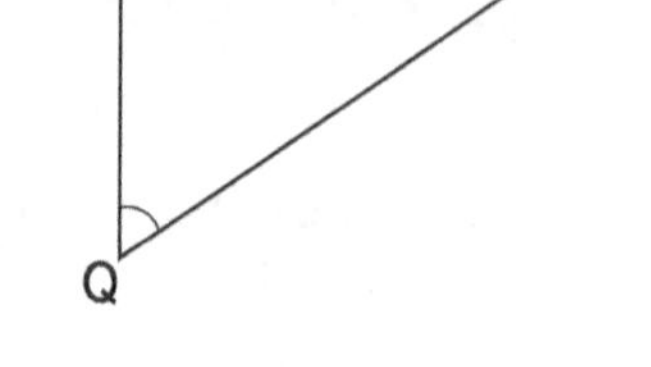

(1) What side is the hypotenuse?

〈Ans.〉

(2) What side are opposite sides of ∠Q and ∠R?

〈Ans.〉 ∠Q ⇨ , ∠R ⇨

(3) Find the measure of the opposite angle of the hypotenuse.

〈Ans.〉

(4) Find the measure of the opposite angle of side PR.

〈Ans.〉

3 **Answer the following questions below about the hypotenuse, the opposite sides and the opposite angles in right triangle DEF with ∠D = 90°.** 5 points per question

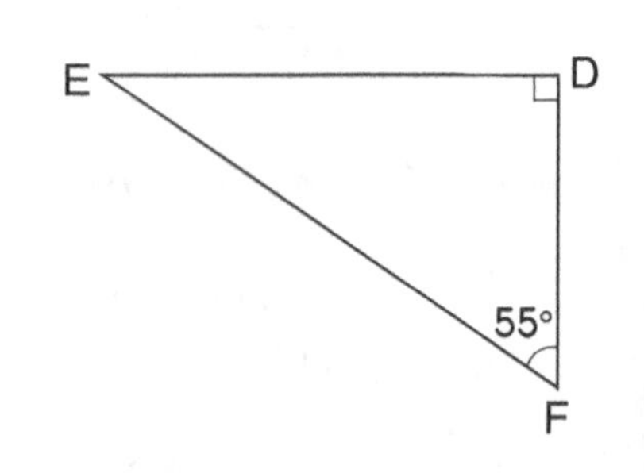

(1) What side is the hypotenuse?

〈Ans.〉

(2) What side are opposite sides of ∠E and ∠F?

〈Ans.〉 ∠E ⇨ , ∠F ⇨

(3) Find the measure of the opposite angle of the hypotenuse.

〈Ans.〉

(4) Find the measure of the angle opposite side DF.

〈Ans.〉

4 Draw the following right triangles, $\triangle ABC$, $\triangle DEF$, and $\triangle PQR$.

5 points per question

(1) Given : $\angle A = 90°$, side $AB = 2$ cm, side $AC = 3$ cm

(2) Given : side $DE = 3$ cm, $\angle D = 90°$, $\angle E = 30°$,

(3) Given : hypotenuse $PQ = 3$ cm, $\angle P = 45°$

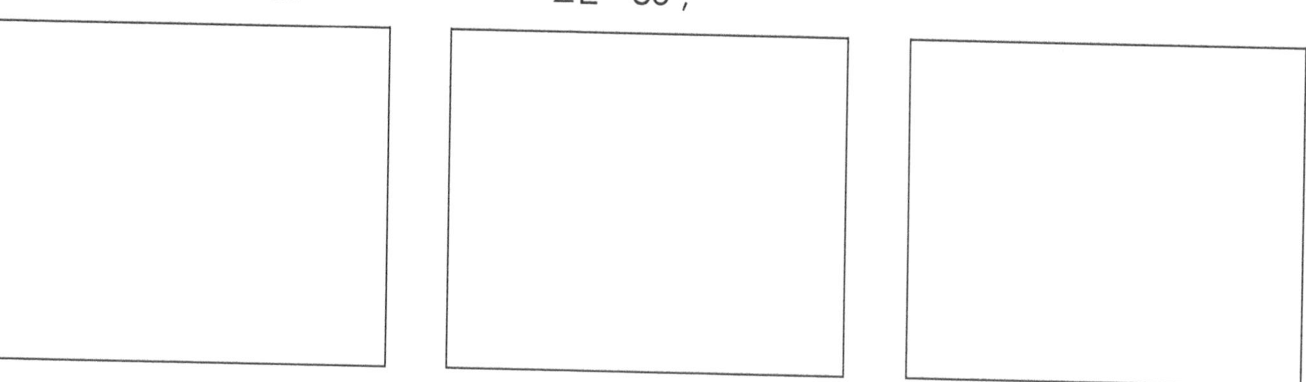

 Draw the right triangle ABC with hypotenuse AB = 5 cm and side BC = 3 cm and fill in the blank.

5 points per question

(1) Draw a circle with a radius of 2.5 cm around middle point O of hypotenuse AB.

(2) Draw an arc with a radius of 3 cm around point B. Then, let one of intersection points with circle O be C.

(3) Connect points A, B, and C to form $\triangle ABC$.
$\triangle ABC$ is a right triangle with $\angle$ ☐ $= 90°$.

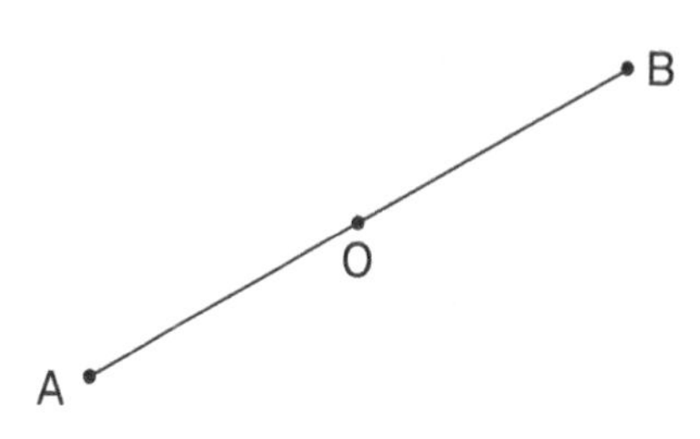

6 Draw right triangle PQR with hypotenuse PQ = 4 cm and $\angle Q = 40°$ and fill in the blank.

5 points per question

(1) Draw a circle with a radius of 2 cm around middle point M of hypotenuse PQ.

(2) Draw the ray QS from the point Q so that $\angle PQS = 40°$. Then let R be the intersection point with circle M.

(3) Connect points P, Q, and R to form $\triangle PQR$.
$\triangle PQR$ is a right triangle with $\angle$ ☐ $= 90°$.

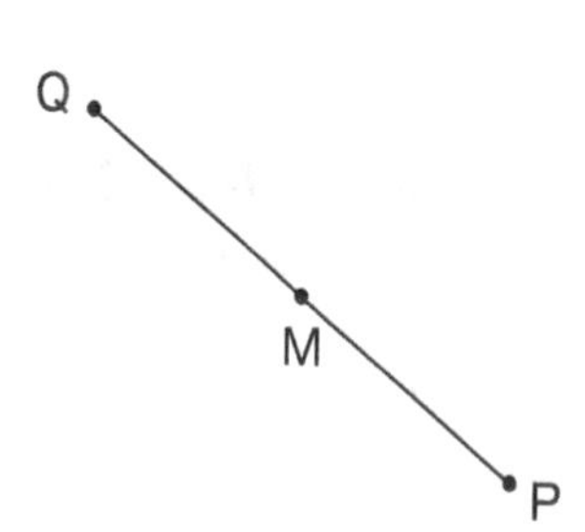

39 Properties of Triangles 1

Level ★★

Date / /

Name

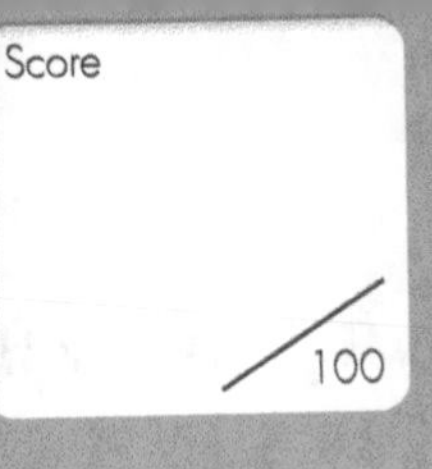

■ The Answer Key is on page 191.

1 **Use a compass to draw the triangle with a side length of 5 cm, 4 cm and 3 cm in the following order.**

4 points per question

(1) Draw line segment AB, which is 5 cm, on the right.

(2) Draw a semicircle with a radius of 4 cm centered on point A on right side of A.

(3) Draw a semicircle with a radius of 3 cm centered on point B on the left side of B. Let two intersection points of semicircles be points C and D.

A • •B

(4) Connect points A, B, and C to form △ABC.

(5) Likewise, connect points A, B, and D to form △ABD. Are △ABC and △ABD congruent?

〈Ans.〉

2 **Draw a triangle with the side length of 5 cm, 3 cm, and 2 cm in the order of question ① above.**

4 points per question

(1) Were you able to draw a triangle? Mark ○ for yes and × for no.

〈Ans.〉

A •————• B

(2) When the length of two sides are 5 cm and 3 cm, in order to make a triangle, how many centimeters should other side be at least?

〈Ans.〉

Don't forget!

●Let a, b and c be the length of triangle's three sides in descending order. ($a \geqq b \geqq c$)
In this case, the following relationship holds.

$a < b + c$

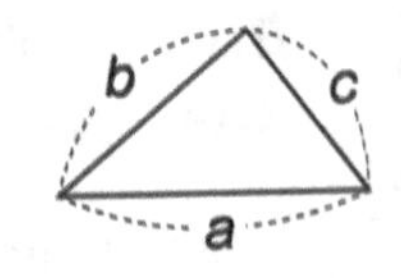

3 **Based on the length of three sides listed below, mark ○ for triangles and × for figures that are not triangles.**

4 points per question

(1) 10 cm · 3 cm · 7 cm 〈Ans.〉

(2) 10 cm · 3 cm · 10 cm 〈Ans.〉

(3) 10 cm · 7 cm · 2 cm 〈Ans.〉

(4) 9 cm · 4 cm · 4 cm 〈Ans.〉

(5) 4 cm · 8 cm · 5 cm 〈Ans.〉

(6) 8 cm · 5 cm · 3 cm 〈Ans.〉

4 **Draw the triangle with two sides of 5 cm and 3 cm and an angle of 30° in the following order. Then answer the questions below.** 4 points per question

(1) Draw $\overline{AB}$ which is 5 cm.

(2) Write ray AP so that ∠BAP from point A is 30°.

(3) Draw the arc with a radius of 3 cm centered on point B to intersect with ray AP.

(4) In this case, how many intersection points of ray AP and arc can be made?

〈Ans.〉

A • • B

(5) How many triangles fit the conditions of the question?

〈Ans.〉

(6) Are these two triangles congruent?

〈Ans.〉

5 **In the following order, draw a triangle in which the length of two sides is 5 cm and 3 cm, and the measure of angle between them is 30°.** 4 points per question

(1) Draw $\overline{AB}$ which is 5 cm.

(2) Draw $\overline{AC}$ with 3 cm in length so that ∠BAC is 30°. In this case, make sure that there is only one point C.

(3) Connect point B and point C with a line in △ABC. In this case, make sure that only △ABC is the triangle that fits the condition of the question.

A • • B

6 **Below are the conditions for a triangle. Answer the correct words and numbers in the blanks.** 3 points per question

(1) The sum of two sides of a triangle is ☐ than the other side.

〈Ans.〉

(2) When three sides matching the condition of (1) are given, there is ☐ triangle that can be formed.

〈Ans.〉

(3) When the length of two sides and one angle are given, triangles can be made in ☐ ways.

〈Ans.〉

(4) When the length of two sides and the angle between them are given, the triangle is ☐ .

〈Ans.〉

40 Properties of Triangles 2

Date / / Name

■ The Answer Key is on page 191.

1 **In the following order, draw a triangle that fits the conditions where the length of one side is 4 cm and the two angles at both ends are 60° and 70°. Then, answer the questions below.** 5 points per question

(1) Find the measure of the three angles of this triangle.

〈Ans.〉 , ,

(2) Draw $\overline{AB}$ of 4 cm on the right. Next, draw ray AP so that ∠BAP = 60°. In the same way, draw the ray BR so that ∠ABR = 70°. When the intersection point between ray AP and ray BR is C, make sure that connecting points A, B, and C with a line will be △ABC.

(3) For this problem, let ∠ABR be 50°. Place intersection C of AP and BR by intersectinging ray AP of ∠PAB = 60° and ray BR of ∠ABR = 50° below line AB in (2). Connecting points A, B, and C with a line will be another triangle ABC.

A • 60° 70° • B

(4) Does △ABC of (2) and △ABC of (3) map onto? Answer with ○ for yes or × for no.

〈Ans.〉

2 **Draw the triangle with a side length of 4 cm and angles at both ends of 50° and 60° in the following order.** 6 points per question

(1) Draw $\overline{AB}$ of 4 cm on the right.

(2) Draw ray AP so that ∠BAP = 60°. In the same way, draw ray BR so that ∠ABR = 50°.

(3) Let C be the intersection point of ray AP and ray BR. In this case, make sure there is only one intersection point.

(4) Connect points A, B, and C to form △ABC. At this time, make sure that △ABC is the only triangle that fits the conditions of the question.

A • 60° 50° • B

Don't forget!

●If only one side and two angles (not at each end point) are given, you can draw two triangles. Once you know the angles at each end of a line segment, you can draw only one triangle.

3 **Below are triangles with the following conditions. Among them, what is $\triangle ABC$ with only one size and shape? Mark ○ for one shape or × for more than one shape.**

6 points per question

(1) $AB = 3$ cm · $AC = 6$ cm · $\angle A = 70°$

〈Ans.〉

(2) $AB = 6$ cm · $BC = 4$ cm · $\angle A = 25°$

〈Ans.〉

(3) $\angle A = 80°$ · $\angle B = 30°$ · $\angle C = 70°$

〈Ans.〉

(4) $AB = 5$ cm · $BC = 6$ cm · $CA = 7$ cm

〈Ans.〉

(5) $\angle C = 50°$ · $\angle B = 30°$ · $BC = 5$ cm

〈Ans.〉

4 **The following are sentences about triangles. Find the appropriate words and numbers in the blanks below.**

(1) (2) 6 points per question
(3) (4) 7 points per question

(1) When $AB = 5$ cm, $BC = 6$ cm, $CA = 7$ cm in $\triangle ABC$, the largest angle is $\angle$☐.

〈Ans.〉

(2) When $\angle A = \angle B = 70°$, the shortest side is side ☐.

〈Ans.〉

(3) The shape of a triangle remains the same when the measure of its three angles are fixed, but the length of the ☐ can vary.

〈Ans.〉

(4) $\triangle ABC$ has the shortest side length of 5 cm, $\angle A = 60°$ and $\angle B = 70°$. The area of $\triangle ABC$ is the largest when side ☐ is 5 cm.

〈Ans.〉

Don't forget!

●When drawing a triangle, if one of the following conditions is known, one fixed triangle can be drawn.

① The length of three sides
② The length of two sides and the measure of the angle between them
③ The length of one side and the measure of the angles at both ends

①

②

③

41 Review 1

Level ★★★

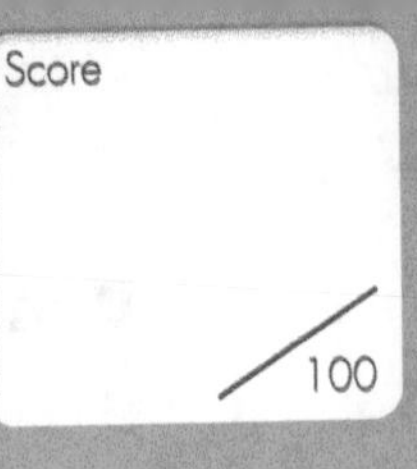

Date / /

Name

■ The Answer Key is on page 191.

1 **Find the surface the area of the prism and the cylinder below.** 6 points per question

(1)

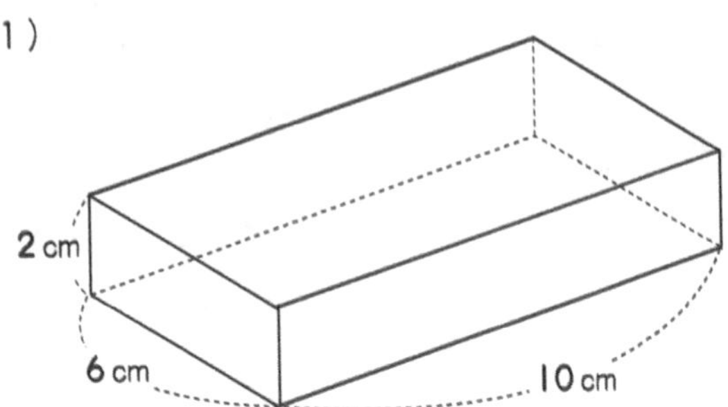

⟨Ans.⟩ ______

(2) (Pi = π)

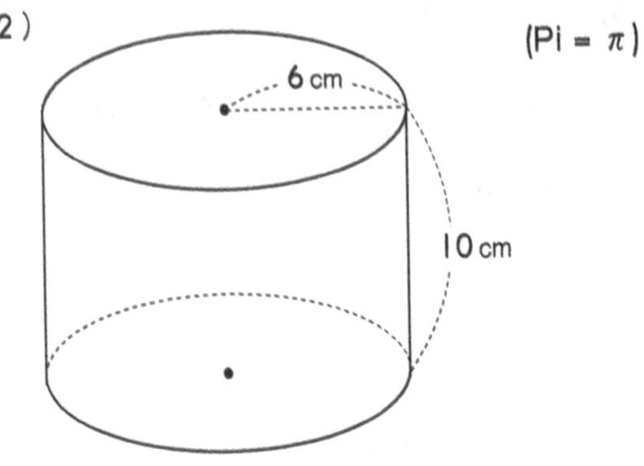

⟨Ans.⟩ ______

2 **Rotate △ABC to the right by one turn around side AB and it becomes a cone. Use this information to answer the questions below.** 6 points per question

p.192

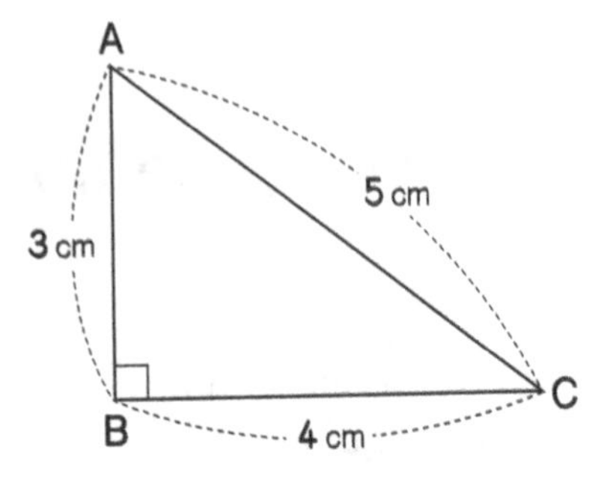

(1) Find the volume of the cone.
(Pi = π)

⟨Ans.⟩ ______

(2) Next, rotate △ABC one time around the side BC. Find the volume of the cone.
(Pi = π)

⟨Ans.⟩ ______

3 **On the right, is a cube. Answer the questions below.** 6 points per question

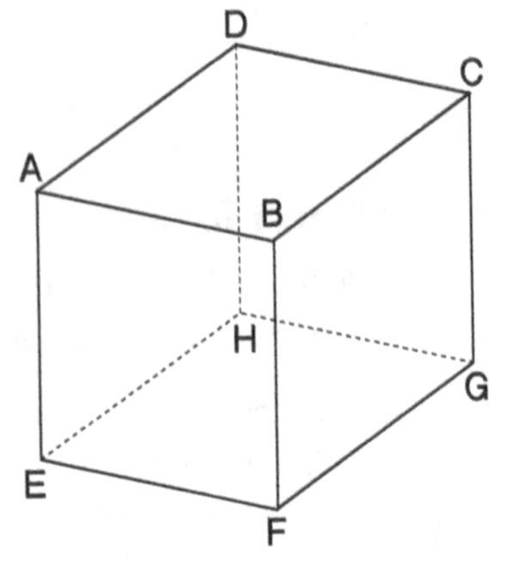

(1) Which surfaces are parallel to edge HG?

⟨Ans.⟩ ______, ______, ______

(2) How many edges are in a skewed position with respect to edge AD.

⟨Ans.⟩ ______

(3) Which surfaces are perpendicular to edge BF?

⟨Ans.⟩ ______, ______

4 **Answer the questions below about a triangular prism on the right.** 6 points per question

(1) Find the surface area.

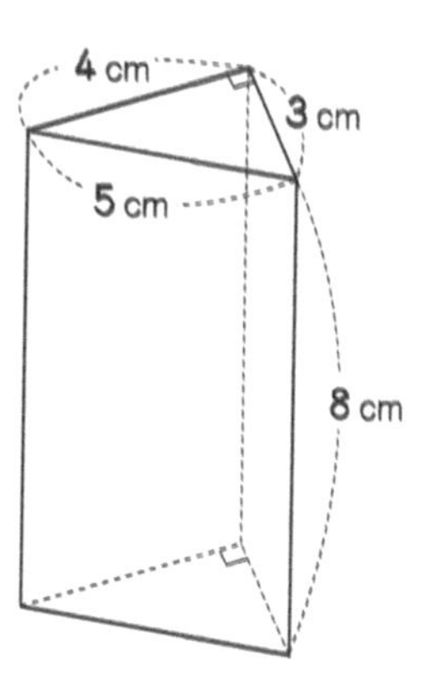

(2) Find the volume.

⟨Ans.⟩ ______________

5 **Answer the questions below about the net of the cone on the right.** 7 points per question

(1) Find the length of the arc in the sector.

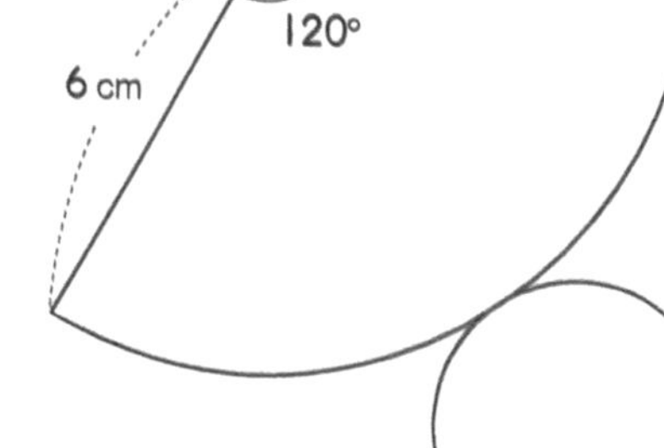

⟨Ans.⟩ ______________

(2) Find the radius of the base.

⟨Ans.⟩ ______________

(3) Find the area of the sector.

⟨Ans.⟩ ______________

(4) Find the surface area of the cone.

⟨Ans.⟩ ______________

6 **On the right is a cylinder with radius of 3 cm and height of 6 cm, and a cone that fits comfortably inside the cylinder. Answer the questions below.** 6 points per question

p.192

(1) Find the volume of the cylinder.

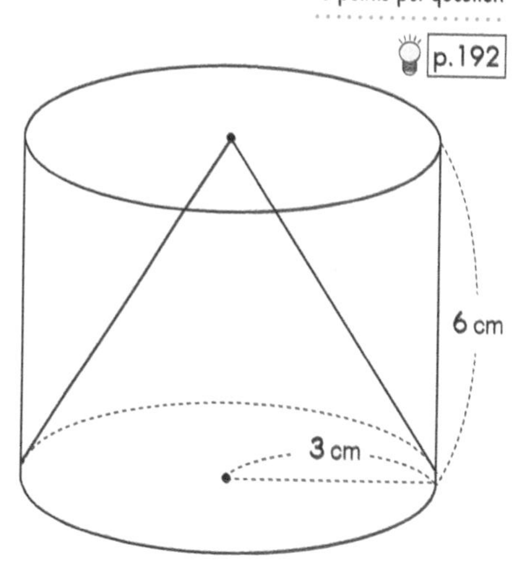

⟨Ans.⟩ ______________

(2) Find the volume of the cone.

⟨Ans.⟩ ______________

(3) Show the ratio of the volume of the cylinder to the volume of the cone. Show your answer in simplest terms.

⟨Ans.⟩ ______________

Review 2

■ The Answer Key is on page 191.

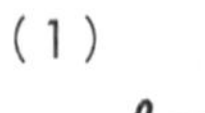

Find the measure of $\angle x$ and $\angle y$ below, given $\ell \parallel m$.

8 points per question

(1)

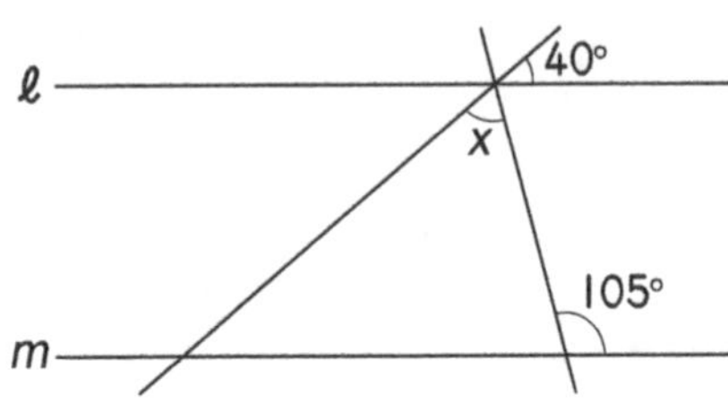

⟨**Ans.**⟩ $\angle x =$

(2)

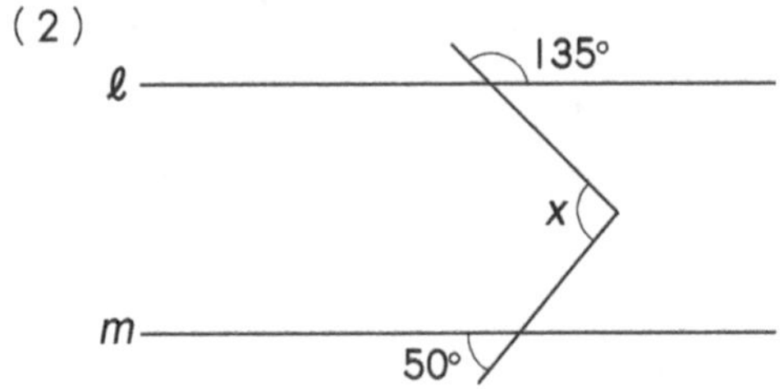

⟨**Ans.**⟩ $\angle x =$

(3)

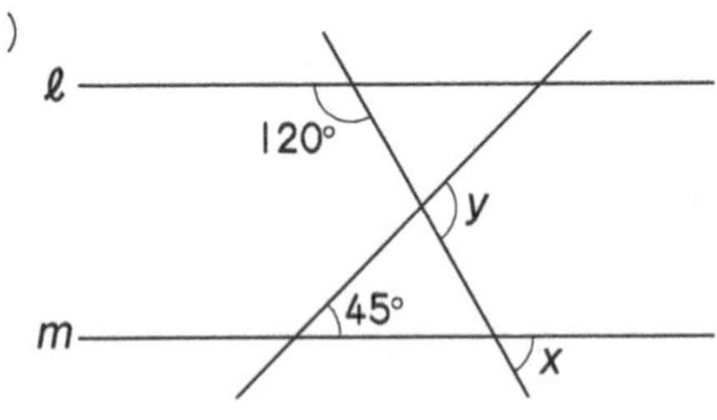

⟨**Ans.**⟩ $\angle x =$, $\angle y =$

(4)

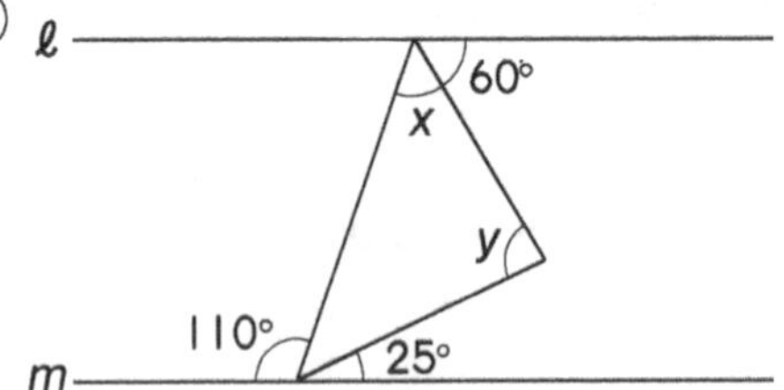

⟨**Ans.**⟩ $\angle x =$, $\angle y =$

(5)

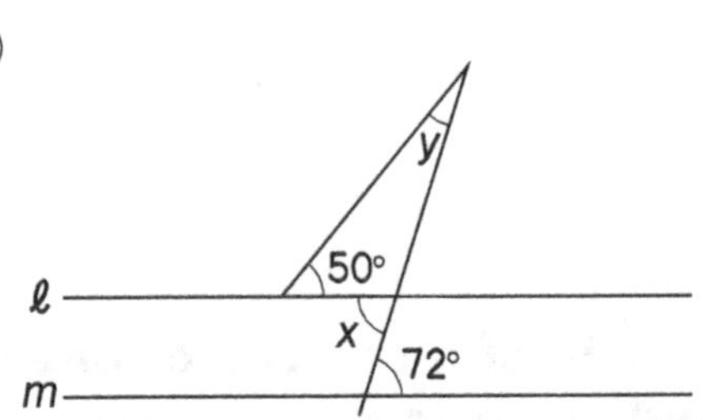

⟨**Ans.**⟩ $\angle x =$, $\angle y =$

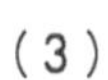

Answer the questions below.

6 points per question

(1) Combine the sum of the exterior angles and the sum of the interior angles of a triangle. Find the sum.

⟨**Ans.**⟩

(2) Find the sum of the interior angles of a hexagon.

⟨**Ans.**⟩

(3) Find the sum of the exterior angles of an octagon.

⟨**Ans.**⟩

(4) Find the measure of one interior angle of a regular octagon.

⟨**Ans.**⟩

3 Find the measure of $\angle x$ and $\angle y$ below.

6 points per question

(1)

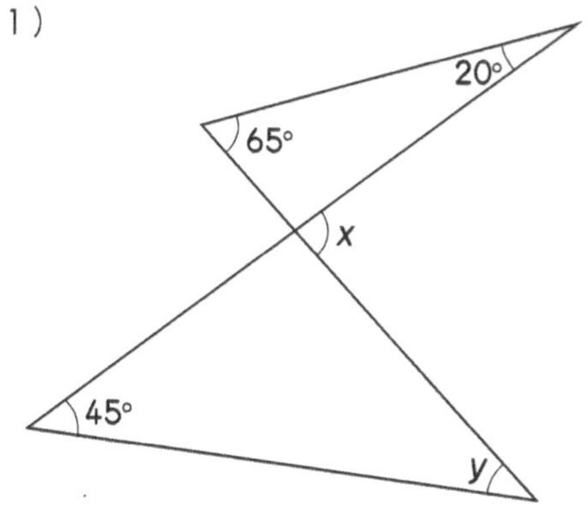

〈Ans.〉 $\angle x =$, $\angle y =$

(2)

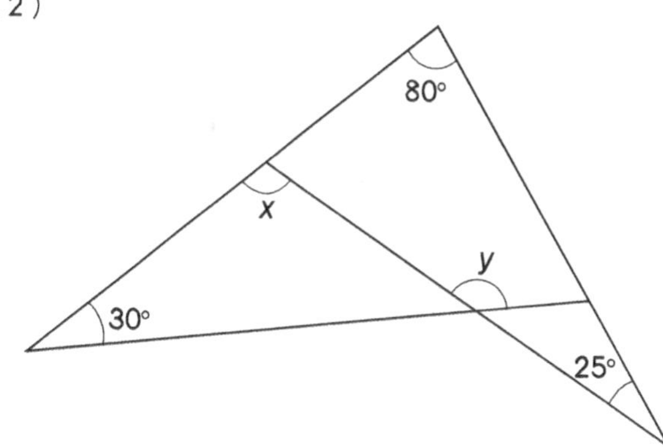

〈Ans.〉 $\angle x =$, $\angle y =$

(3)

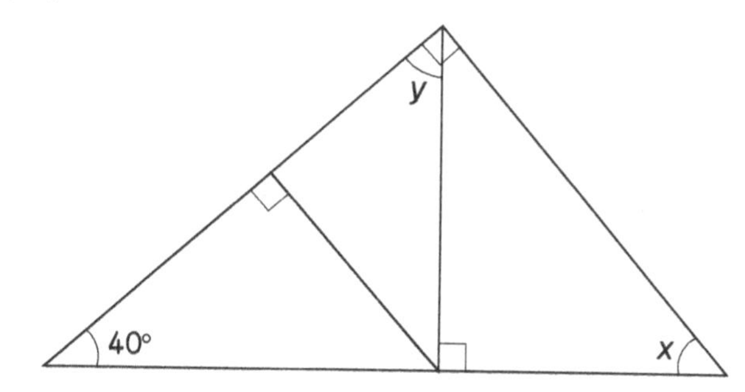

〈Ans.〉 $\angle x =$, $\angle y =$

(4)

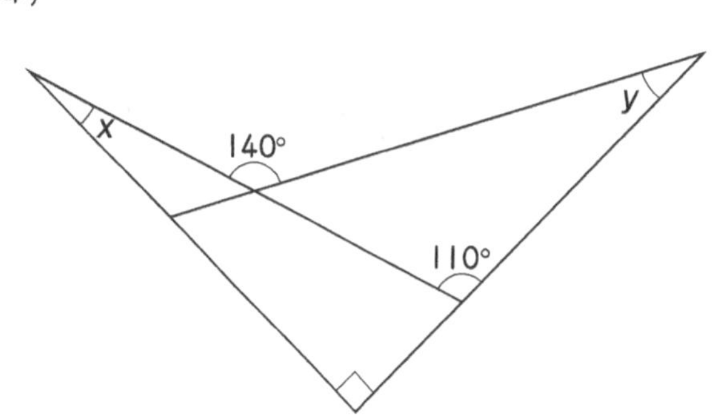

〈Ans.〉 $\angle x =$, $\angle y =$

4 Find the measure of $\angle x$ and $\angle y$ below.

6 points per question

(1)

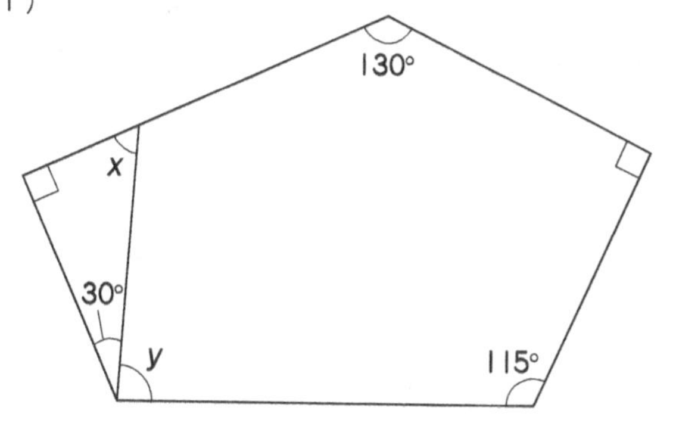

〈Ans.〉 $\angle x =$, $\angle y =$

(2)

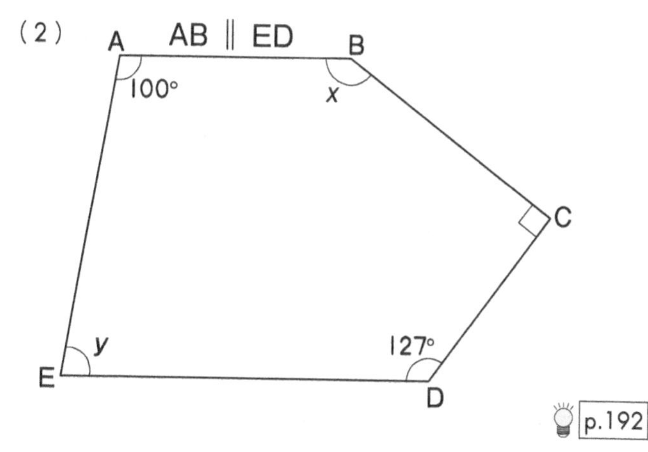

p.192

〈Ans.〉 $\angle x =$, $\angle y =$

43 Review 3

Level 

Date / / Name

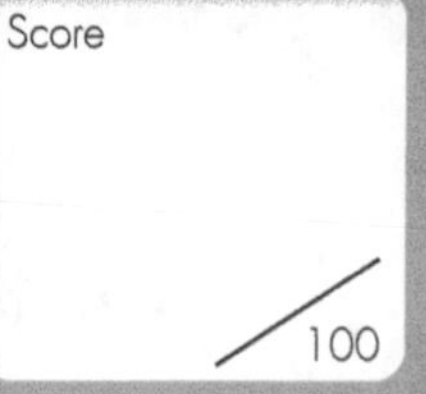

■ The Answer Key is on page 191.

1 **On the right, circle O circumscribes square ABCD. Answer the questions below.** 3 points per question

(1) On diagonal AC, which line segment has the same length as CO?

⟨Ans.⟩

(2) Find the measure of central angle BOC of arc BC.

⟨Ans.⟩

(3) Find the measure of inscribed angle BDC of arc BC.

⟨Ans.⟩

(4) There is another inscribed angle of arc BC. What is the inscribed angle?

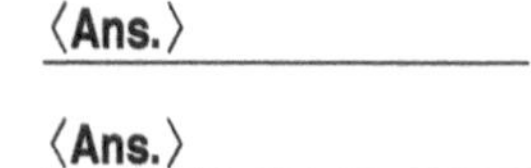

⟨Ans.⟩

(5) Find the measure of central angle AOC of arc ABC. ⟨Ans.⟩

(6) Find the measure of the inscribed angle of arc ABC. ⟨Ans.⟩

2 **Find the measure of $\angle x$.**

5 points per question

(1)

⟨Ans.⟩ $\angle x =$

(2)

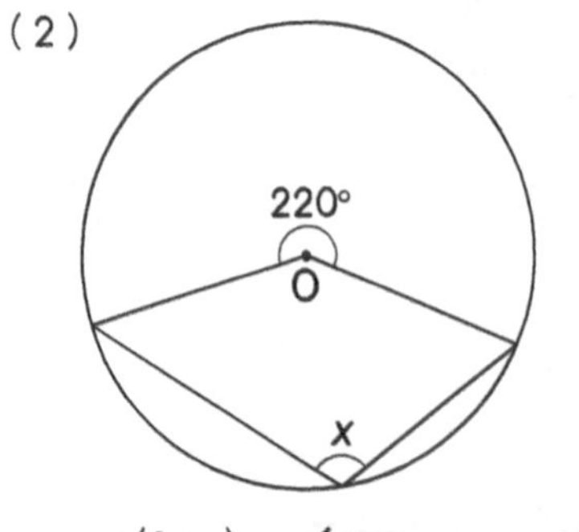

⟨Ans.⟩ $\angle x =$

(3)

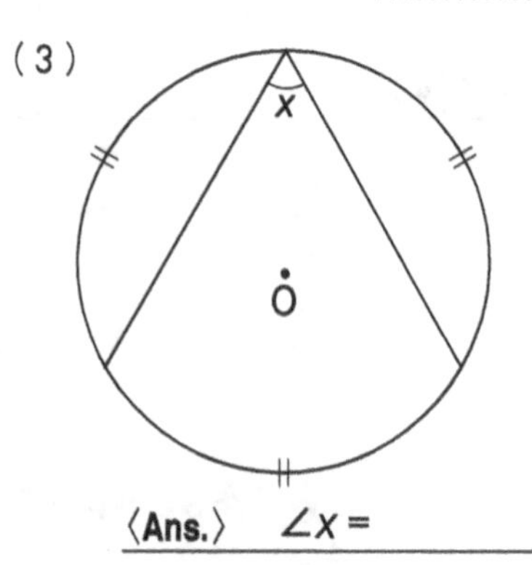

⟨Ans.⟩ $\angle x =$

3 **There is point P on the circumference of semicircle O with diameter AB. Given $\angle$PBA = 35°, answer the questions below.** 3 points per question

(1) Find the measure of $\angle$OPB.

⟨Ans.⟩

(2) Find the measure of $\angle$POA.

⟨Ans.⟩

(3) Find the measure of $\angle$APO.

⟨Ans.⟩

(4) Find the measure of $\angle$APB. ⟨Ans.⟩

(5) Given $\angle$PBA = 40°, find the measure of $\angle$APO. ⟨Ans.⟩

4 **The following is a method to draw equilateral triangle ABC inscribed in circle O. Fill the appropriate words and numbers in the blanks below.** 4 points per question

(1) Draw circle P with the same ☐ as circle O around point P on the circumference of circle O.

(2) Let A and B be the ☐ points of circle O and circle P.

(3) Draw chord ☐ by connecting point A and point B.

(4) Draw circle A with radius AB around point A. Let C be the intersection point of circle A and circle O. Connecting Point A, Point B, and Point C will result in $\triangle ABC$. Since $AB = AC = CB$, $\triangle ABC$ is ☐ triangle.

O

P

(5) Since $\triangle ABC$ is an equilateral triangle, $\angle ACB$ is 60°. $\angle ACB$ is the ☐ angle of arc APB.

(6) Since $\angle AOB$ is the center angle of arc APB, it is ☐°.

(7) Next, let D be the intersection point with circle O when CO is extended. In this case, point D overlaps with point ☐. CD passing through point O is ☐ of circle O.

(8) Since side CD is the diameter of circle O, $\angle DBC$ is ☐°.

(9) Put a point E on arc BC. $\angle AEB$ is the ☐ angle of arc AB like $\angle ACB$.

(10) $\angle AEB$ is ☐°.

5 **The followings are descriptions of triangles. Fill the appropriate words and numbers in the blanks below.** 3 points per question

(1) $\triangle ABC$ has different lengths for all three sides. In order for side AB to be the longest, opposite angle ☐ can be set to ☐° or more.

(2) Given $\angle A = 30°$ and $\angle B = 80°$ in $\triangle ABC$, the second longest side is side ☐.

(3) When the perimeter of a triangle is 20 cm long, the longest side cannot exceed ☐ cm.

(4) Make a triangle with three lines. If the two lines are 5 m and 3 m, the length of the remaining line must be longer than ☐ m and shorter than ☐ m.

1 Line & Plane 1 pp 98, 99

1 (1) AB, BC, CD, DA (2) BC, CG, GF, FB
(3) CD, DH, HG, GC (4) CG, DH, AE, BF
(5) BA, FE, GH, CD (6) DA, CB, GF, HE
(7) CD (8) CG
(9) a (ABCD) (10) ABFE

2 (1) *f* (FGHE) (2) *d* (DAEH)
(3) *e* (EABF) (4) FG, AD, EH
(5) CG, DH, AE (6) GH, FE, BA
(7) *d*, *e*, *b*, *c* (8) *e*, *f*, *c*, *a*
(9) *d*, *f*, *b*, *a* (10) AE, BF, CG, DH
(11) DA, CB, GF, HE (12) FG, EH, EF, HG
(13) AB, BF, FE, EA (14) AD, EH, CD, GH
(15) AE, DH, EF, HG

2 Line & Plane 2 pp 100, 101

1 (1) AB ∥ Y (2) BC ∥ Y (3) BF ⊥ Y (4) DH ⊥ X
(5) FG ∥ X (6) CD ∥ Y (7) AE ⊥ X (8) AB ⊥ Z
(9) EH ∥ Z (10) GH ⊥ Z

2 (1) X ⊥ Z (2) Y ⊥ Z (3) AEHD ⊥ Y (4) BC
(5) FG (6) CD (7) BC ∥ W (8) AD
(9) FG (10) X (11) Z (12) CDHG
(13) EF, GH, CG, BF

3 Line & Plane 3 pp 102, 103

1 (1) $\ell \parallel$ Y (2) always (yes)
(3) $m \perp$ Y (4) not (no)

2 (1) $m \perp$ X (2) $n \perp \ell$
(3) $n \parallel$ Y (4) always (yes)

3 (1) $n \parallel$ X (2) $\ell \perp m$
(3) $n \parallel k$

4 (1) always (yes) (2) always (yes)
(3) always (yes) (4) not (no)
(5) $k \perp n$ (6) always (yes)

4 Line & Plane 4 pp 104, 105

1 (1) 1 ○ 2 ○ 3 ○ 4 ×
5 × 6 ○ 7 ○

(2) 1 parallel 2 intersect
3 parallel 4 skewed position
5 skewed position 6 parallel
7 intersect

2 (1) AD, BE, AC, BC
(2) DE
(3) CF, FE, DF

3 (1) CD, EF, GH (2) BF, AE, BC, AD
(3) CG, DH, FG, EH (4) CD, DH, HG, GC
(5) DCGH (6) AE, BF, CG, DH
(7) BA, FE, GH, CD

4 (1) DE, EF, BE (2) AD
(3) AD, BE, CF (4) DEF
(5) BC, EF

5 Distance in Space pp 106, 107

1 (1) 5 cm (2) 3 cm (3) 3 cm (4) 3 cm
(5) 3 cm (6) 8 cm (7) FG (8) BF

2 (1) 5 cm (2) 4 cm (3) 3 cm (4) 6 cm
(5) BC, EF (6) AB, AC, AD

3 (1) 5 cm (2) 5 cm

6 Solids 1 pp 108, 109

1 (1) regular quadrangular prism
(2) regular pentagonal prism
(3) regular triangular prism
(4) regular hexagonal prism

2 (1) equilateral triangle (2) 2
(3) rectangle (4) 3

3 (1) regular pentagon (2) 2
(3) rectangle (4) 5

4 (1) regular hexagon (2) 2
(3) rectangle (4) 6

5 (1) AF, EJ, DI (2) ABCDE, FGHIJ
(3) ABCDE, FGHIJ (4) FGHIJ
(5) AF, BG, DI, EJ (6) 7

6 (1) DJ, EK, FL, AG, FE, LK
(2) ABCDEF, GHIJKL
(3) 4
(4) GHIJKL, EFLK
(5) IJ, JK, GL, GH, DJ, EK, FL, AG

7 Solids 2 pp 110, 111

1 (1) × (2) ○ (3) ○ (4) ×
(5) ○ (6) ×

2 (1) square (2) 1
(3) isosceles triangle (4) 4

3 (1) regular hexagon (2) I
(3) isosceles triangle (4) 6

4 (1) AD, BC, OD, OC (2) DC
(3) OD, OC (4) AD, DC
(5) OA, OB, AB

5 (1) ODA, ABCD (2) 3 cm
(3) 4 cm (4) 5 cm
(5) AB, BC

8 Net 1

pp 112, 113

1 (1) 9 cm (2) IH
(3) HG (4) G
(5) M, I (6) ④
(7) ①
(8) ②, ③, ④, ⑥
(9) 5 cm (10) 9 cm

2 (1) ⑤ (2) ED
(3) ①, ⑤ (4) intersect
(5) skewed position

3 (1)

(2) ①, ③ (3) J
(4) ④ (5) ED
(6) BC (DC), IH (GH) (7) CH
(8) ⑤

9 Net 2

pp 114, 115

1 (1) H (2) AE, CG (3) BF, DH (4) FH

2 (1) ⑤ (2) ED
(3) ③, ⑥ (4) skewed position
(5)

3 (1) *b*, *f* (2) *a*, *d* (3) *f* (4) *c*, *e*
(5) *d* (6) *f*

4 (1) $\ell \parallel Y$ (2) $m \perp X$ (3) $\ell \perp Y$ (4) $\ell \parallel m$

5 (1) × (2) ○ (3) × (4) ○

10 Cross Section of a Solid

pp 116, 117

1 (1) isosceles triangle (2) $X \perp Y$
(3) $\ell \perp$ ABCD (4) The vertex O and the point G
(5) isosceles triangle (6) O
(7) ④ (8) square

2 (1) equilateral triangle (2) isosceles triangle
(3) ② (4) square

3 (1) rectangle (2) regular hexagon

11 Solids 3 (Cylinder & Cone)

pp 118, 119

1 (1) ○ (2) △ (3) ○ (4) △
(5) △ (6) ○

2 (1)

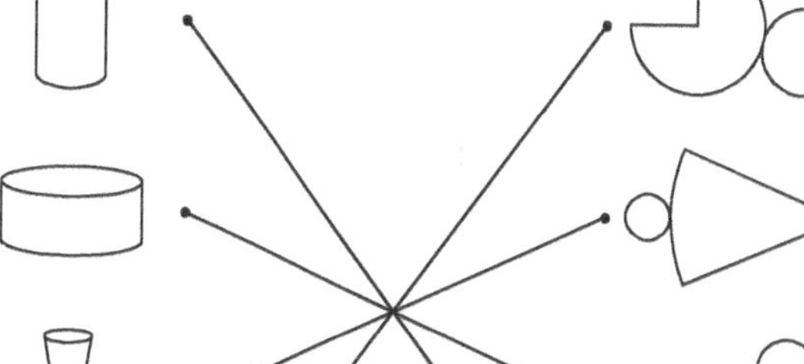

(2)

(3)

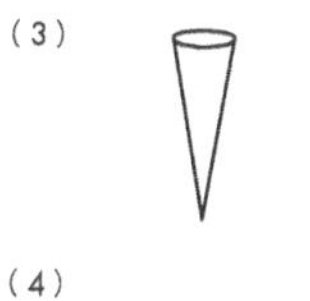

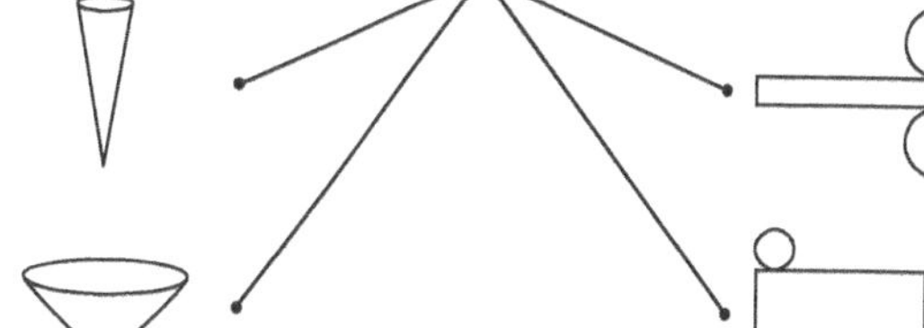

(4)

3 (1) 8π cm (2) rectangle
(3) *a* : 6 cm, *b* : 8π cm

4 (1) *c*, *b*, *a* (2) ① B, ② C, ③ A

12 Projection of a Solid

pp 120, 121

1 (1)

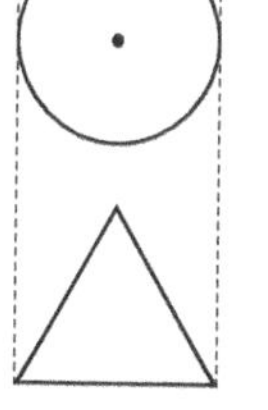

(2)

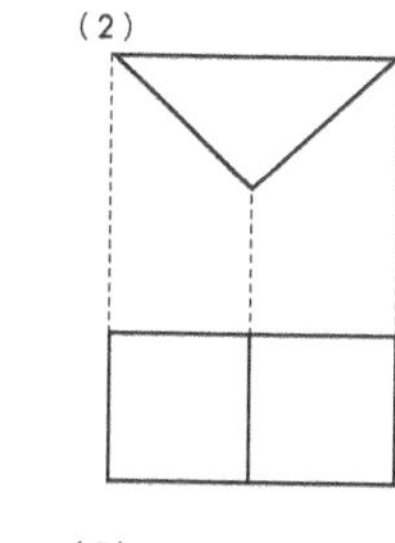

(3)

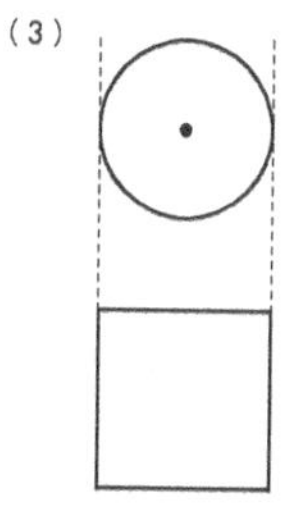

(4)

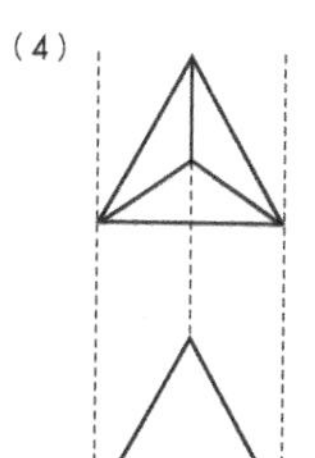

(5)

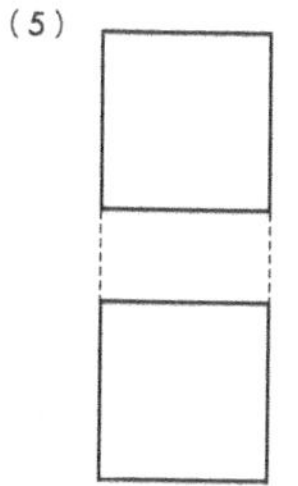

2 (1) sphere

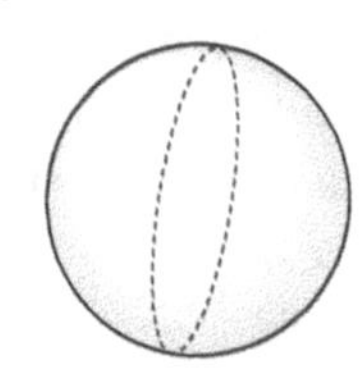

(2) regular hexagonal prism

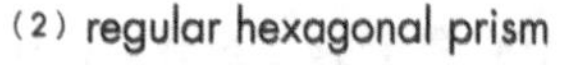

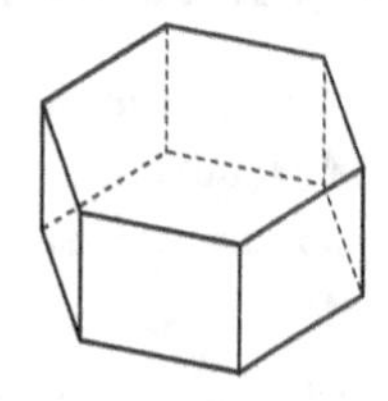

(3) triangular pyramid

(4) regular pentagonal pyramid

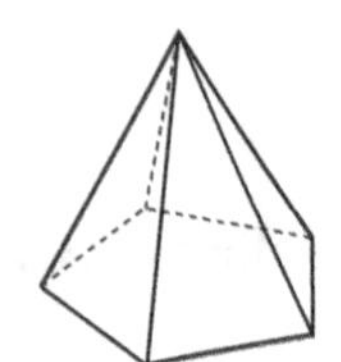

(5) regular rectangular prism

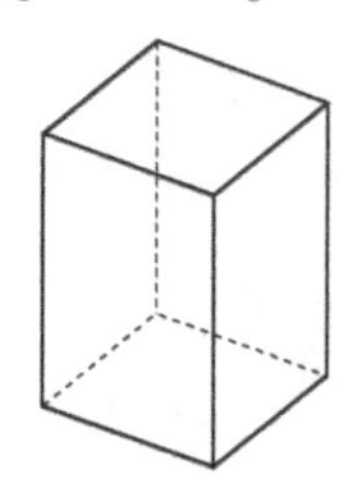

13 Volume 1 (Prism & Cylinder) pp 122, 123

1 (1) 15 in.^{2} (2) 45 in.^{3}

2 (1) 24 in.^{2} (2) 48 in.^{3}

3 (1) 100π in.^{2} (2) 1200π in.^{3}

4 (1) 140 in.^{3} (2) 1000 cm^3 (3) 240 in.^{3}
(4) 120 in.^{3} (5) 200π cm^3 (6) 320π cm^3
(7) 1600 cm^3 (8) 240 cm^3

14 Volume 2 (Pyramid & Cone) pp 124, 125

1 (1) 24 cm^2 (2) 80 cm^3

2 (1) 36 cm^2 (2) 96 cm^3

3 (1) 25π cm^2 (2) 75π cm^3

4 (1) 84 in.^{3} (2) 30 in.^{3} (3) 80 in.^{3} (4) 80 in.^{3}
(5) 200π cm^3 (6) 80π cm^3 (7) 120 cm^3 (8) 30 cm^3

15 Nets of Prisms & Cylinders pp 126, 127

1 (1) 4 cm (2) 6 cm^2
(3) 12 cm (4) 48 cm^2

2 (1) 10 cm (2) 8 cm^2
(3) 12 cm (4) 120 cm^2
(5) 136 cm^2

3 (1) 6 cm (2) 6π cm
(3) 6π cm (4) 4 cm
(5) 24π cm^2

4 (1) 4π cm^2 (2) 4π cm
(3) 4π cm (4) 32π cm^2
(5) 40π cm^2

16 Surface Area 1 (Prism & Cylinder) pp 128, 129

1 (1) area of the base: 6 in.^{2}
surface area: 84 in.^{2}
(2) area of the base: 18 in.^{2}
surface area: 216 in.^{2}
(3) area of the base: 9π in.^{2}
surface area: 78π in.^{2}
(4) area of the base: 16 in.^{2}
surface area: 224 in.^{2}
(5) area of the base: 6 in.^{2}
surface area: 108 in.^{2}
(6) area of the base: 25π in.^{2}
surface area: 170π in.^{2}

2 (1) lateral area: 120 cm^2
surface area: 168 cm^2
(2) lateral area: 128 cm^2
surface area: 160 cm^2
(3) lateral area: 100π cm^2
surface area: 150π cm^2
(4) lateral area: 96π cm^2
surface area: 128π cm^2
(5) lateral area: 160 cm^2
surface area: 184 cm^2
(6) lateral area: 144 cm^2
surface area: 192 cm^2

17 Net of a Pyramid pp 130, 131

1 ③, ⑤

2 (1) isosceles triangle (2) 25 cm^2
(3) 100 cm^2 (4) 25 cm^2
(5) 125 cm^2

3 (1) lateral area: 300 cm^2
surface area: 400 cm^2
(2) lateral area: 140 cm^2
surface area: 189 cm^2

4 (1) 96 cm^2 (2) 132 cm^2
(3) 81 cm^2 (4) 105 cm^2

18 Net of a Cone

pp 132, 133

1 (1) 6π cm (2) circumference of the base
(3) 18π cm (4) $\frac{1}{3}$
(5) 120° (6) 27π cm²
(7) 9π cm² (8) 36π cm²

2 (1) 90° (2) 60°

3 (1) 2 cm (2) 4π cm²
(3) 8π cm²

4 (1) 16 cm (2) 64π cm²

5 (1) 48π cm² (2) 24π cm²
(3) 96π cm² (4) 60π cm²

19 Surface Area 2 (Cone & Pyramid)

pp 134, 135

1 (1) 6 cm² (2) 72 cm²
(3) 84 cm²

2 (1) 16 cm² (2) 64 cm²
(3) 80 cm²

3 (1) 36π cm² (2) 144π cm²
(3) 216π cm²

4 (1) 16π cm² (2) 32π cm²
(3) 48π cm²

5 (1) 105 in.² (2) 260 in.²
(3) 156 in.² (4) 80π in.²
(5) 108π in.² (6) 100π in.²

20 Volume & Surface Area (Sphere)

pp 136, 137

1 (1) $\frac{32}{3}\pi$ cm³ (2) $\frac{256}{3}\pi$ cm³
(3) 288π cm³

2 (1) 36π cm² (2) 144π cm²
(3) 324π cm²

3 (1) circle (2) $\frac{128}{3}\pi$ cm³
(3) 48π cm²

4 (1) sphere (2) $\frac{500}{3}\pi$ cm³
(3) 100π cm²

5 (1) $\frac{4}{3}\pi$ cm³ (2) $\frac{\pi}{6}$

21 Parallel Lines & Angles 1

pp 138, 139

1 (1) 60° (2) 110°
(3) a: 40°, b: 40° (4) a: 120°, b: 60°

2 (1) $\angle d$ (2) $\angle e$
(3) $\angle C = 180° - (\angle b + \angle d)$

3 (1) 60° (2) 80°
(3) 80°

4 (1) $\angle e$ (2) $\angle f$
(3) $\angle g$ (4) $\angle h$

5 (1) $\angle g$, 120° (2) $\angle e$, 120°
(3) $\angle e$, 120° (4) $\angle g$, 120°
(5) $\angle f$, 60° (6) $\angle g$, 120°

22 Parallel Lines & Angles 2

pp 140, 141

1 (1) corresponding angle (2) alternate interior angle
(3) $\angle a = \angle e = \angle g$, $\angle b = \angle f = \angle h$
(4) 50°

2 (1) 30° (2) 150°
(3) 30°

3 (1) $j \parallel m$, $k \parallel \ell$ (2) $\angle a = \angle d$, $\angle b = \angle c$

4 (1) corresponding angle (2) alternate exterior angle
(3) $\angle a = \angle e = \angle g$, $\angle b = \angle f = \angle h$
(4) 60°

5 (1) 125° (2) 125°

6 (1) $j \parallel \ell$, $k \parallel m$ (2) $\angle a = \angle c$, $\angle b = \angle d$

23 Parallel Lines & Angles 3

pp 142, 143

1 (1) $\angle e$, $\angle p$ (2) $\angle g$, $\angle r$
(3) $\angle b$, $\angle u$ (4) $\angle g$, $\angle r$
(5) $\angle d = 75°$, $\angle g = 105°$, $\angle q = 75°$, $\angle v = 105°$

2 (1) $\angle x = 60°$, $\angle y = 120°$
(2) $\angle x = 115°$, $\angle y = 65°$
(3) $\angle x = 110°$, $\angle y = 60°$
(4) $\angle x = 70°$, $\angle y = 110°$

3 (1) 50° (2) 130°
(3) 80° (4) 100°
(5) 120° (6) 125°

4 (1) $\angle x = 40°$, $\angle y = 50°$
(2) $\angle x = 45°$, $\angle y = 110°$
(3) 95° (4) 130°
(5) 35° (6) 75°

24 Parallel Lines & Angles 4 pp 144, 145

1 (1) ○ (2) ○ (3) ×

2 (1) $\angle x = 65°$, $\angle y = 65°$
(2) $\angle x = 45°$, $\angle y = 135°$
(3) $\angle x = 140°$, $\angle y = 40°$
(4) $\angle x = 65°$, $\angle y = 65°$

3 (1) $\angle a = \angle b$ (2) $\angle a + \angle c = 180°$
(3) 125°

4 (1) $\angle x = 130°$, $\angle y = 110°$
(2) $\angle x = 70°$, $\angle y = 65°$
(3) $\angle x = 130°$, $\angle y = 40°$
(4) $\angle x = 115°$, $\angle y = 25°$

25 Opposite Angle & Opposite Side pp 146, 147

1 (1) ∠A ⇨ BC, ∠B ⇨ AC, ∠C ⇨ AB
(2) AB ⇨ ∠C, AC ⇨ ∠B, BC ⇨ ∠A

2 (1) ∠P ⇨ QR, ∠Q ⇨ PR, ∠R ⇨ PQ
(2) PQ ⇨ ∠R, QR ⇨ ∠P, PR ⇨ ∠Q

3 (1) EF (2) DF (3) DE
(4) ∠E (5) ∠F (6) ∠D

4 right triangle: b, f, h
acute triangle: a, c, g
obtuse triangle: d, e

5 (1) side: a [3], b [2], c [1]
angle: A [3], B [2], C [1]
(2) side: a [3], b [1], c [2]
angle: A [3], B [1], C [2]

26 Interior & Exterior Angles of a Triangle 1 pp 148, 149

1 ∠ACE, ∠ECD, ∠ECD, ∠BCD

2 (1) 70° (2) 110°
(3) 135° (4) 115°

3 ∠ACB, ∠ACD, ∠ACD

4 (1) 125° (2) 130°
(3) 55° (4) 60°
(5) 60° (6) 35°

27 Interior & Exterior Angles of a Triangle 2 pp 150, 151

1 (1) 126° (2) 65°
(3) 135°

2 ∠B, ∠D, ∠C, ∠D

3 (1) 63° (2) 60°
(3) $\angle x = 65°$, $\angle y = 25°$

4 (1) $\angle x = 90°$, $\angle y = 130°$
(2) $\angle x = 70°$, $\angle y = 60°$

5 (1) 73° (2) 55°

6 (1) 64° (2) 25°

28 The Sum of the Interior Angles of a Polygon pp 152, 153

1 (1) 2, 2 (2) 3, 3
(3) 4, 4 (4) 2, $n-2$
(5) 1080° (6) 1260°
(7) 1800°

2 (1) 5, isosceles, 72, 108, 540
(2) 36, 144, 1440

3 (1) 1080° (2) 135°
(3) 150° (4) 7
(5) 6

4 the sum = 720°, $\angle x = 58°$

29 The Sum of the Exterior Angles of a Polygon pp 154, 155

1 5, 2, 3, 5, 5, 3, 360

2 $n-2$, n, 2, 360

3 (1) 360° (2) 45°
(3) 20

4 (1) 95° (2) 80°
(3) 75° (4) 60°

30 Summary of Parallel Lines & Angles 1 pp 156, 157

1 (1) 55° (2) 50°
(3) 105° (4) 130°
(5) 50°

2 (1) $\angle x = 60°$, $\angle y = 120°$
(2) $\angle x = 90°$, $\angle y = 50°$
(3) $\angle x = 115°$, $\angle y = 145°$
(4) $\angle x = 30°$, $\angle y = 55°$
(5) $\angle x = 95°$, $\angle y = 115°$

3 (1) $\angle x = 72°$, $\angle y = 32°$
(2) $\angle x = 75°$, $\angle y = 30°$
(3) $\angle x = 130°$, $\angle y = 50°$
(4) $\angle x = 90°$, $\angle y = 55°$
(5) $\angle x = 120°$, $\angle y = 55°$

4 (1) 85° (2) 105°
(3) 58° (4) 145°

31 Summary of Parallel Lines & Angles 2 pp 158, 159

1 (1) 85° (2) 100°
(3) 90°

2 (1) 75° (2) 45°

3 (1) 60° (2) 35°

4 (1) 360° (2) 360°
(3) 1440° (4) 144°
(5) regular octadecagon

5 (1) 45° (2) isosceles triangle
(3) 67.5° (4) 135°

6 (1) $\angle x = 110°$, $\angle y = 100°$
(2) 115°

32 Circles & Triangles 1 (Tangent) pp 160, 161

1 (1) ①0, ②2, ③1, ④2
(2) ③ (3) 5

2 (1) (2)

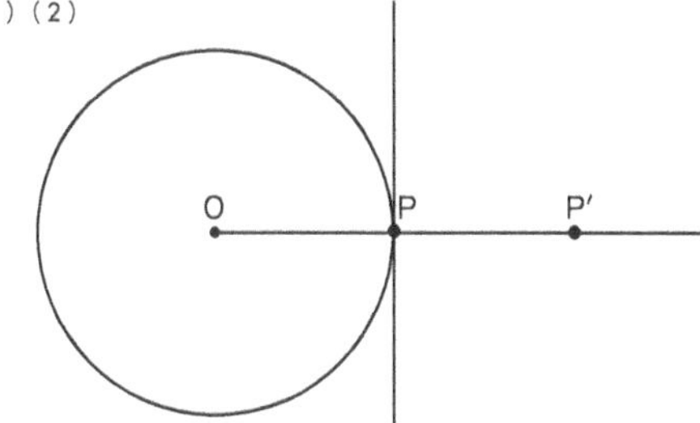

3

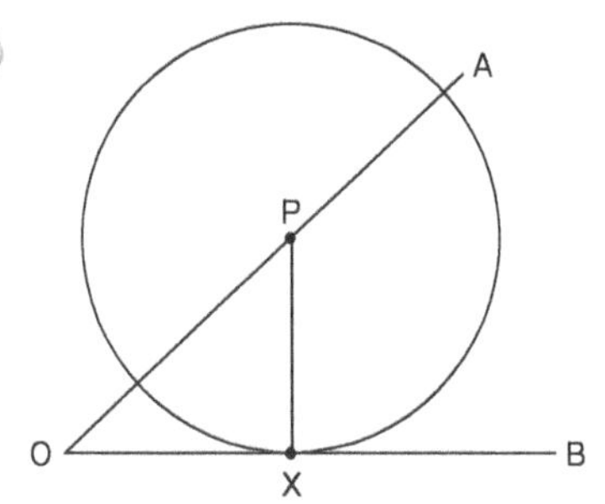

4

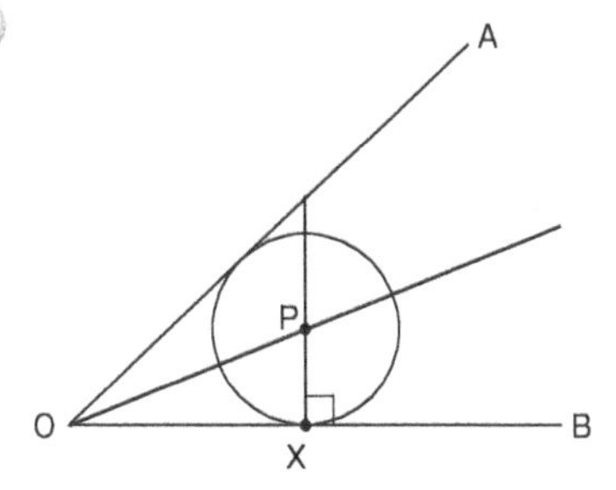

33 Circles & Triangles 2 (Inscribed in a Circle) pp 162, 163

1 (1) two (2) P
(3) Q (4) O
(5) P (6) six
(7) hexagon (8) equilateral triangle
(9) six

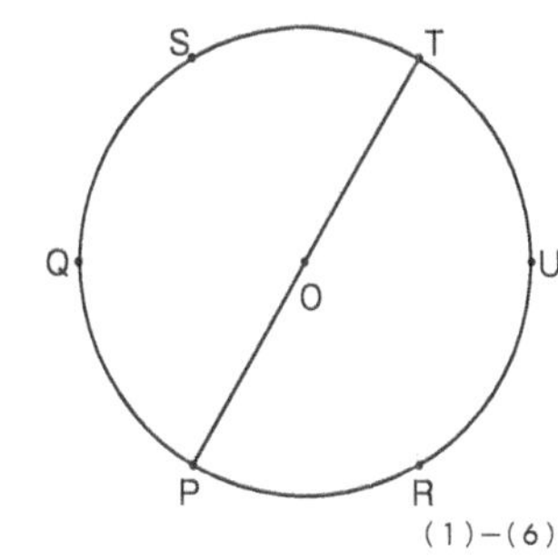

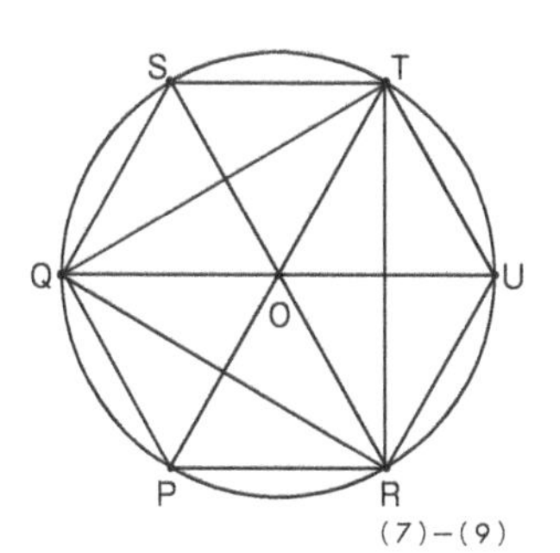

2 (1)–(6)

(2) two (4) A
(6) equilateral triangle

3 (1) 120° (2) 60°
(3) $CD \perp AB$
(4) $\angle AOD = 60°$, $\angle ACD = 30°$
(5) $\angle BOD = 60°$, $\angle BCD = 30°$
(6) BOD, DB, $\frac{1}{2}$

34 Circles & Triangles 3 (Inscribed Angles 1) pp 164, 165

1 (1) 5 cm (2) 50°
(3) 12 cm (4) 135°

2 (1) 20° (2) 50°

3 (1) $\angle CAO$, 2 (2) x, x
(3) 90

4 (1) $\angle AOB$ (2) $2a$
(3) $2b$ (4) $a + b$
(5) $2\ (a + b)$ (6) $\frac{1}{2}\angle AOB$

5 (1) 30° (2) 100°
(3) 40° (4) 70°

35 Circles & Triangles 4 (Inscribed Angles 2) pp 166, 167

1 (1) 60° (2) 50°
(3) 50° (4) 130°

2 (1) 30° (2) 5 cm
(3) 50° (4) 5 cm
(5) 80° (6) 20 cm

3 (1) 180° (2) 90°
(3) right triangle

4 (1) 90° (2) 40°
(3) 30° (4) 70°

5 (1) 105° (2) 260°
(3) 220° (4) 105°
(5) 110° (6) 100°
(7) 55° (8) 120°

36 Isosceles Triangles 1 pp 168, 169

1 (1) 6 cm (2) 75°
(3) 30° (4) 80°

2 (1) 8 cm (2) 6 cm
(3) 5 cm

3 (1) AC, C, 55, 55, 70
(2) 40°
(3) $180° - 2a°$

4 (1) 73° (2) 80°
(3) $2a°$

5 (1) 65° (2) 110°

37 Isosceles Triangles 2 pp 170, 171

1 (1)–(3)

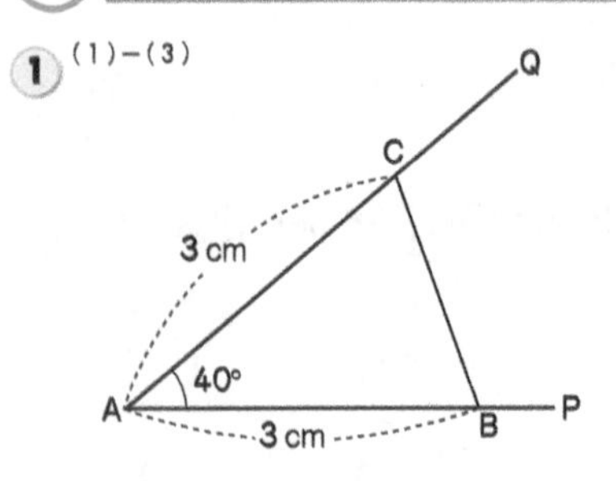

(4) 70°

2 (1)–(4)

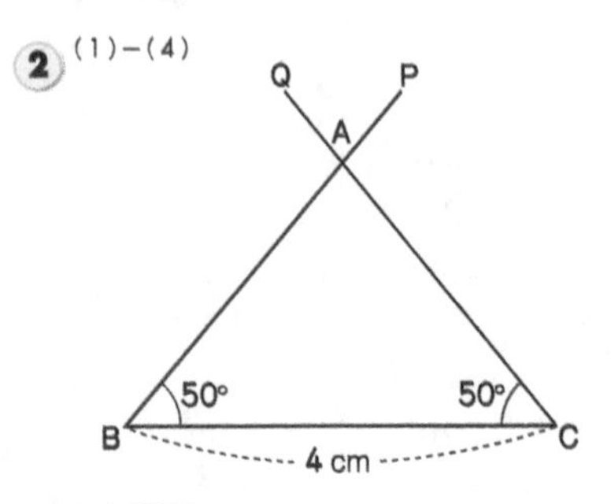

(5) 80°

3 (1)–(4)

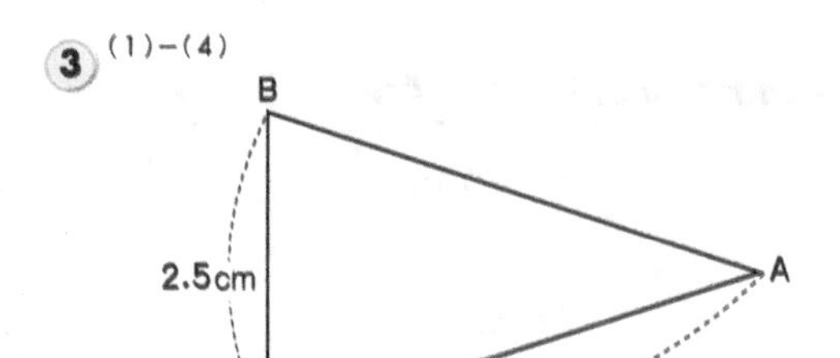

4 (1)–(4)

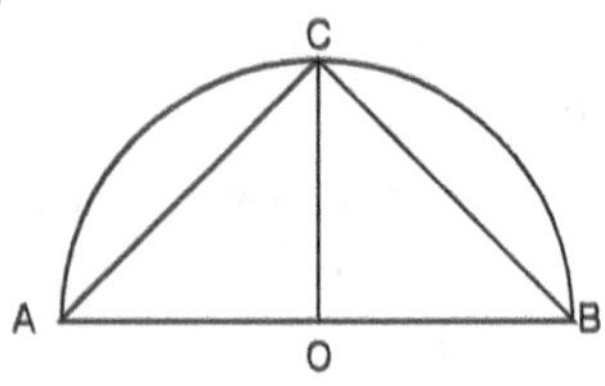

(5) $\overline{OB}$, isosceles (6) OCB, 45
(7) 45 (8) 90

5 (1)–(3)

C
A B
D

(4) $\overline{AD}$
(5) 30
(6) $\overline{BD}$
(7) 30
(8) 120

38 Right Triangles pp 172, 173

1 (1) ∠A ⇨ BC, ∠B ⇨ AC, ∠C ⇨ AB
(2) AB ⇨ ∠C, AC ⇨ ∠B, BC ⇨ ∠A
(3) BC

2 (1) QR (2) ∠Q ⇨ PR, ∠R ⇨ PQ
(3) 90° (4) 55°

3 (1) EF (2) ∠E ⇨ DF, ∠F ⇨ ED
(3) 90° (4) 35°

4 (1)

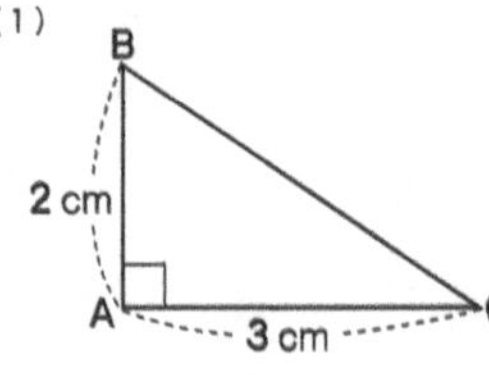

(2)

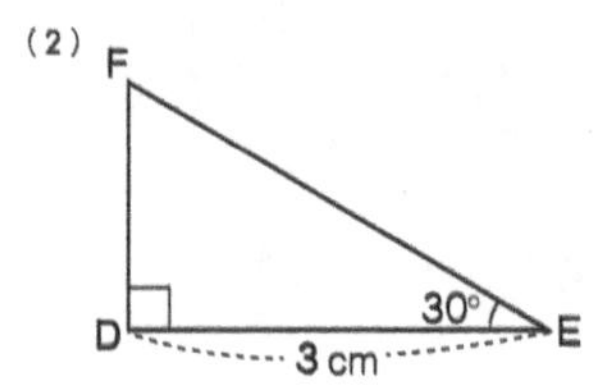

(3)

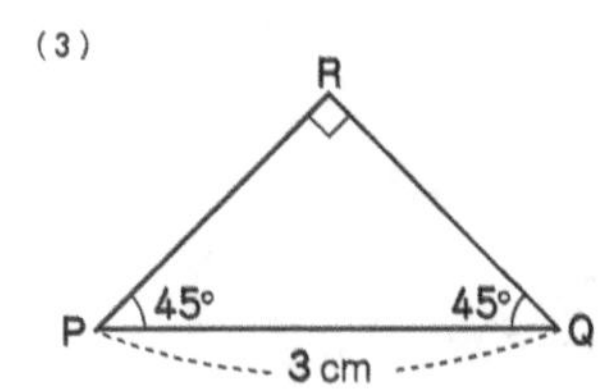

5 (1) (2)

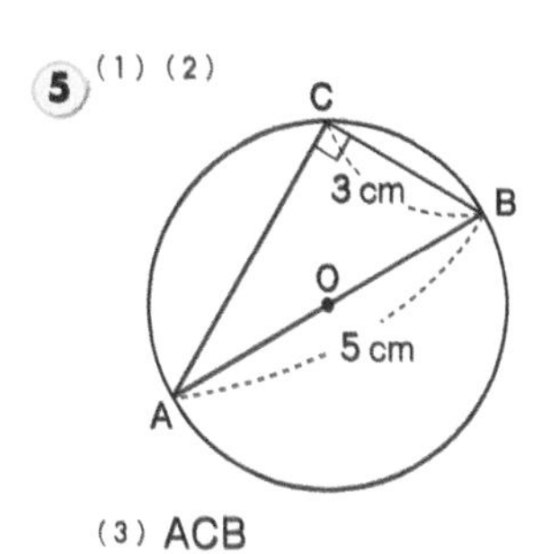

(3) ACB

6 (1) (2)

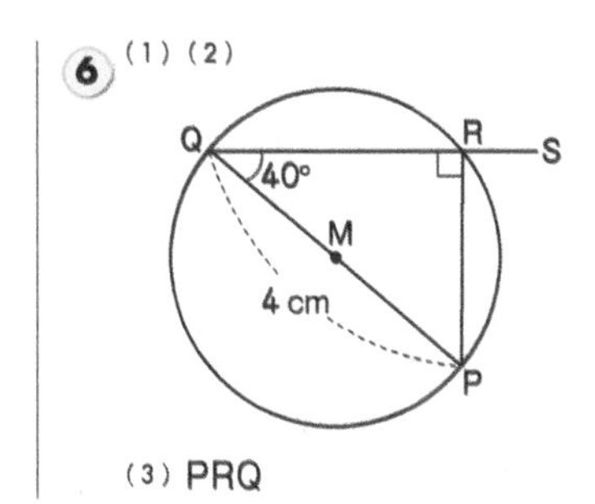

(3) PRQ

39 Properties of Triangles 1

pp 174, 175

1 (1)–(4)

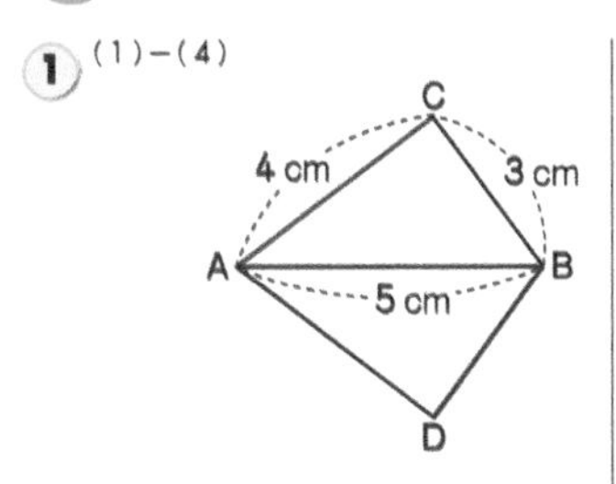

(5) congruent (yes)

2

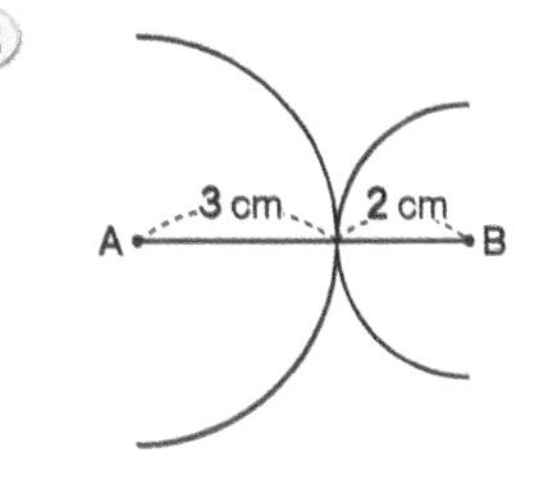

(1) ×　(2) more than 2 cm

3 (1) ×　(2) ○　(3) ×　(4) ×　(5) ○　(6) ×

4 (1)–(3)

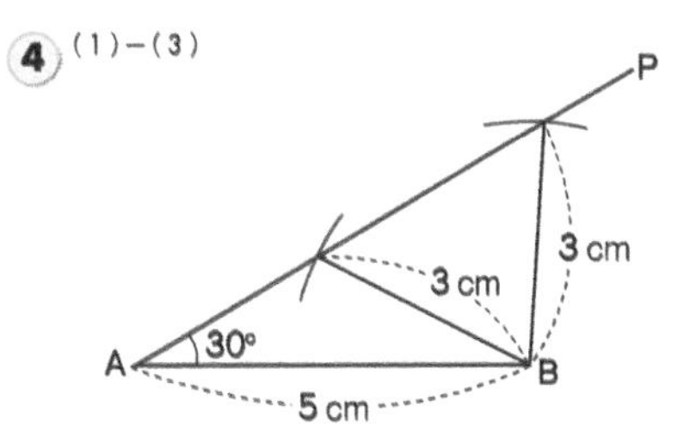

(4) two　(5) two　(6) not congruent (no)

5

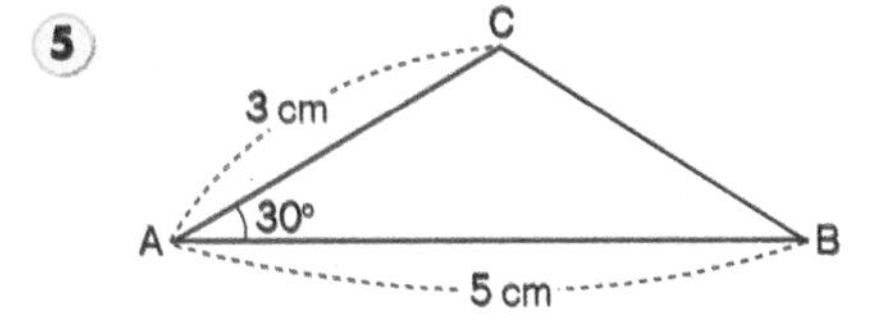

6 (1) larger　(2) one　(3) two　(4) one

40 Properties of Triangles 2

pp 176, 177

1 (1) 60°, 70°, 50°

(2) (3)

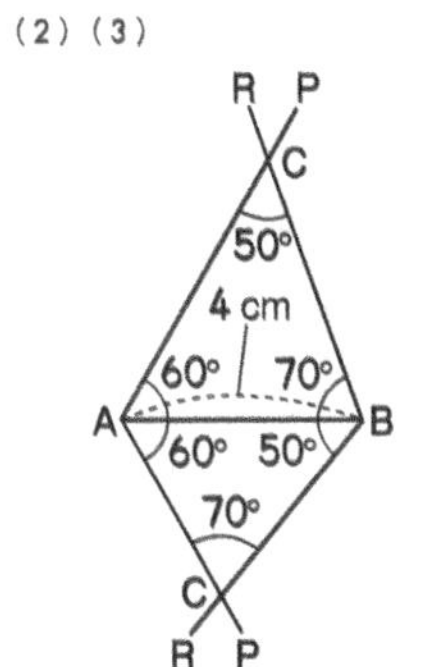

(4) ×

2 (1)–(4)

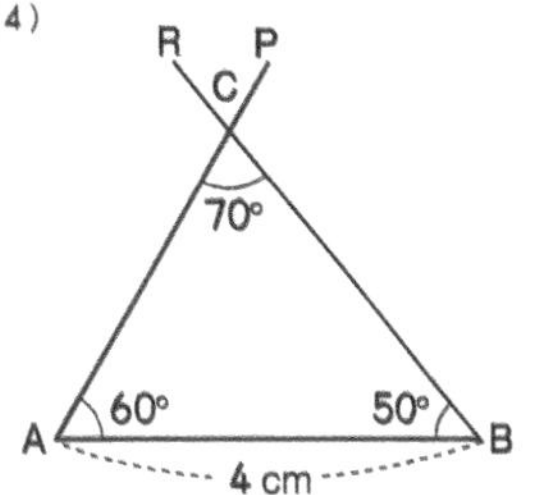

3 (1) ○　(2) ×　(3) ×　(4) ○　(5) ○

4 (1) ∠B　(2) AB　(3) sides　(4) AB

41 Review 1

pp 178, 179

1 (1) 184 cm²　(2) 192 π cm²

2 (1) 16 π cm³　(2) 12 π cm³

3 (1) AB, DC, EF　(2) 4　(3) ABCD, EFGH

4 (1) 108 cm²　(2) 48 cm³

5 (1) 4 π cm　(2) 2 cm　(3) 12 π cm²　(4) 16 π cm²

6 (1) 54 π cm³　(2) 18 π cm³　(3) 3 : 1

42 Review 2

pp 180, 181

1 (1) 65°　(2) 95°
(3) $\angle x = 60°$, $\angle y = 105°$　(4) $\angle x = 50°$, $\angle y = 85°$
(5) $\angle x = 72°$, $\angle y = 22°$

2 (1) 540°　(2) 720°　(3) 360°　(4) 135°

3 (1) $\angle x = 85°$, $\angle y = 40°$
(2) $\angle x = 105°$, $\angle y = 135°$
(3) $\angle x = 50°$, $\angle y = 50°$
(4) $\angle x = 20°$, $\angle y = 30°$

4 (1) $\angle x = 60°$, $\angle y = 85°$
(2) $\angle x = 143°$, $\angle y = 80°$

43 Review 3

pp 182, 183

1 (1) OA　(2) 90°　(3) 45°　(4) ∠BAC
(5) 180°　(6) 90°

2 (1) 32°　(2) 110°　(3) 60°

3 (1) 35°　(2) 70°　(3) 55°　(4) 90°　(5) 50°

4 (1) radius　(2) intersection　(3) AB
(4) equilateral　(5) inscribed　(6) 120
(7) P, diameter　(8) 90　(9) inscribed
(10) 60

5 (1) ACB(C), 90　(2) AB　(3) 10　(4) 2, 8

How to Solve

P.102 3 – 1 (4)

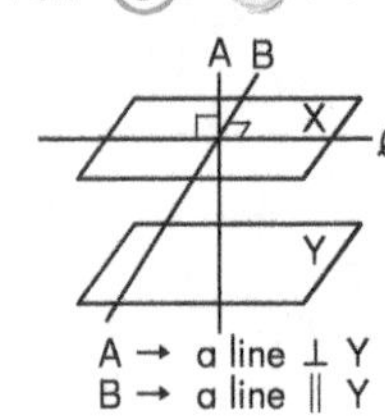

A → a line ⊥ Y
B → a line ∥ Y

P.103 3 – 3 (3)

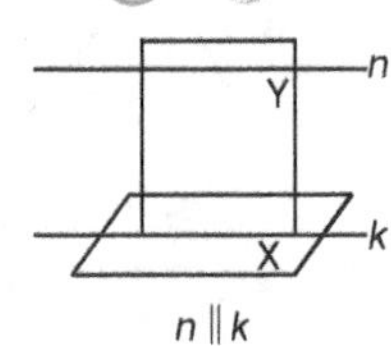

$n \parallel k$

P.103 3 – 4 (5)

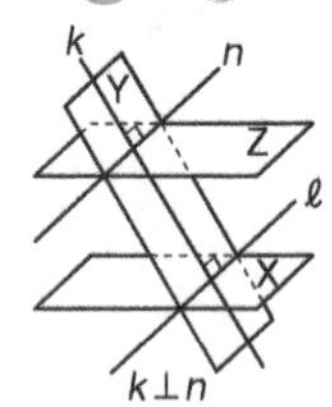

$k \perp n$

P.107 5 – 3 (2)

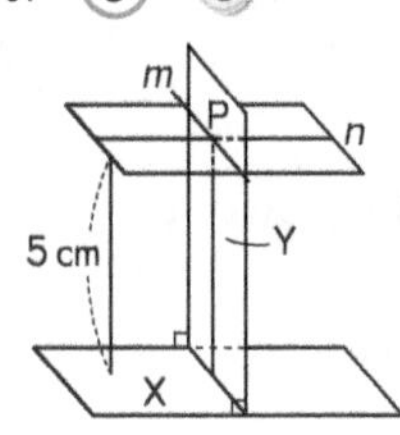

P.115 9 – 5 (1)

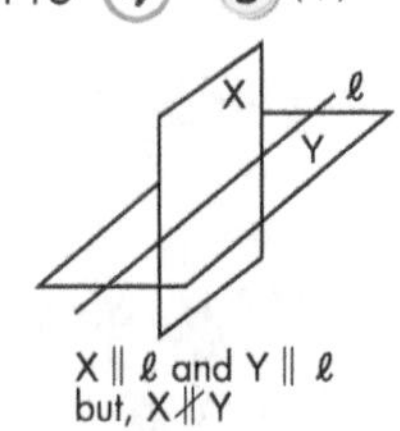

X ∥ ℓ and Y ∥ ℓ
but, X ∦ Y

(3)

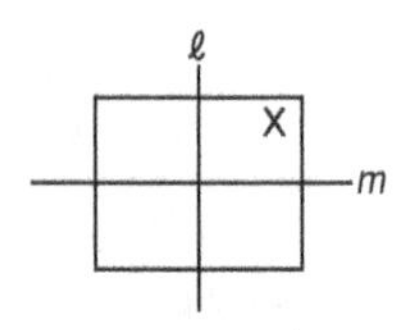

X ∥ ℓ and X ∥ m
but, ℓ ∦ m

P.117 10 – 2 (2)

●Draw a front view.

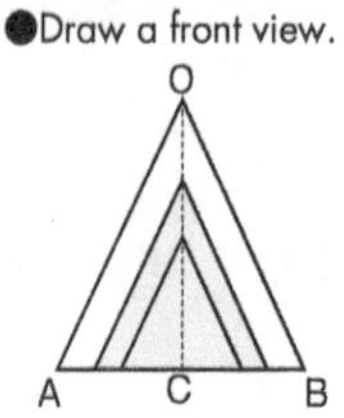

(3)

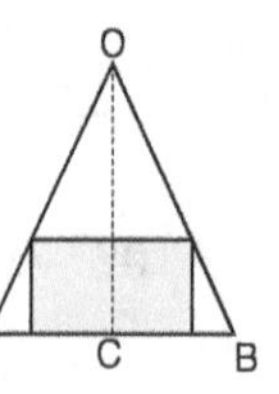

(4)

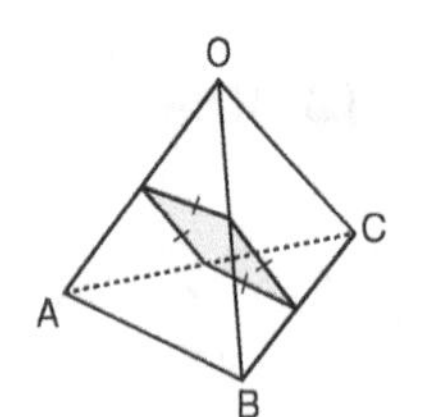

P.117 10 – 3 (2)

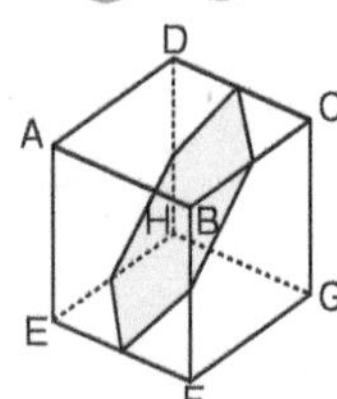

passing through the midpoint of the sides BF and DH, too.

P.126 15 – 2 (4)

Since 'a' is the base, ABCD is the lateral area.

P.128 16 – 1 (1)

(surface area) = (lateral area) + (area of the base) × 2

Therefore, $(3+4+5) \times 6 + \left(3 \times 4 \times \frac{1}{2}\right) \times 2 = 84$ in².

P.129 16 – 2 (3)

lateral area : $10\pi \times 10 = 100\pi$ cm²

surface area : $100\pi + (5 \times 5 \times \pi) \times 2 = 150\pi$ cm².

P.131 17 – 4 (1)

(surface area) = (lateral area) + (area of the base)

Therefore, $\left(4 \times 10 \times \frac{1}{2}\right) \times 4 + 4 \times 4 = 96$ cm².

P.132 18 – 2 (1)

(the length of the arc) = (the circumference of the base) = 6π cm

(the circumference of the circle) ÷ (the length of the arc) =

$24\pi \div 6\pi = 4$

(the central angle) = $360° \div 4 = 90°$

P.133 18 – 5 (1)

(the lateral area of the cone) = (the erea of the sector)

(the circumference of the circle) ÷ (the length of the arc) =

$24\pi \div 8\pi = 3$

(the area of the circle) ÷ 3 = (the area of the sector) =

$144\pi \div 3 = 48\pi$ cm² = (lateral area of the cone)

P.137 20 – 5 (2)

(the volume of the sphere) = $\frac{4}{3} \times 1 \times 1 \times \pi = \frac{4}{3}\pi$ cm³

(the volume of the cube) = $2 \times 2 \times 2 = 8$ cm³

(How many times larger) = $\frac{4}{3}\pi \div 8 = \frac{\pi}{6}$

P.159 31 – 6 (2)

$360 - 100 - 90 - (50 + a) -$
$(55 + b) = 65 - (a + b) = 0$

Since, $a + b = 65°$

$x = 180 - 65 = 115°$

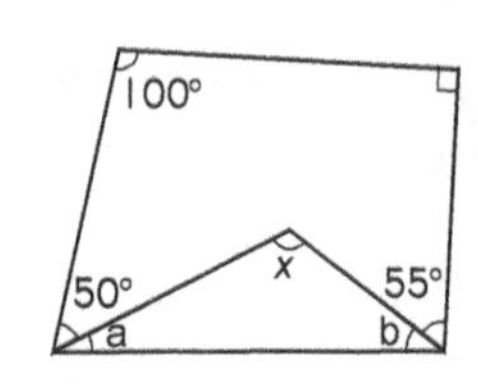

P.167 35 – 4 (4)

Connecting AD, ∠DAB = 20°.

Since △ABD is a right triangle,

∠ADB = 90°.

$x = 180 - 90 - 20 = 70°$.

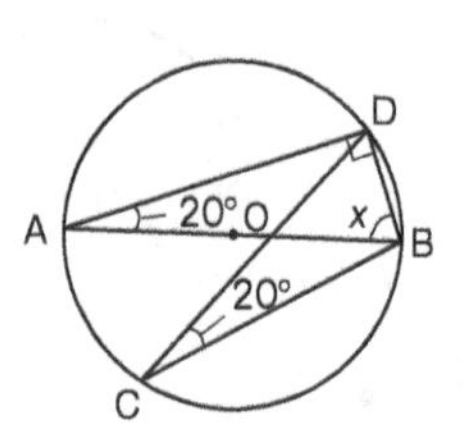

P.178 41 – 2

(1) the volume of the cone = $16\pi \times 3 \times \frac{1}{3} = 16\pi$ cm³.

(2) the volume of the cone = $9\pi \times 4 \times \frac{1}{3} = 12\pi$ cm³.

P.179 41 – 6

(1) the volume of the cylinder = $9\pi \times 6 = 54\pi$ cm³.

(2) the volume of the cone = $9\pi \times 6 \times \frac{1}{3} = 18\pi$ cm³.

(3) the ratio = $54\pi : 18\pi = 3 : 1$

P.181 42 – 4 (2)

$x = 180 - 37 = 143°$

$y = 180 \times (5-2) - (100 + 143 + 90 + 127) = 80°$

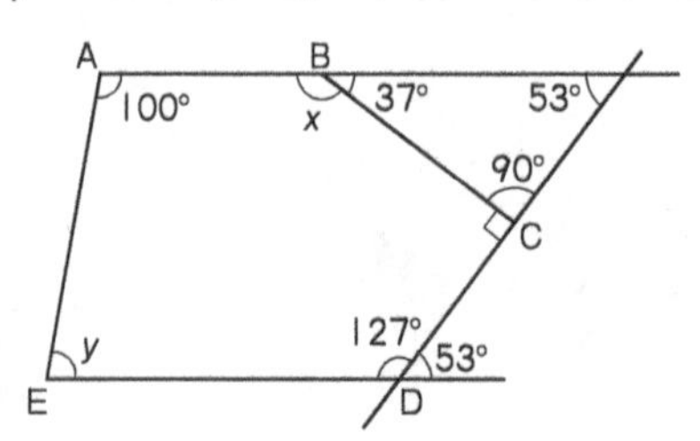

